Praise for *Applied Embedded Electronics*

Applied Embedded Electronics is a must-have reference for every engineer working on embedded systems. The author does a fantastic job covering the essential components and how to ensure you take everything into account to design a robust electronic system.

—*Jacob Beningo, Embedded Systems Consultant,*
Beningo Embedded Group

This is the best book in the industry for the aspiring design engineer wanting to create products for real-world applications. This should be required reading for all design engineers.

—*John Catsoulis, Design Engineer, Author of* Designing Embedded Hardware, *Founder of Udamonic.com*

As an embedded engineer with 25 years of experience, I can tell you this book helps demystify the hardware and system considerations of an embedded design. This book provides information to design correctly the first time without having to debug much later in the process.

—*Mark Kraeling, Embedded Systems Engineer*

Applied Embedded Electronics
Design Essentials for Robust Systems

Jerry Twomey

Beijing · Boston · Farnham · Sebastopol · Tokyo

Applied Embedded Electronics

by Jerry Twomey

Copyright © 2024 Gerald Twomey. All rights reserved.

Published by O'Reilly Media, Inc., 1005 Gravenstein Highway North, Sebastopol, CA 95472.

O'Reilly books may be purchased for educational, business, or sales promotional use. Online editions are also available for most titles (*http://oreilly.com*). For more information, contact our corporate/institutional sales department: 800-998-9938 or *corporate@oreilly.com*.

Acquisitions Editor: Brian Guerin	**Indexer:** Potomac Indexing, LLC
Development Editor: Angela Rufino	**Interior Designer:** David Futato
Production Editor: Christopher Faucher	**Cover Designer:** Karen Montgomery
Copyeditor: Audrey Doyle	**Illustrator:** Jerry Twomey
Proofreader: Kim Cofer	

November 2023:   First Edition

Revision History for the First Edition

2023-11-14:   First Release

See *http://oreilly.com/catalog/errata.csp?isbn=9781098144791* for release details.

978-1-098-14479-1

[LSI]

This book is dedicated to the memory of:

*James "Eddie" Curnyn, W1IZB,
who taught a young and
impressionable kid that you can
point an antenna at the moon and
actually listen to the Apollo
astronauts.*

*That "one small step" lit a fire
under me that has lasted a lifetime.*

Table of Contents

Preface

Why I Wrote This Book

Whereas most academic textbooks cover the ideal theory of electronics, the approach here is centered on design methods that work in the challenging environment of real-world devices.

The topics were motivated by my experiences troubleshooting unreliable and problematic devices for many clients. The book sprang from my revisions of deficient electronics and my coaching sessions with designers on important issues that must be addressed to develop reliable products. Many such incidents motivated me to publish a series of trade magazine articles dealing with related topics, and the need for this book became evident.

Who This Book Is For

This book is a reference for engineers, scientists, and other designers who want to create an electronic system. The term *applied embedded electronics* covers a wide swath of devices that includes embedded controllers, cell phones, medical instruments, computers and tablets, all consumer electronics, industrial robots, automation systems, and countless others. All electronic devices that use digital control methods also fall under this broad heading.

The topics covered here are common to all of these devices. The emphasis here is on hardware design—the actual circuits and systems—with a brief look at coding and software issues.

The reader is presumed to have a background in engineering or science and a grasp of fundamental electronics.

Evolving Design Methods: A Different Approach

Electronics textbooks tend to fit into three groups. The first group includes academic textbooks that introduce engineering students to specific topics. These books use a theoretical approach and, for simplicity, focus on ideal situations that are free of second-order effects and problems.

The second group includes books explaining electronics to hobbyists and nonprofessional experimenters. Although these books have a significant following, they tend to be very global on the basics and concentrate on methods not used in commercial products.

The final group consists of books for practicing engineers. These books take a deep technical dive into a specific aspect of technology, without delving into the surrounding issues needed to utilize that technology in a functional product.

This text takes a different approach. Modern technology has evolved to the "system on a chip" (SoC) era, where a large amount of the "deep technical dive" has been integrated into an application-specific integrated circuit (ASIC). Consequently, the IC designer requires technical depth, but the system and printed circuit board (PCB) designers need a different approach.

Here the focus is on putting the system together, paying special attention to the important items that make the system reliable and functional under all conditions. This book takes the perspective that modern electronics use SoC methods wherever possible, and creators need the "design essentials" to bring those devices together into a larger system. To do this, the emphasis here is on explaining the important concepts that are common to all electronic systems and illustrating these concepts using extensive graphics and minimal equations.

The evolution of electronic systems has seen two distinct trends. First, wherever possible, analog methods have yielded to digital signal processing and digital control methods. In a modern system, real-world analog information is converted to digital data equivalents as quickly as possible.

Second, the pursuit of progressively higher levels of integration has been relentless. Historically, what was once a myriad of discrete transistors and circuits evolved to more modular op-amps, wideband amplifiers, ADCs, DACs, and logic gates. That modular level then progressed to higher levels of integration with specific applications being targeted.

Wherever product volumes make it cost-effective, dedicated ICs are created to cover entire systems in a single chip. For those single-chip products, you just add a power source and external connections, and (in theory!) you should be good to go.

More complex systems utilize a collection of application-specific analog front ends (AFEs) that connect to the outside world and provide a data interface to a digital controller. Ideally, designers simply select a suitable digital controller and write the code to process and control the data streams, and they have a product.

Problems with these modern systems tend to be in areas like noise-corrupted signals, data link reliability, power system stability, improper battery capacity, radiated EMI, driver and sensor circuitry failures, improperly configured feedback control, regulatory testing failures, and poorly written code. With the methods laid out here, designers can avoid these issues and make robust and reliable products.

With the evolution of electronics, the designer's skill set has also changed. Transistor-level circuits are now the realm of IC designers, and modern system designers have become more skilled in coding and digital methods.

How This Book Is Organized

Many of these topics are interwoven and need to be discussed from different perspectives. In this book, when I provide a "Further Discussed In" (FDI) cross-reference (e.g., "FDI: Essentials," "FDI: Architecture," "FDI: EMI & ESD," "FDI: Power," etc.) I'm letting you know that supplemental information is available elsewhere in the book. As well, each chapter includes a "Further Reading" section for those who seek additional sources.

The book covers the following topics:

Chapter 1, "Essential Concepts" (Essentials)
> Starting off, this chapter makes sure all readers have a common foundational knowledge of real-world electronics. Readers are introduced to the nonideal nature of electronics, how academic simplifications can mislead, how impedance of connections affects performance, parasitic coupling issues, nonideal grounding and ground bounce, and nonideal components, among many other issues.

Chapter 2, "Architecting the System" (Architecture)
> How does it all fit together? What processor should you use? Every system has things to monitor, things to control, and signals to process. In this chapter, big-picture considerations are addressed, including understanding digital controller features, picking appropriate digital control methods, and partitioning a system. Also discussed are signal processing methods, the use of chip sets, and creating a "mostly digital" architecture.

Chapter 3, "Robust Digital Communication" (Digital)
> Moving digital information between devices can require many different strategies, depending on data rates, distance traveled, and the application environment.

Both wired and wireless digital interfaces are reviewed for capabilities and limitations.

Chapter 4, "Power Systems" (Power)

All systems need reliable, safe power and energy efficiency. Topics include AC power safety, ground-protected and double-insulated AC safety, weakest-link protection methods, AC/DC converters, linear voltage regulators, low dropout regulators, buck switching regulators, boost switching regulators, configuring multiregulator power systems, bypass and decoupling methods, noise optimization for power converters, special challenges of digital current transients, high-frequency stabilization of power grids, and controlled power sequencing for the entire system.

Chapter 5, "Battery Power" (Battery)

Battery power is essential for modern devices, including plugged-in devices, which frequently have internal batteries to preserve critical systems during power outages. Battery chemistry types, chargeable and single-use options, current capability, projected battery life, calculating time and discharge profiles, and how to design a battery pack and charging system are covered here.

Chapter 6, "Electromagnetic Interference and Electrostatic Discharge" (EMI & ESD)

Most electronic systems create electronic noise and radiate noise energy. Any commercial product must meet legally imposed limits on that radiated noise energy. Techniques to make a system quieter and not interfere with other devices are presented. Also, systems will be exposed to electronic noise and ESD events, so designing a system to be noise immune and to survive ESD without damage is covered as well.

Chapter 7, "Data Converters: ADCs and DACs" (ADC & DAC)

Important performance parameters for data converters are discussed. The limitations of commonly used PWM DACs are examined, and circuit techniques to improve performance are presented.

Chapter 8, "Driving Peripheral Devices" (Drive)

Methods used to drive motors, actuators, lights, visual displays, speakers, and other devices are covered. Using power transistors and selecting application-appropriate devices are also discussed.

Chapter 9, "Sensing Peripheral Devices" (Sense)

Ways to measure parameters such as temperature, air pressure, force, acceleration, speed, location, and others are examined. Sensor information can be noisy or inconsistent due to the application environment. Methods for processing raw sensor information so that it's both usable and reliable to a digital controller are covered.

Chapter 10, "Digital Feedback Control" (Control)

Driving and sensing peripherals can be used as part of a feedback control loop. Classic analog control system methods are largely obsolete, and most modern devices use digital methods. Industry-prevalent methods of digital PID control and trajectory management are examined.

Chapter 11, "Schematic to PCB" (PCB)

Implementing a PCB design requires a suitable grounding and power plane strategy, good signal integrity and noise isolation, and a built-in strategy for testing. Schematic organization, component selection, footprint creation, mechanical mounting, PCB layers, physical design rules, placement strategy, interconnect RLC, between-layer connection vias, transmission lines, fabrication notes, and manufacturing files are all discussed.

Chapter 12, "Software and Coding" (Code)

The topics of real-time operating systems, port and processor configuration, developing device drivers, code for peripheral communication, defensive coding methods, techniques for self-recovery from faults, watchdog timers, and multi-controller coding are covered.

Chapter 13, "Special Systems and Applications" (Special)

Not all electronic systems are designed with the same methods and priorities. Regulatory restrictions and compliance drive the design in many sectors. Special requirements in avionics, astrionics, military, medical, automotive, consumer, and industrial automation are explored.

Chapter 14, "Creating Great Products" (Great Products)

With solutions to technical challenges explored in prior chapters, in this chapter a brief sampling of nontechnical issues is provided, including marketing and product demand, understanding the target market, products that solve the customer's problem, viable product pricing, markets limited by time windows, product ease of use, design team considerations, minimal design to avoid feature creep, ease of manufacturing, and getting customer feedback.

All of these topics are essential to a successful, robust, and reliable electronic system. A designer can connect a collection of peripherals to a digital controller, but the system won't function reliably if the designer doesn't address the important issues explored here. Frequently, designers learn this the hard way: when a new design is nonfunctional, failures crop up in the field or a product launch is delayed due to regulatory testing failures.

Discussions of electronics technology use a large number of acronyms, and this book is no exception. Most acronyms here are in common use and won't need to be defined for many readers. When used as a reference book for answers on specific topics, the book may not be read front to back. Therefore, acronyms are included in a glossary for convenient reference.

Let's get started!

Jerry Twomey
San Diego, October 2023
effectiveelectrons.com

Conventions Used in This Book

The following typographical conventions are used in this book:

Italic
Indicates new terms, URLs, email addresses, filenames, and file extensions.

`Constant width`
Used for program listings, as well as within paragraphs to refer to program elements such as variable or function names, databases, data types, environment variables, statements, and keywords.

`Constant width bold`
Shows commands or other text that should be typed literally by the user.

`Constant width italic`
Shows text that should be replaced with user-supplied values or by values determined by context.

This element signifies a tip or suggestion.

This element signifies a general note.

This element indicates a warning or caution.

O'Reilly Online Learning

 For more than 40 years, *O'Reilly Media* has provided technology and business training, knowledge, and insight to help companies succeed.

Our unique network of experts and innovators share their knowledge and expertise through books, articles, and our online learning platform. O'Reilly's online learning platform gives you on-demand access to live training courses, in-depth learning paths, interactive coding environments, and a vast collection of text and video from O'Reilly and 200+ other publishers. For more information, visit *https://oreilly.com*.

How to Contact Us

Please address comments and questions concerning this book to the publisher:

O'Reilly Media, Inc.
1005 Gravenstein Highway North
Sebastopol, CA 95472
800-889-8969 (in the United States or Canada)
707-829-7019 (international or local)
707-829-0104 (fax)
support@oreilly.com
https://www.oreilly.com/about/contact.html

We have a web page for this book, where we list errata, examples, and any additional information. You can access this page at *https://oreil.ly/applied-embedded-electronics*.

For news and information about our books and courses, visit *https://oreilly.com*.

Find us on LinkedIn: *https://linkedin.com/company/oreilly-media*

Follow us on Twitter: *https://twitter.com/oreillymedia*

Watch us on YouTube: *https://youtube.com/oreillymedia*

Acknowledgments

The first half of this book was written during the COVID-19 lockdowns, so a special thanks to the healthcare workers, first responders, and frontline workers who could not stay home. All of you helped keep us alive and safe. Thank you!

Much appreciation goes to the O'Reilly team and the technical reviewers who helped bring it all together. I could not have gotten it done without all your efforts. Thank you!

Much gratitude is owed to all my friends and family who have tolerated my "limited participation in life" during my extended efforts to get this book done. It's been a long grind, and your patience and support is greatly appreciated. Thank you!

CHAPTER 1

Essential Concepts

An introduction to electronics gives the basics of Ohm's law; Kirchhoff's current and voltage laws; current and voltage sources; how resistors, capacitors, and inductors function; and similar topics. A typical curriculum will cover the ideal versions of these topics to clearly convey the basics. Unfortunately, academic simplifications can be misleading. Real-world applications frequently run into problems when ideal concepts fall short. Nothing is ideal.

This chapter takes a brief look at some of the "less than ideal" components seen in real-world applications, how they differ from ideal devices, and where designers need to pay attention to the limitations of nonideal devices. This chapter is about building an awareness of problems that are commonly encountered; the rest of the book develops solutions or strategies to avoid these issues.

To solve any problem, the designer needs to be aware of the problem first. Again, nothing is ideal.

Basic Electronics

There are many textbooks that cover fundamental electronic design from scientific, engineering, and hobbyist perspectives. The focus here is not on basic electronics; rather, it is on the important things that are necessary to design a reliable electronic system. Figures 1-1 and 1-2 serve as a quick checkpoint. If you are reasonably familiar with the topics listed in these figures, you should have the background required to work with the material presented here.

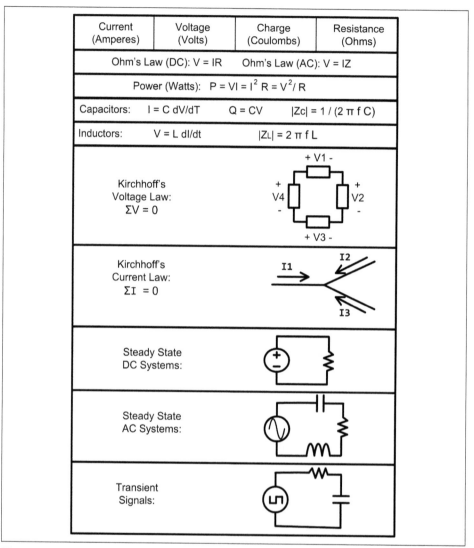

Current (Amperes)	Voltage (Volts)	Charge (Coulombs)	Resistance (Ohms)

Ohm's Law (DC): $V = IR$ Ohm's Law (AC): $V = IZ$

Power (Watts): $P = VI = I^2 R = V^2 / R$

Capacitors: $I = C \, dV/dT$ $Q = CV$ $|Z_C| = 1 / (2 \pi f C)$

Inductors: $V = L \, dI/dt$ $|Z_L| = 2 \pi f L$

Kirchhoff's Voltage Law: $\Sigma V = 0$

Kirchhoff's Current Law: $\Sigma I = 0$

Steady State DC Systems:

Steady State AC Systems:

Transient Signals:

Figure 1-1. Basic concepts, part I

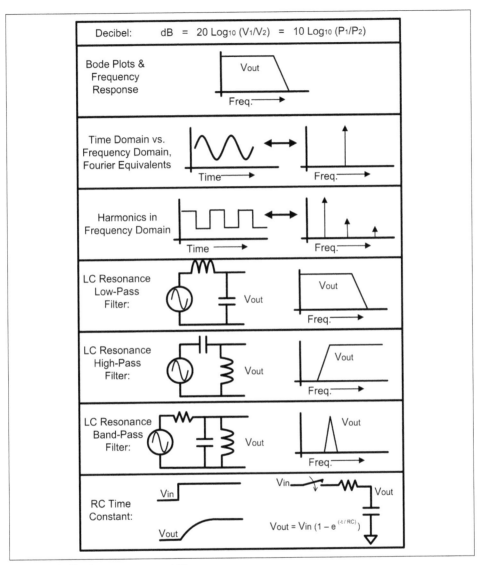

Figure 1-2. Basic concepts, part II

If some items in the figures are unfamiliar to you, there are many reference books, online tutorials, and websites that you can explore. Many of those sources cover the basics.

Ideal Simplifications of Academia

Modern electronic systems are predominantly digital, but most design problems are analog in origin. Noise, signal integrity, power stability, electromagnetic interference (EMI), and connection impedance are common problems. These issues can quickly degrade any electronic system, or render it nonfunctional. Digital systems are more tolerant of these issues, but fully digital devices can be broken by analog limitations.

When explaining electronics, a simplified representation gives a quick idea of how something should work. Such a "first-order" model is useful for simplicity and illustration but frequently leaves out important details. That simple model is often an incomplete model, and many of the "second-order effects" that are omitted can affect device performance significantly.

Becoming aware of a more detailed model helps lay the foundation for better design. The techniques presented here all pay attention to second-order effects and provide methods to deal with them.

Interconnections

As a start, the idea of connecting things together needs a closer look.

As shown in Figure 1-3, a short piece of wire can have significant impedance. A piece of 24 AWG wire, 10 cm long, will have approximately 100 nH of inductance, 10 milliohms of resistance, and capacitive coupling to the environment around it.

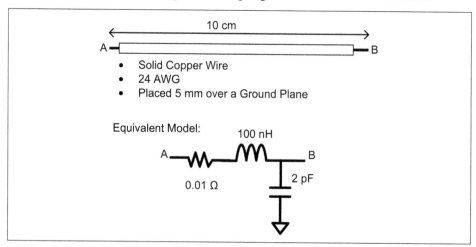

Figure 1-3. Wire impedance

In this example, the wire impedance and capacitance to the outside environment will act as a low-pass filter (LPF) above 300–400 MHz and be sensitive to motion, due to capacitance changing as a function of wire placement relative to ground. Larger wire can reduce resistive losses, and parallel wire connections can reduce inductance some, but the inductance is not easily removed.

A similar 10 cm connection on a printed circuit board (PCB) shows many of the same characteristics. In Figure 1-4, the inductance is about the same as the wire, and the capacitance has increased due to the connection being tightly spaced over a ground plane. For this situation, the PCB trace will start to perform as an LPF around 80–90 MHz due to the increased capacitance. One advantage of the PCB trace over the wire is that the characteristics of the impedance will not change due to the fixed environment that the PCB creates.

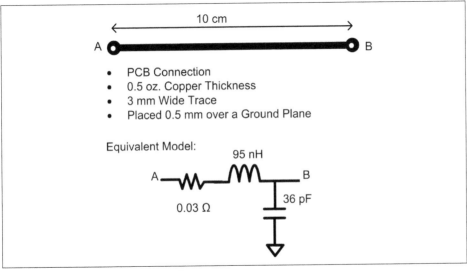

Figure 1-4. PCB connection impedance

The takeaway here is that every connection has impedance, and some coupling to an outside environment.

When connecting things together in Figure 1-5, the impedance of the source, connection, and load plays a part in how much signal loss occurs. Signal loss and distortion get progressively worse with both longer connections and higher frequencies. Designers working under 50 MHz and on a small (10 cm × 10 cm) PCB generally can ignore much of this and survive. High bandwidth, long distances, and connections off the PCB make interconnect issues a significant part of the design problem.

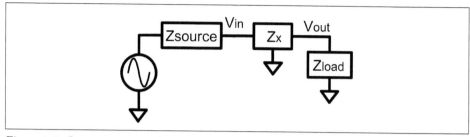

Figure 1-5. Connection impedance

In addition to loading and losses associated with connection impedance, high-frequency signals with a lengthy connection also exhibit transmission line character-istics (Figure 1-6). For the example of the 10 cm connection, current takes about 0.7 ns to transit the wire. Depending on the connection length and the frequency of the signals involved, improperly terminated transmission lines can affect signal integrity.

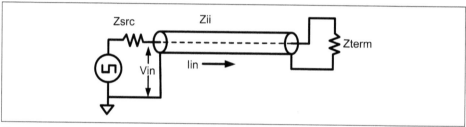

Figure 1-6. Transmission line

Consequently, designers need to consider transmission line issues when the signal wavelength becomes a significant part of the connection distance. For the 10 cm wire example, keeping under 0.1 wavelength on the wire would limit this at a 280 MHz sinusoid. A digital signal has multiple harmonics that need to be included, further limiting the connection's capability.

The termination impedance, Z_{term}, should match the characteristic input impedance, Z_{ii}, of the line to minimize reflections on the transmission line. Practical applications terminate at both ends of the line (Z_{src} and Z_{term}) to minimize both the initial reflection and any residual reflection that went back to the driven end. The use of impedance matching, striplines, and creating data paths with high signal integrity are topics covered in Chapters 3 and 11 (FDI: Digital, FDI: PCB).

As shown in Figure 1-7, even a solid copper sheet used as a PCB ground plane will exhibit resistance and inductance between locations. A current surge anywhere on a dedicated ground plane will cause that location to have a voltage transition relative to other locations on the ground plane. A surge current into the ground is analogous to a person bouncing on a trampoline. That ground bounce can be kept low by minimizing ground impedance and the magnitude of current surges into the ground.

It is important to recognize the concept that there is not a singular ideal ground, but rather that the voltage of ground can vary, with both the proximity between points and the characteristics of the current dynamics passing through the ground plane.

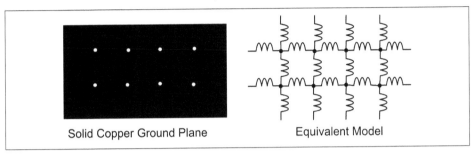

| Solid Copper Ground Plane | Equivalent Model |

Figure 1-7. Planar PCB connections

Figure 1-8 shows power and ground (P&G) connections exhibiting impedance. The typical electronic device presents switched loads between power and ground, causing the power voltage to drop and the ground to rise. The magnitude of P&G variance is influenced by the connection impedance and the transient rate at which the current changes (di/dt). The transient current can be changed through the load resistance variance and by how much capacitance is present across the load. Proper system design adjusts these variables to give P&G stability that is good enough to maintain proper functionality.

In most situations, P&G bounce can be kept manageable with proper circuit techniques, power bypass filtering, and PCB layout.

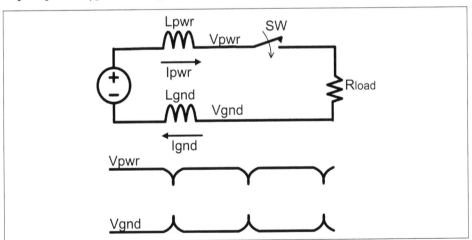

Figure 1-8. Power and ground impedance, ground bounce, and power instability

As shown in Figure 1-9, connections between circuit boards will have signal integrity problems on all connections. Current surges on P&G wires will create voltage variances between the separated boards. That variance makes noise on any ground-referenced signal, relative to the local ground. An external stimulus like EMI or electrostatic discharge (ESD) can further corrupt all signals.

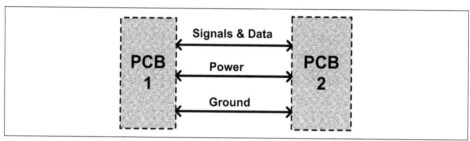

Figure 1-9. Impedance between circuit boards

Special efforts need to be made to "clean up" the noisy raw power coming into the board and create clean power on the PCB. Also, data and signals need to be passed between boards in a manner that is not dependent on local grounds or power supplies. Chapters 3, 4, and 6 deal with these issues (FDI: Power, FDI: EMC & ESD, FDI: Digital).

Distributing a common signal to multiple receivers can result in phase errors between the received signals. Figure 1-10 shows one signal going to five locations. Present in the connection are the distributed connection impedance, path length, and capacitive receiver load. All of these result in five different phases of the original signal. This is a common problem in clock tree distribution, especially with high clock frequencies, multiple destinations, and longer distances.

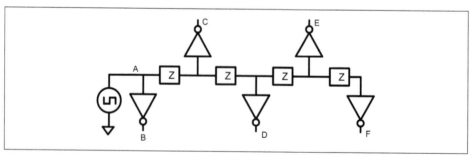

Figure 1-10. Phase errors in digital signals

Frequently, designers don't consider interconnect issues until problems arise. At that point, an expensive redesign is necessary. When high currents, long connection paths, or high frequencies are involved, they need to be carefully considered in the design.

Basic Components

The evolution of electronics includes a diverse collection of devices, many of which are now obsolete. The emphasis here is on modern systems that will be commercially manufactured in volume. Axial lead devices and other through-hole components are minimized in most high-volume manufactured devices. Consequently, surface mount circuit components are the focus here.

Capacitors

A distributed element model of a surface mount multilayer ceramic capacitor (SMT MLCC) is shown in Figure 1-11. This type of capacitor has multiple interleaved conducting and insulating layers. The actual capacitance is between adjacent plates (C), all plates have a small amount of resistance (R), and the interconnection has inductance (L). A distributed element model can be awkward to work with, and an equivalent model is generally sufficiently accurate.

The *equivalent model*, also known as a *lumped element model*, includes elements that model externally observed behavior. This device includes a single Equivalent Series Inductance or ESL (L), a single Equivalent Series Resistance or ESR, a single capacitor, and a leakage resistance (R_{leak}). The principal capacitor element can vary, primarily as a function of temperature and the voltage across it. The variability is due to the insulating dielectric characteristics used between the plates.

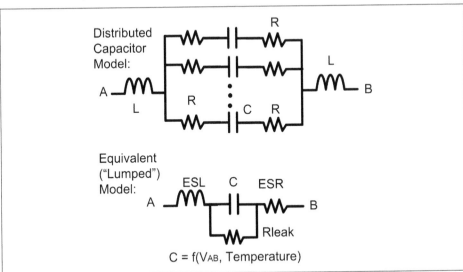

Figure 1-11. Surface mount multilayer ceramic capacitor

Depending upon the application, some elements can be ignored or, in other scenarios, can limit the device's performance. ESL is important for RF circuit performance and high-frequency power filters but does not affect low-frequency performance. ESL with C creates a self-resonant tuned circuit (Figure 1-12), which limits the high-frequency response of the capacitor.

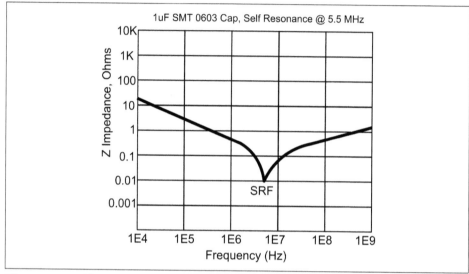

Figure 1-12. Self-resonance of surface mount multilayer ceramic capacitor

Leakage resistance becomes an issue if a capacitor is used to hold a static charge for an extended time. ESR becomes apparent in circuits using high-surge currents through the capacitor. The SMT MLCC makes up the majority of PCB capacitors due to excellent reliability, low cost, and the wide selection of options available.

Modern capacitors make trade-offs in:

- Capacitance per unit volume
- Maximum applied voltage (breakdown voltage)
- Min/max operating temperatures
- Temperature variation from nominal value
- Applied voltage variation from nominal value
- Aging variation from nominal value
- Nominal value accuracy
- Total overall life

These parameters interact with each other. If the manufacturer changes one parameter, the change can affect another parameter. For example, creating a higher breakdown voltage implies a larger body size for the same value of capacitance.

SMT MLCCs have different dielectrics, which leads to different performance parameters. Looking at the more common devices, some brief comments are useful to better understand the differences.

C0G and *NP0* are known as Class 1 capacitors, which are designed for minimal thermal variance and minimal change from voltage bias. Class 1 devices sacrifice capacitance per unit volume to achieve higher accuracy and stability. For a given body size, the total capacitance is limited. These devices are useful for tuned circuits and other applications where accuracy and stability are needed.

X5R, X7R, Y5V, and others are known as Class 2 capacitors (Table 1-1). Their design sacrifices voltage bias accuracy and thermal stability to achieve more capacitance within a given volume. The notation used for these devices is a little confusing but actually simple to decode. The three characters decode to a minimum temperature, maximum temperature, and value variance over temperature. An X7R capacitor is designed for use from $-55°C$ to $125°C$ and will have $+/-15\%$ variance over temperature. The Y5V capacitor is designed for use from $-30°C$ to $85°C$ and will have a value ranging from $+22\%$ to -82%. Picking the proper capacitor dielectric can significantly affect accuracy.

Table 1-1. EIA RS-198 code for Class 2 capacitors

Low temperature (°C)	High temperature (°C)	Temperature variance
X = −55	4 = 65	P = +/−10%
Y = −30	5 = 85	R = +/−15%
Z = 10	6 = 105	S = +/−22%
	7 = 125	T = +22%, −33%
	8 = 150	U = +22%, −56%
	9 = 200	V = +22%, −82%

Class 2 devices also exhibit *DC bias effect*, also known as the *DC voltage characteristic*, which consists of changes to the capacitance value as a function of the static DC voltage across the device. Generally, the value of the capacitance decreases as the bias voltage increases. This can be significant, with as much as 60% of the capacitance value changing depending on the specifics of the device.

Devices with a higher breakdown voltage tend to have less DC bias effect for the same voltage change. This can be useful if a designer needs to reduce this effect. If a specific capacitor needs high accuracy, a Class 1 capacitor may be required.

Entire books have been dedicated to capacitors. A closer look at the limitations of the SMT MLCC is warranted because it is the predominant capacitor used in modern designs.

Several other capacitors are frequently used in modern designs. The aluminum electrolytic capacitor (AEC) is widely used in DC power supply filters and other applications where large amounts of capacitance in a small package, coupled with low cost, are needed. All AECs suffer from poor high-frequency response due to high ESL, so AECs are not suitable for RF situations. High-frequency response of the AEC can be supplemented by parallel SMT MLCC devices if needed. The AEC also has a limited life, fussy temperature restrictions, and significant ESR. Check device specifications for aging information and lifetime versus temperature data. Most AEC devices have performance and lifetime parameters that are unique to a specific manufacturer and product line.

Both tantalum and aluminum-polymer capacitors are available in high-reliability and long-life variants. Component selection here needs to be done on a case-by-case basis because these devices also have short life variants as well.

Following are important things to look for in a capacitor:

- Nominal component value and fabrication tolerance
- Breakdown voltage
- Package size
- Dielectric type, temperature range, and variance over temperature
- DC bias effect and device variation due to bias voltage
- ESL characteristics and self-resonant frequency (SRF) for high-frequency applications
- ESR characteristics where series resistance can affect performance
- Reliability, lifetime, and aging data

Resistors

A typical resistor model is shown in Figure 1-13. In addition to the ideal resistance (R), some additional elements need consideration. These are ESL, Equivalent Parallel Capacitance (EPC), and internally generated noise (V_n) created by the resistance itself.

Ideally, a resistance has flat impedance over frequency. Because of EPC and ESL, the impedance may not be flat at high frequencies. For DC bias applications, the ESL and EPC can be ignored. Multiwatt resistors are commonly fabricated with wire-wound methods and have significant ESL and EPC components.

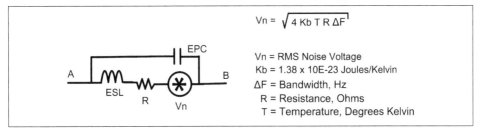

Figure 1-13. High-frequency resistor model

Most modern surface mount resistors are some type of film resistor: thick film, thin film, and metal film are common descriptions of devices available. Multiple vendors have found many different approaches to making resistors. Generally, metal film resistors exhibit lower noise than carbon and thick film devices, but check specific vendors for noise data.

Thermal noise, also known as *Johnson-Nyquist noise*, is created by the thermal agitation of electrons in a resistive material. For most large-signal applications, this noise is not a significant issue. It becomes important in RF frontends, communication channels, and low-amplitude scenarios, where signals are in the nanovolts.

Gigaohm resistors are available from manufacturers, but special considerations have to be made to use them. Dust and humidity (Figure 1-14) in the environment can have a lower impedance path than the resistor. As a general rule, anything over 100 K ohms needs special environmental attention to avoid alternative current paths around the resistor.

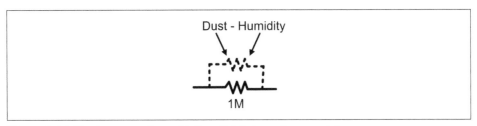

Figure 1-14. Environmental sensitivity of high-impedance resistors

Thankfully, most modern designs don't need high-value resistors, because the analog circuits that used them in the past have been replaced with more reliable digital methods.

The following are important things to look for in a resistor:

- Nominal component value and fabrication tolerance
- Material composition; film resistors are preferred, older carbon-based resistors should be avoided
- Power rating
- Min–max temperature range
- Maximum voltage rating for high-voltage applications
- Thermal variance, sometimes specified as temperature coefficient of resistance
- ESL for high-frequency situations
- Thermal noise, carbon contact noise
- Stability, repeatability, and aging data

Inductors

A high-frequency model of an inductor is shown in Figure 1-15. Some EPC exists, which limits the useful frequency range. With EPC and L in parallel, the device will resonate as a tank circuit and look like an open circuit at the SRF. Below the SRF, the device functions as an inductor; above the SRF, the device performs like a capacitor.

The ESR defines the quality factor (Q) of the inductor when used as part of a filter. When used in a switching power converter, the ESR will limit power efficiency.

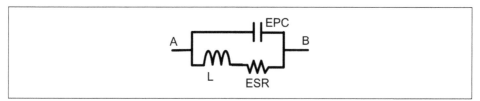

Figure 1-15. High-frequency inductor model

Inductors are commonly used in power conversion, RF signal processing, and EMI limiting. Inductors are used in switched-mode power conversion, in both AC/DC and DC/DC converters. Inductor priorities for power conversion are maximum current, saturation current, thermal range, and resistive losses. Switching converters have high current transients, and suitable devices are wire-wound on a magnetic core, with high current capability and low resistive losses (FDI: Power).

Inductors in RF signal processing are used in tuned circuits and need consistent accuracy for reproducible frequency response. Priorities are a high quality factor, component tolerance/accuracy, and ensuring that the SRF does not affect the signals being processed. Lower currents are used in signal processing inductors, and many components exist in surface mount variants.

Inductors used for EMI limiting are commonly known as *chokes*. Their purpose is to pass DC current while reducing transient currents in the connection. High-current chokes are commonly implemented with a wire-wound power inductor. Low-current chokes can be implemented with smaller surface mount inductors (FDI: EMC & ESD).

Following are important things to consider in an inductor:

- Whether the component value and tolerance will meet the design's required accuracy
- SRF and whether it affects the signals of interest
- Current rating, peak and average
- Temperature range
- Core material and saturation current
- DC resistance and device Q
- Whether it is shielded or unshielded

Voltage Sources and Batteries

The ideal voltage source doesn't exist. An infinite source of current at a fixed voltage is an academic concept. Practical implementations of a voltage source can be the output of a voltage regulator, a battery, or another generation source. Invariably, these all have limitations.

Any detailed voltage source model (Figure 1-16) includes source resistance (R_{src}), a noise component (V_{noise}), and the voltage source being a dependent function of the output current (I_{load}). The goal of a voltage source design is to minimize the R_{src}, the V_{noise}, and the dependency on I_{load}.

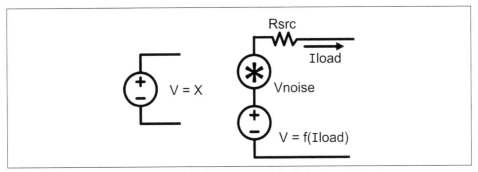

Figure 1-16. Voltage sources, ideal (left) versus real (right)

Batteries have a limited capacity, so voltage variation over discharge has to be added to the model. Also, batteries have current limitations around charging and discharging, reduced performance with age, and a myriad of special considerations depending on the particular battery type. Chapter 5 is devoted to the design, care, and feeding of battery supplies (FDI: Battery).

The characteristics of the V_{noise} signal will be highly dependent upon the methods used to create the voltage source. Linear power supplies and regulators that have been carefully optimized for noise are available when low-noise power is needed. Batteries are very low noise, but they are not perfect, and they can generate electrochemical voltage noise dependent on charge state, or the battery charger can superimpose noise onto the power.

Switching power supplies will achieve high efficiency while generating commutation noise associated with the voltage regulation function. Digital electronics don't need ultra-low-noise power, but any chip designed into a system needs to be investigated for its "noise on the power" limitations.

Current Sources

Similar to ideal voltage source, an ideal current source doesn't exist. A number of circuit designs come close but have restrictions and limitations on impedance and voltage range.

A practical current source (Figure 1-17) includes some source resistance (R_{src}), a noise component (I_{noise}), and an output current that is a dependent function of the output voltage (V_{out}).

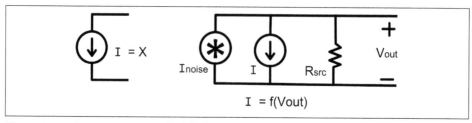

Figure 1-17. Current sources, ideal (left) versus real (right)

Most current sources are limited in voltage range to maintain appropriate, fixed-current behavior. Discrete current sources are not commonly used in board-level system on a chip (SoC) design, but they are an important building block within analog and mixed-signal integrated circuit (IC) design. Readers interested in this topic can research "Cascode current sources" for further information.

Switches and Relays

Mechanical switches and relays can be problematic, especially if they are used as control inputs on a logic port. Both exhibit similar problems due to their mechanical implementations.

For instance, mechanical switches and relays exhibit contact bounce (Figure 1-18), where the process of opening and shutting contacts creates multiple brief open/shut states. If they are used as digital logic inputs, opening and closing contacts are seen as an erratic string of data states due to imprecise mechanical contacts. Using a switch on a logic port requires software polling the port multiple times and determining state change after remaining stable for 50–100 ms.

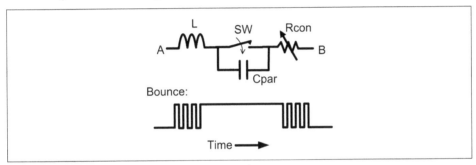

Figure 1-18. Switches and relays with contact bounce

In addition to contact bounce, switches have inductance (L) that adds impedance to a closed switch. Parasitic capacitance (C_{par}) allows high-frequency signals to pass through an open switch. Also, contacts have highly variable resistance (R_{con}) and can change with every switch cycle. This variability gets worse with age as contacts become dirty or damaged.

Operational Amplifiers

The ideal operational amplifier (op-amp) claims to have infinite gain, infinite bandwidth, infinite voltage range, infinite current output, zero input current, zero noise, zero offset voltage, and, best of all, no need for a power supply. As a math model, it's an interesting idea, but reality is something else.

The real op-amp (Figure 1-19) has limitations that affect performance. First, the device requires power and ground (V_{power}, V_{gnd}), and the output (V_{out}) can be sensitive to noise present (NP, NG) on these connections. The output has impedance (R_{out}), and the output voltage (V_{out}) is restricted by the power supply voltage. The gain is not infinite; rather, an open loop gain of 80 dB is typical.

The gain response over frequency will have both bandwidth limitations and an additive phase response. Also, the output response speed is limited by a maximum slew rate at which the device can respond. Since an op-amp is designed to function within a closed feedback system, the high-frequency gain of the device must be internally limited to keep it stable within a feedback configuration. Due to this frequency compensation, op-amps may not be the best devices to use in high-frequency designs.

Input capacitance (C_{inp}) creates loading that can affect high-frequency performance. Op-amps designed with bipolar transistors will have input bias current (I_{bias}). The input transistors used within the op-amp will not be perfectly matched and will create an equivalent input offset voltage (V_{off}) that is typically 1–10 mV depending on the specific device. That offset voltage becomes a problem when dealing with small signals or high-gain closed loop configurations. The internal circuitry will create noise, which is modeled as an input referred noise (IRN) source.

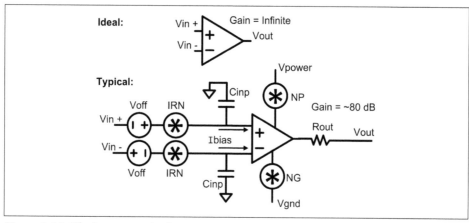

Figure 1-19. Ideal versus typical op-amp

Many of these limitations are not a problem when working with larger signals, but pushing for high gain and high bandwidth or using sub-mV signals will show performance limitations. Performance specifics will depend on the op-amp selected, and many offerings are commercially available.

Voltage Comparators

A voltage comparator has many things in common with the op-amp model. The comparator (Figure 1-20) also has sensitivity to power and ground noise (NP, NG), input loading capacitance (C_{inp}), IRN, and input offset (V_{off}), similar to an op-amp. Since the typical application is against a fixed reference voltage (V_{in} −), a simple one-sided model will suffice.

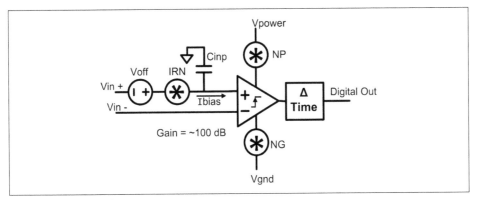

Figure 1-20. Typical comparator

The difference between an op-amp and a comparator centers on the fact that an op-amp is designed to function within an analog feedback loop and a comparator is not. Comparator gain is often higher (~100 dB), no internal frequency compensation is used, and the output is digital.

A variable time response is part of a comparator model since input/output time can be dependent on the input signal amplitude. Small signals with a minimal crossover voltage are slower to propagate through the circuit than a larger voltage transition.

Nonideal Digital Devices

Digital devices are generally more resilient to noise than analog signals. However, even digital systems have nonideal problems and limitations.

Figure 1-21 shows communication between two digital devices. In modern designs, discrete gates are uncommon, but this illustrates communication between any two digital devices, such as field-programmable gate arrays (FPGAs), microcontrollers, or other devices that share a common P&G.

As illustrated, the P&G between devices includes connections with distributed inductance. Depending on what else is connected to the P&G network, currents drawn from or injected into the network can cause the connection to reactively dip or rise. As well, the magnitude of that "bounce" can be different at various locations on the network.

In Figure 1-21, the P&G network at one location (V_{PWR1}, V_{GND1}) has different characteristics than at another location (V_{PWR2}, V_{GND2}). How much of this bounce is tolerable depends on the transistor technology used in the digital gates, but keeping the P&G stable has to be a priority.

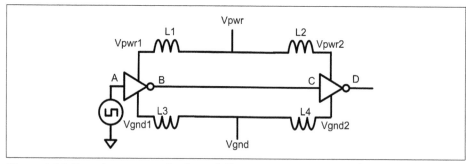

Figure 1-21. Impedance-created voltage differences across power and ground

Generally, CMOS semiconductors are designed with transistor threshold voltages that are about one-third of the power supply voltage and can tolerate P&G instability of 20% of the nominal power before digital devices start to generate false states. Any commercial product will specify acceptable voltage ranges for both high and low logic states. That specification is a quick guideline to how much P&G noise is tolerable.

As shown in Figure 1-22, a clean two-state signal at A can be corrupted by unstable P&G (V_{PWR1}, V_{GND1}) at its output (B), which is then further disturbed by the variation in P&G (V_{PWR2}, V_{GND2}) at the receiving end (C), leaving an output (D) that no longer represents the original signal.

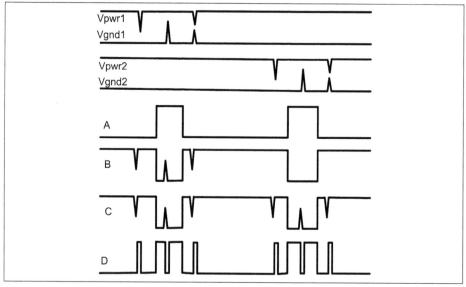

Figure 1-22. Digital sensitivity to power and ground noise

False state generation is just one of many problems seen in digital systems. Consider the clocked system shown in Figure 1-23, with two devices communicating at the same frequency but using uncorrelated clocks. Although the clocks are the same frequency, there is no fixed-phase relationship between devices. As phase changes, the data–clock relationship periodically violates setup and hold times, and data errors occur. Chapter 3 (FDI: Digital) covers techniques to deal with reliable unsynchronized data transfer.

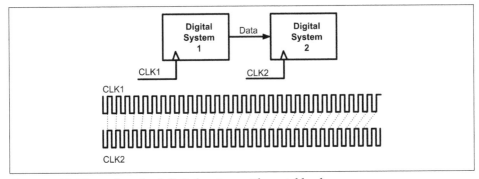

Figure 1-23. Unsynchronized digital system with variable phase

Asynchronous logic should be avoided due to unpredictable outputs and digital glitches. Although Figure 1-24 is a bit contrived, the Boolean analysis says the output (C) should never assert. Reality shows that the propagation delay of the inverter affects output. This illustrates the motivation for synchronous logic where data resynchronization to a clock is used to avoid glitches.

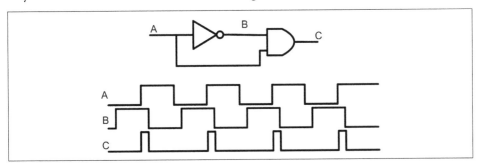

Figure 1-24. Race conditions in asynchronous logic

In addition to problems created by timing relationships, digital functionality can suffer due to the analog characteristics of digital signals. Several factors can contribute to this: higher data rates, resistance of the transmission path, or a transmission path that is loaded with capacitance.

A simple driver and receiver pair is shown in Figure 1-25. The middle diagram in the figure shows the interconnect impedance with resistance of the PCB (R_{pcb}), inductance (L_{pcb}), and capacitance (C_{pcb}). The bottom diagram in the figure includes an equivalent model for the output driver and receiver. For CMOS logic, the output driver can be modeled as a pair of switches and a resistor. R_{out}, depending on the transistors used, will typically be 10 Ω to 80 Ω. In addition, the receiver input behaves as a capacitance for most situations. This input capacitance is due to large, high-current ESD protection circuits, and 2 pF to 10 pF is common for each logic gate. Ignoring interconnect inductance, this model shows that an RC low-pass filter is present in every digital interconnect.

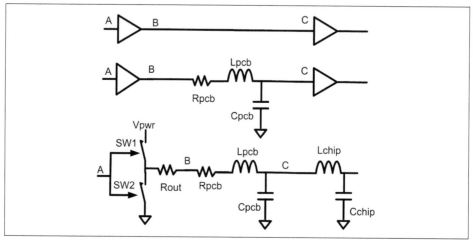

Figure 1-25. RC loading in digital logic interconnects

With this more detailed model, the frequency limitations of logic are apparent. Figure 1-26 illustrates the limitations created by the RC circuits due to the output driver resistance and attached capacitance. On the left, at lower frequencies, enough time is available for the transient behavior to settle out before the next transition. On the right, as time is reduced between transitions, the device has insufficient time to settle out. The RC time constant of the system remains the same, but no well-defined logic states are achieved.

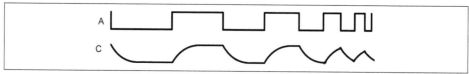

Figure 1-26. RC loading and signal degeneration

The system fails when V_{high} or V_{low} limits are violated, or when setup and hold times on flip-flops are not met. Ground referenced logic frequently fails when used for high-frequency communication. Differential signals are commonly used at high data rates to minimize this limitation (FDI: Digital).

Signal Integrity

Undesired effects can be created by a circuit in addition to its normal functionality. Many electronic systems create unwanted electronic noise by creating signals that radiate externally. Also, one part of a system can create noise that interferes with the proper functionality of a different part of the same system. In some cases, a poorly designed system can be sensitive to externally created electromagnetic interference. Signal integrity, radio frequency interference (RFI), EMI, electromagnetic compatibility (EMC), electronic noise, RF immunity, electromagnetic shielding, radiated emissions, crosstalk, and others are all related topics in this area.

Using the 10 cm connection discussed earlier, Figure 1-27 creates radiated emissions with an ideal quarter-wavelength antenna at about 700 MHz. For this example, a 100 MHz clock would have its seventh harmonic have an optimal antenna, making the connection a radio transmitter. All frequencies would radiate from that connection, albeit with varying amounts of efficiency.

Radiated emissions are especially problematic from computers and from any device with large amounts of high-frequency digital circuitry. The Federal Communications Commission (FCC) and other worldwide regulatory bodies restrict the EMI that any device is allowed to produce to minimize interactive problems between electronic devices (FDI: EMC & ESD).

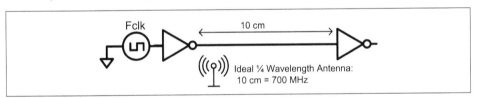

Figure 1-27. Radiated emissions

As illustrated in Figure 1-28, as well as radiating EMI to the outside world, devices can be internally sensitive to unintended communication between circuits, commonly called *crosstalk*. This noise coupling can be capacitive or magnetic and often takes both paths. This illustrates a common problem with digital signals, especially high-frequency signals, as these can be coupled to an unconnected circuit path (FDI: Digital, FDI: EMC & ESD, FDI: PCB).

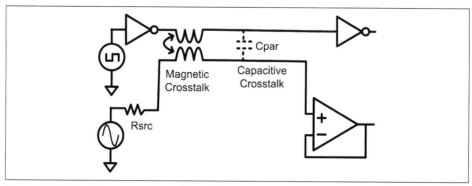

Figure 1-28. Sensitivity to external signals

Summary and Conclusions

The important points of this chapter are as follows:

- Academic simplifications often omit important details for the sake of clarity.
- All connections have impedance, which can cause voltage variance across P&G.
- Phase errors between receivers can exist even when driven by a common signal.
- Capacitors have ESL, ESR, self-resonance, leakage, and voltage variance issues.
- Resistors have issues with ESL, EPC, and thermally generated noise.
- High-impedance resistors are susceptible to error due to environmental issues.
- Inductors have ESR, EPC, and self-resonance issues.
- Voltage and current sources can have both impedance and noise components.
- Switches have contact bounce, inductance, and variable contact resistance.
- Op-amps have limitations including input/output voltage range, bandwidth, offset, output current, additive noise, and others.
- Comparators have limitations in voltage range, input offset, response time, and others.
- Power–ground instability can corrupt logic.
- Asynchronous logic can create unpredictable behavior.
- Ground-based logic can be limited by RC loading that degenerates signal rise time and amplitude.
- Digital devices often create radiated emissions.
- Signals in one circuit can couple to another circuit without a direct connection.
- Nothing is ideal.

The intent of this chapter is to build awareness of the limitations and nonideal nature of electronics. Many engineers build awareness due to device failures from many of the issues described here. Learning from experience is invaluable, but failure analysis is not the most efficient use of time or effort.

Moving forward, the techniques presented in this book address these limitations while developing design solutions proven to work in real-world applications.

Further Reading

- "Capacitor Guide – Dielectric Materials," EETech Media, LLC, *www.capacitor-guide.com/dielectric-materials*.

- "Understanding Ceramic Capacitors: types – MLCC, C0G, X7R, Y5V, NP0, etc.," Electronics Notes, *https://www.electronics-notes.com/articles/electronic_components/capacitors/ceramic-dielectric-types-c0g-x7r-z5u-y5v.php*.

- "Here's What Makes MLCC Dielectrics Different," Kemet Corporation, *https://ec.kemet.com/blog/mlcc-dielectric-differences*.

- "Types of Capacitors: An Essential Overview," Electronics Notes, *https://www.electronics-notes.com/articles/electronic_components/capacitors/capacitor-types.php*.

- "High-Reliability Capacitors: When the Mission Just Can't Fail," Kemet Corporation, *https://www.aerodefensetech.com/component/content/article/adt/features/articles/27962*.

- C.K. Boggs, A.D. Doak, and F.L. Walls, "Measurement of Voltage Noise in Chemical Batteries," *Proceedings of the 1995 IEEE International Frequency Control Symposium* (49th Annual Symposium), May 31–June 2, 1995.

- *Analysis and Design of Analog Integrated Circuits, 5th Edition*, by Paul R. Gray, Paul J. Hurst, Stephen H. Lewis, and Robert G. Meyer, 2009, ISBN 978-0-470-24599-6, John Wiley & Sons.

CHAPTER 2

Architecting the System

This chapter defines the key parts of a system at the box level, discusses how to implement digital control at the core, and takes a detailed look at the criteria for a digital control unit (DCU). A procedure for selecting a microcontroller (MCU) is outlined as well. Later chapters detail the internal circuitry of those other boxes.

Topics covered here include:

- What a "mostly digital" system is
- Where DCUs are used
- Where multiple DCU devices are needed
- Differences between DCUs, MCUs, and microprocessors (MPUs)
- Alternative DCU methods, including field-programmable gate arrays (FPGAs), complex programmable logic devices (CPLDs), and application-specific integrated circuits (ASICs)
- Specialty methods for digital signal processing (DSP) and streaming data
- How to avoid data port bottlenecks

This chapter touches upon many topics that later chapters investigate in detail. Since these topics affect the architecture discussion, they need to be briefly defined for the reader to understand certain architecture strategies. We will start with some preliminary ideas and methods that need to be applied to all designs.

Preliminary Ideas

Defining a successful approach to a commercial product is the emphasis here. Certain common ideas need to be clarified at the start.

Simulate or Build

Generally there is no need to create a full simulation of an entire embedded controller system. The digital controller code will be simulated within the integrated design environment (IDE) many times to get the control signals functioning properly, but trying to simulate the peripheral functions created by those control signals probably isn't necessary.

If the system is composed of system on a chip (SoC) devices, simulating the peripheral circuits is generally not required to create a first build. That work will have been done by the integrated circuit (IC) designers of the SoC devices. Priorities and methods are very different in IC versus printed circuit board (PCB) design. For ICs, it makes sense to pursue accurate design simulations. IC fabrication is costly in both time and money (FDI: Twomey, 2014).

Building out a first-generation PCB, however, is low cost and quickly accomplished. A real product can rapidly get into the hands of designers for debugging, so simulating the device is unnecessary. For an MCU with a simple set of peripherals, another option is a development/demo board. These are available from most MCU manufacturers for minimal cost, and they allow designers to go "hands-on" quickly.

Through-Hole/Leaded Components (Obsolete)

Circuit components with pins or wires that go through holes in the PCB are suitable for use by hand assemblers and hobbyists. However, these *through-hole devices*, also called *leaded components*, are not suitable for automated assembly or high-frequency circuits. Also, leaded components tend to be much larger than their surface mount equivalents.

Unless there is no other option, the use of leaded components is strongly discouraged. Any commercial design should use surface mount components for lower assembly cost, better high-frequency performance, and smaller physical implementation.

Discrete Gate Logic (Obsolete)

The 7400 logic series, typically with four to six logic gates per chip (but also available in many variants), was developed in the mid-1960s and is still manufactured and widely sold. However, this chapter focuses on modern methods using programmed devices and code-driven logic functions. Singular digital devices should be seen only as level shifters and buffer drivers.

Modern Design Strategies

Modern design strategies need to eliminate some older and outdated practices. First, as a general rule, analog circuits and signal processing should be avoided or minimized wherever possible. Figure 2-1 illustrates why this is a good idea.

As shown in Figure 2-1, implementing an analog band-pass filter using operational amplifiers (op-amps) can have many issues. The capacitors need to be of high quality and have good absolute accuracy and linearity. The accuracy of both the resistors and capacitors will affect the accuracy of the frequency response. Nonideal op-amps also contribute noise, distortion, and DC signal offset.

In this example, the op-amp band-pass filter requires four high-accuracy plastic film capacitors, which are expensive. Using an analog-to-digital converter (ADC) followed by a DSP-defined filter is a low-cost and very repeatable solution. Frequently, an ADC built into the MCU and a software (SW) defined filter will get it done. Independent ADCs are also available with a wide range of performance criteria and low unit cost.

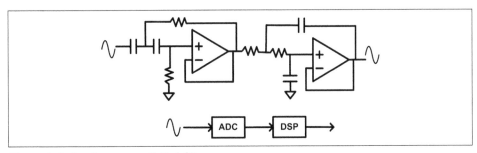

Figure 2-1. Analog versus digital filters

There will always be analog signals to deal with, however. The "analog world" is readily interfaced with through a multitude of sensors. Getting those signals reliably processed is important.

As a general rule, quickly transforming any analog signal into a digital equivalent is a good strategy (Figure 2-2). Analog signals have issues with noise, distortion, and other signal integrity concerns. Digital data streams are less problematic to connect over distance and can remain functional in a hostile electromagnetic interference (EMI) environment.

In Figure 2-2, Case A illustrates a sensor with a signal output that traverses a long interconnect before encountering a gain amplifier. The signal amplitude output of the sensor, depending on the device, could be anything from microvolts to volts. With small amplitude signals, the sensitivity to external EMI can be high.

Placing an amplifier in close proximity to the sensor, as in Case B, improves this situation. Ideally, to eliminate long-path analog signals, Case C moves the ADC close to the sensor as well.

More and more sensor devices are being developed where the sensor, amplifier, and ADC are implemented in a single package with a serial port digital interface.

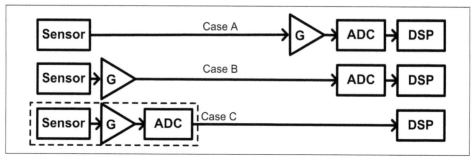

Figure 2-2. Quantizing signals near the source

Control systems is another area where analog methods are outdated. A classic analog control feedback loop is shown in Figure 2-3. A device under control (DUC) is controlled through a device driver and a linear system of forward transfer function, H(S); a sense system; a feedback transfer function, G(S); and an error differencing amplifier. Although this type of control feedback system exists internally within many analog and mixed-signal ICs, it is rarely used anymore at the system level in modern designs.

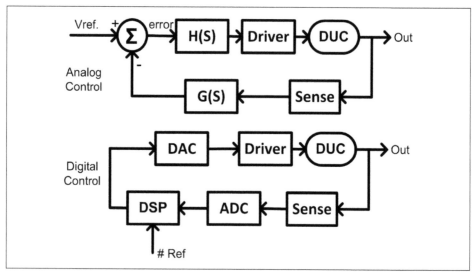

Figure 2-3. Control systems: classic analog versus digital methods

Modern control feedback systems are mixed signal in nature, with some analog characteristics within the driver, DUC, and sense blocks. However, the DSP functionality can take many forms to achieve optimum control.

Communication and control between PCBs (Figure 2-4) in a multiboard system needs to be structured for EMI immunity.

Controls and signals between PCBs share ground, power, and data. The data communication should use a differential digital system with error-checking capability. This approach avoids most noise problems and signal integrity issues. Also, this method tolerates variation between the grounds of each PCB.

Also, power on each PCB should be locally regulated, thus allowing a large amount of variance in the raw DC power before on-board functionality is degraded. Remember that all connections between PCBs have impedance and will cause voltages to vary (FDI: Power, FDI: PCB).

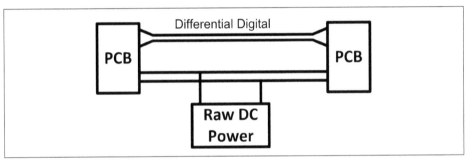

Figure 2-4. Communication and power between PCBs

Next, we will discuss what a mostly digital approach looks like and what its limitations are.

Mostly Digital Design

The mostly digital system is a simple concept:

- Convert any analog signal (control voltage, sounds, video information, etc.) to a digital-equivalent data stream.
- Process that data stream using digital methods only.
- Convert the data stream to an analog format only if needed as an output.

Figure 2-5 illustrates this idea. The industry frequently refers to this as a *mixed-signal* approach. Any form of signal processing that can be defined as a math function or some type of input/output relationship can be created in a DSP system. Following are some examples:

- Mixer (multiply), divide, sum, difference, integral, derivative
- Nonlinear transfer
- Closed loop control—proportional-integral-derivative (PID) controller
- Closed loop control—fitted trajectory
- Encoder or decoder
- Amplitude qualification—threshold or comparator
- Modulator or demodulator
- Filters—FIR or IIR (LPF, HPF, BPF, and others)
- Averaging and smoothing (AVG)
- Time domain to frequency domain (Fourier) transform
- Viterbi detection

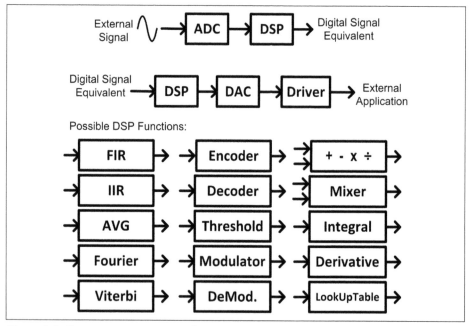

Figure 2-5. Common structures used in a DSP strategy

If you can accurately quantize a signal, the DSP possibilities are endless. However, ADC performance can limit this method in some cases.

DSP Methods: Versatility and Limits

Improving signal-to-noise performance requires increasing the resolution of the ADC, ergo an ADC with a higher number of bits. To accurately sample a signal, you need to convert at a rate faster than twice the signal bandwidth (FDI: ADC & DAC).

The internal circuitry of an ADC has performance limitations. Figure 2-6 illustrates that power consumption, the number of bits, and the conversion rate interact in an ADC design. Increasing the conversion rate or increasing the number of bits increases power consumption. Designing a "fast" converter with a "large" bit count and "low" power consumption can be challenging, but a steady improvement on "fast – large – low" performance is still ongoing. What was difficult to achieve a few years back is now readily available.

As of 2023, ADCs are available with resolutions from 4 to 32 bits and sampling rates from 4 samples/sec to 6.4 G samples/sec. Many of the very fastest devices are expensive, power hungry, and not suitable for use in consumer applications. However, many cost-effective options are available at under 500 M samples/sec (FDI: ADC & DAC).

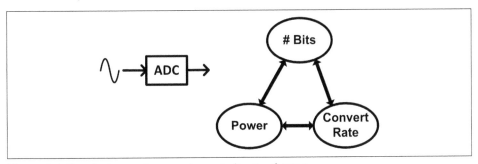

Figure 2-6. Resolution and conversion rate limits of ADC

A simplified case study (Figure 2-7) of a radio system illustrates why analog methods are still needed in some situations. The "Software Defined Radio" is a simple idea: connect an ADC to an antenna and perform all the signal processing in a DSP block. Case 1 in Figure 2-7 illustrates that idea.

Although Case 1 looks easy on paper, the specifics of implementation are a different story. The dynamic range of a typical radio receiver is 90 dB. To support that dynamic range would require 16 bits or more of resolution. State-of-the-art ADCs with 16 bits of resolution top out at 500 M samples/sec, so this approach may be usable up to a 250 MHz carrier frequency. Most cell phones operate from 900 MHz to 2 GHz, so it's not going to work for cellular.

Case 2 introduces a low-noise amplifier (LNA) between the antenna and ADC. Adding 20 dB of gain drops the needed resolution to 12 bits. As of 2023, there are options available up to 6 G samples/sec, but they are expensive, very power hungry, and more suitable to rack-mounted instruments than handheld cellular devices. Case 2 could be implemented, however, for RF signals up to 3 GHz, with a power cord connected and a high price to implement.

Case 3 and Case 4 introduce a mixer, which converts the RF signal down to a low-frequency baseband. The high frequency of the RF carrier is no longer an issue, so only the bandwidth of the signal modulation needs to be converted. For these situations, the signal of interest is under 100 MHz, and frequently is much lower than that.

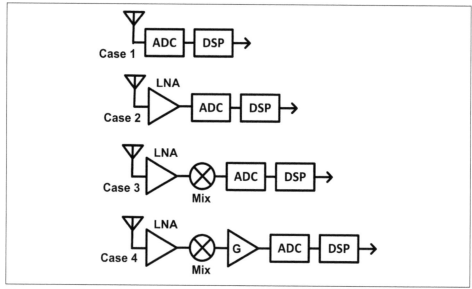

Figure 2-7. Software-defined radio, with antenna connected to ADC

At lower conversion rates, a multitude of converters are available with battery-friendly power levels and suitable bit resolutions. Most cell phones now use methods similar to Case 3 and Case 4. This is a simplified example, but it illustrates the need for analog methods in some situations.

Audio systems (Figure 2-8) are amenable to the mostly digital approach. Modern audio, on the inside, is largely digital circuitry. Many microphone signals are fed directly to an ADC and turned into a serial data stream. High-quality recording microphones are now available with digital USB outputs that connect directly to a computer.

Some recording studios use analog differential signal microphones and analog differential connectors and cables, largely due to legacy infrastructure. Modern mixer boards immediately convert the analog microphone input and digitally process the audio data stream.

As shown in Figure 2-8, audio output can be kept digital up to the speaker's power amplifier that uses a digital input. The encoder driving the amplifier converts the binary data to a pattern suitable for driving a switched amplifier.

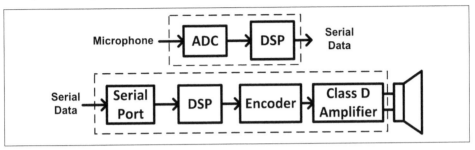

Figure 2-8. Mostly digital audio

Sensors that interface to the analog world come in many forms. Figure 2-9 shows that a sensor output may require amplification to use the full range of the ADC. Also, the characteristics of the sensor output may need to be filtered by analog methods before connecting to the ADC to avoid aliasing (FDI: Sense, FDI: ADC & DAC).

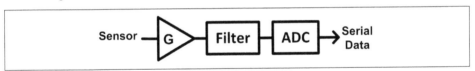

Figure 2-9. Analog signal processing: signals and sensors

Digital Control Methods: DCU, MCU, MPU, FPGA, CPLD, and ASIC

An *embedded system* or *embedded controller system* can be loosely defined as an electronic or electromechanical system that uses a digital controller to control and sense things in the system. The DCU is a generic concept in that it can be an MCU or an MPU. Alternatives include more specialized digital methods like FPGA, CPLD, or digital ASIC.

Digital control can be either hardware based, as in a Boolean state machine, or software based, as in code written for a computer. Compiled code that was written for and downloaded into an MCU is known as *firmware*.

An embedded system is single purpose, unlike a multipurpose computer that has the capability to change programming. Figure 2-10 shows a motor driver controller PCB using an MCU to control three stepper motors. The MCU issues step and direction commands to the motors and has a sensor for the home position. Also, the MCU communicates to a remote manager over a serial port bus to receive commands and send status information.

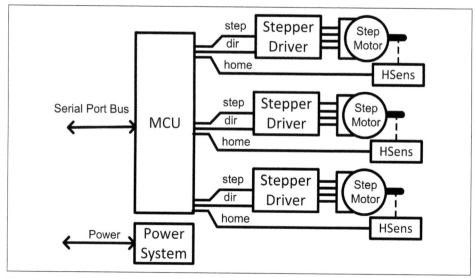

Figure 2-10. *Typical example of an embedded controller system*

The structure in Figure 2-10 is typical of that used in a robot for motion control where dedicated MCUs control groups of motors.

Terminology in MCU and MPU Specifications

As of 2023, there are over 40 semiconductor manufacturers of microcontroller chips, and there are thousands of MCU and MPU variants available. While sorting through specifications, the specialized jargon can be confusing. A brief terminology primer is a good start:

Harvard architecture
    A Harvard architecture CPU uses separate data-address buses for instruction memory and data memory. Most MCU devices are Harvard architecture.

Princeton architecture (aka von Neuman architecture)
    A Princeton architecture CPU uses common address-data buses for both instruction memory and data memory.

RISC (Reduced Instruction Set Computer)
    Typical RISC CPUs limit the instruction set to reduce the complexity of the physical design. This also keeps the execution time of an instruction within a

single clock cycle. Instructions and resulting data are often pipelined in a RISC device. Most MCU devices are RISC devices.

CISC (Complex Instruction Set Computer)

A CISC processor executes more complex instructions than a RISC device. The compiler for a RISC device breaks those complex instructions into simpler instructions suitable to the RISC instruction set. The internal electronics of a CISC machine need to be more complex. Most present-day processors use RISC architecture, except for Intel's X86 devices.

X86 processor

The X86 architecture includes the Intel 8086, 80186, 80286, 80386, 80486, and others. The X86 series is a CISC architecture that has been the foundation of Intel-Microsoft personal computers. Generally, X86 devices are not used in MCUs within embedded systems.

ARM (Advanced RISC Machine or Acorn RISC Machine)

ARM is a series of RISC processors with 32-bit data and 26- or 32-bit addressing. ARM Holdings licenses ARM processors as intellectual property to many semiconductor manufacturers. Chip manufacturers add peripheral support devices, creating a finished MCU.

MIPS (Microprocessor without Interlocked Pipeline Stages)

A MIPS processor is a RISC processor that uses a pipelined architecture. MIPS devices are optimized for streaming data, and they are available from multiple vendors as a standalone MPU or are implemented by semiconductor manufacturers within their chips. MIPS cores are popular in general data processing and in applications that process data streams.

8051 and 6502

The 8051 is an early-generation, 8-bit data, 16-bit address MCU from the early 1980s. The device is still used in many simple controller applications, and is implemented as a digital equivalent on a more modern silicon process. A multitude of variants with different features are available from many vendors. The 6502 was developed around 1975 and is similar to the Motorola 6800 MPU of that era. The 6502 uses 8-bit data and 16-bit addressing and is still available, albeit implemented on more modern silicon. Both the 8051 and 6502 are programmed at a low level, using Assembly language programming.

PIC, AVR, and ATmega

These are several different families of microcontrollers that started as 8-bit Harvard architecture RISC processors. Some of these have expanded their product lines to include higher-performance devices.

Hardware Controllers

Hardware-based (HW) controllers include the PAL, PLD, CPLD, FPGA, digital ASIC (metal layers), and digital ASIC (all layers). Here is a brief look at each of these:

PAL (programmable array logic) and PLD (programmable logic device)
> These are both programmable logic arrays. PAL devices tend to be 28 pins or less, with 5 V compatible interfaces, and most have through-hole mounting. PLD devices are smaller variants of the CPLD, although no industry-standard naming conventions exist, so this can vary. PAL use is being phased out and is discouraged for new designs.

CPLD
> The CPLD ranges in size from 20 I/O ports to over 200. Internally, the CPLD uses a programmable interconnect on a field of gates or macrocells. If a hardware-programmable approach to a DCU is needed, a CPLD is often capable of getting it done. Both CPLD and FPGA devices are generally configured using Verilog or VHDL, to define functionality.

FPGA
> The FPGA ranges in size from 10 I/O ports to over 1,900. The FPGA uses programmable lookup tables to form logic functions. Complex, large array, dedicated logic can be done within an FPGA. Frequently, an FPGA is used in a first-generation product, and a digital ASIC is then created when the design has been proven.

Digital ASIC
> The digital ASIC can't be programmed, so it is not suitable for initial product development. A functional digital design needs to be worked out using another platform before fabricating an ASIC. If the product is high volume and needs fast logic, this is the preferred method. Two variants of the digital ASIC exist. The first uses unique metal layers on a standardized array of fabricated transistors. The second is the all-layer implementation where transistors and metal interconnects are unique to the device. Both are widely used.

HW-based controllers are suitable when fast dedicated logic is needed. The majority of controllers are readily implemented using a software-based controller.

Software Controllers

Software/firmware-based (SW) controllers include MCU and MPU devices. Differences between MPU and MCU features can be fuzzy. An MCU tends to have more features to make it self-sufficient, but there can be a lot of feature overlap.

As shown in Figure 2-11, most MPUs keep the RAM and ROM outside the chip with an emphasis on interfacing to external memory and peripherals. Many state-of-the-art MPUs have advanced to a greater number of internal features than shown here, but this illustrates the concept.

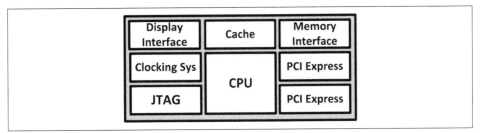

Figure 2-11. Typical MPU feature set

A typical MCU (Figure 2-12) includes all the features necessary to allow the device to function with minimal external support. The working memory RAM and the control program memory ROM are internalized. Serial communication to external peripherals is supported by UART, SPI, I²C, and CAN interface capability (FDI: Digital). Both ADC and DAC capability are common features.

Frequently, a clocking system includes clock multipliers and a real-time clock (RTC) oscillator. Internal timers are commonly used to time or control external functions, and a watchdog timer is often provided as a safety reset feature. General-purpose input/output (GPIO) is also available to control and monitor individual control lines.

SERCOM UART	RAM	GPIO
Clocking Sys & RTC	CPU	Watchdog Timer
JTAG		TIMERS
ADCs In	ROM	PWM DACs
SERCOM SPI	SERCOM I²C	SERCOM CAN

Figure 2-12. Typical MCU feature set

There are thousands of different MCU and MPU devices on the market, ranging from simple to multicore and offering a diverse and extensive set of options. This chapter provides the background knowledge to navigate the options maze.

Computers Versus Controllers

A multitude of single-board computers are readily available, and many vendors also produce development boards for their MCU devices. Off-the-shelf boards are one way to start, but they may not be a good final fit to what the project needs.

Single-board computers are configured to interface with the usual computer peripherals, and an MCU board is typically set up to interface to other ICs and provide stimulus/response through individual ports. A side-by-side comparison is useful.

Raspberry Pi (MPU) Versus Arduino (MCU)

Figure 2-13, which depicts two very popular single-board devices, illustrates the typical differences between an MPU and an MCU. The Raspberry Pi can be considered a single-board computer (MPU) and the Arduino is a microcontroller development board (MCU).

The processor details, and what's included on the board, tell the story:

- The MCU runs on a 16 MHz clock and the MPU uses a 1.5 GHz clock. The MCU is configured with interfaces for peripheral ICs using SPI and I²C interfaces and the MPU is set up with video display, USB, and Ethernet interfaces. Analog-friendly ADC and DAC interfaces are available on the MCU. The MCU has an internal 8-bit structure, and the MPU has 64 bits.

- The MCU is optimized to be fast enough and have enough processing power to control the slower functions of an electromechanical system.

- The MPU is configured for speed and enough processing power to run an operating system, feed graphics to two video ports, and interface with multiple peripherals over multiple I/O interfaces.

The differences in the target applications are clear.

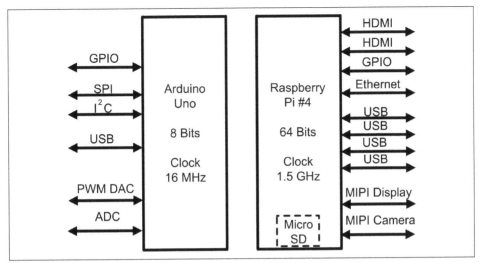

Figure 2-13. Arduino MCU versus Raspberry Pi MPU

Multipurpose and Specialty MCUs

Frequently, MCU cores are bundled with a specialized block of logic. Figure 2-14 shows two of the most commonly bundled implementations as an FPGA gate array or a dedicated DSP engine. The FPGA/DSP core can process data streams at high speed, while the MCU controls and monitors the process. DSP cores are optimized for the floating-point math of DSP algorithms.

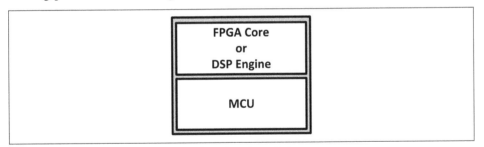

Figure 2-14. DSP or FPGA core combined with MCU

Figure 2-15 shows a small sample of the multitude of other devices that are being bundled with an MCU for specialty applications. If sufficient demand exists, some manufacturer has probably produced a specialty device to fill the market.

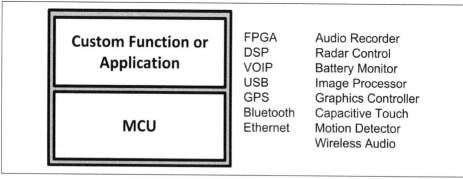

Figure 2-15. Examples of specialty devices

Single-chip solutions are popular for use in consumer products where volumes are high and low cost is important. Chip sets, grouped for use in select products, are popular in highly complex devices.

Chip Set Methods

A *chip set* is a group of ICs designed to function together to create the electronics for a particular product. A chip set needs minimal external support, as it is designed to be plug-together compatible.

Chip sets have been used since the early 1970s. The original Intel 4004 MPU was part of a four-chip set for the electronics of a calculator. A chip set is often developed when a high-volume demand exists or when a small form factor is mandatory.

One of the most widely adopted chip sets (Figure 2-16) is the one used on the personal computer (PC) motherboard. Many early PCs used a four-chip system of MPU, memory controller (Northbridge chip), I/O and peripheral controller (Southbridge chip), and BIOS. This was the heart of a PC motherboard.

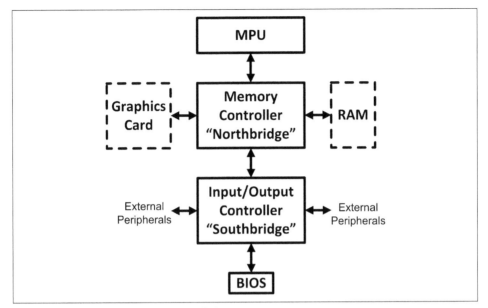

Figure 2-16. Computer motherboard chip set

Depending on the particular architecture, newer designs have reduced the chip count further.

Creating a chip set for the smartphone requires a very high level of integration to achieve both low power and compact size. Most smartphones use a chip set similar to the one shown in Figure 2-17. Most digital signal processing and control exists on a single digital ASIC that uses a silicon process optimized to high-density logic.

Supporting that are chips that provide other needed functions and are often fabricated using a semiconductor process better suited to their application. For example, the PWR MGR (power management chip) depicted in Figure 2-17 supports battery management and charging. The RF Front End chip is fabricated using an "RF-friendly" semiconductor process. And the RF Power Amplifier (RF PA) is often implemented using a specialty process (gallium nitride or gallium arsenide) suitable to the needed RF output power.

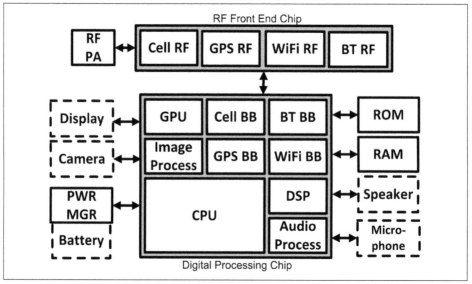

Figure 2-17. Cell phone chip set

Some other products that embrace the chip set approach are digital cameras, HDTV systems, and set-top boxes.

System Architecture Options

MCU and MPU devices see a countless number of applications. Not all are single processor, and frequently it's not evident that there's an MCU in the device. Taking a look at some common applications gives insight into some typical MCU use scenarios.

An MCU with some additional special feature content is frequently implemented as a standalone device. Figure 2-18 is an example with pushbutton switch inputs, a battery, and an infrared LED. This example exists in TV remote controls. A similar implementation that exists in many wireless key devices includes an internal RF transmitter and an antenna instead of the infrared LED.

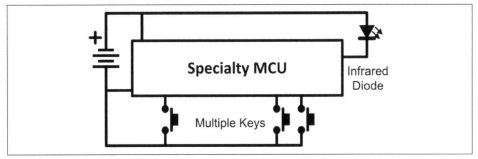

Figure 2-18. All-in-one SoC device (infrared remote control)

Generic MCUs can be used for specialty applications too. Figure 2-19 shows a generic MCU without special internal electronics. Five input switches, an external sensor on a serial port, and several relay controls are attached. Another serial port is set up to drive a liquid crystal display.

This example is taken from an electronic thermostat. The sensor is for temperature, and the three relay controls are for heat, cool, and fan.

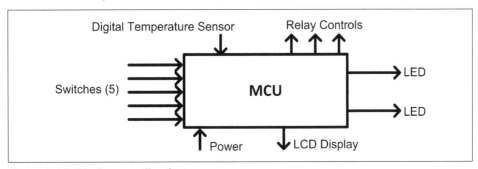

Figure 2-19. Single-controller device

Figure 2-20 shows an industrial application that illustrates multiple MCUs working together but not electronically linked. A conveyor belt system can detect the presence and location of packages along the belt. Each belt section has a separate controller that can turn the belt off when no packages are present (saving energy) and control the belt speed. If the package is not picked up by the next belt, the system can hold packages in the middle, waiting for the next belt to take up the package. Communication and coordination between belts is via the handoff of packages, rather than having adjacent belts in electronic communication. Setup and configuration is minimal since each belt acts as a self-sufficient system.

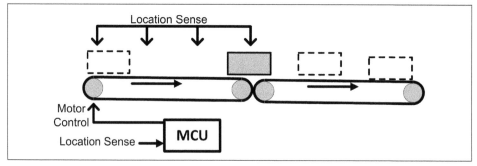

Figure 2-20. Multicontroller system without a manager

More complex industrial control systems or robotic devices frequently make use of a distributed controller scenario (FDI: Special). In Figure 2-21, a manager controller runs the global system functionality while local controllers deal with the detailed control and sense operations of the device they are connected to. This is a common approach in industrial control, multi-axis motion systems, robotics, and other areas.

If communication is lost to the manager MCU, local control MCUs can initiate a safety shutdown protocol. Frequently, this type of system uses the manager MCU to download programming into the other MCU devices, allowing easy code updates to the entire system.

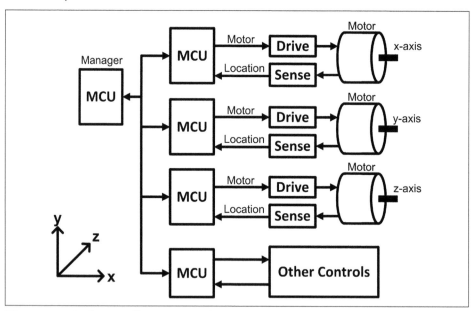

Figure 2-21. Multicontroller system with sequence manager

The examples so far include MCUs used to sense and control things. Frequently, some form of data stream needs to be created from a device, and the resulting data needs to be processed.

If the data rates are suitable, an MCU can be used as part of the data streaming path (Figure 2-22). The viability of this method hinges on the data rate from the ADC, the complexity of the DSP algorithm, and the clocking rate of the MCU.

Figure 2-22. Streaming data processed through the controller

Some configurations may choose to not stream the data through the MCU. Figure 2-23 shows a common alternative that streams data through a dedicated signal processing path, where the MCU controls the process but is not in line with the data stream. Here, the MCU can sample the numbers seen at the ADC output, and can adjust the gain amplifier to maintain an appropriate signal amplitude into the ADC. The MCU can monitor the DSP system and make adjustments there as well.

This particular example is typical of a magnetic disk read channel, where the HW-DSP includes a phase equalization algorithm, bandwidth-limiting LPF, and data detection algorithm.

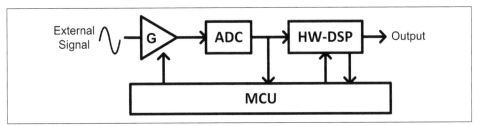

Figure 2-23. Streaming data through a dedicated DSP

SoC devices frequently have MCU/DCU devices within them that are not even mentioned in the specification. A good example is the wireless microphone system shown in Figure 2-24. The microphone has a DCU that controls the Bluetooth (BT) transmitter and provides the control protocol associated with the BT standard. All the microphone electronics exist as a single SoC.

The receiver has four DCU devices. One controls the Bluetooth receiver, one configures the equalizer, and one encodes the binary audio data to control patterns needed by the Class D audio amplifier. The prior three devices may actually be dedicated state machines within the SoC devices. The fourth device is an MCU that orchestrates the overall system.

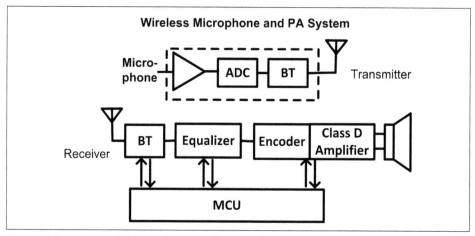

Figure 2-24. SoC modules used as a set for a wireless microphone

A typical controller interface to peripherals does not just turn things on and off. The interconnection can have more complexity. Adding some more detail to a typical controller in Figure 2-25 shows the expected single input/output for switches, LED indicators, and buzzers. In addition, other external peripherals are connected through addressable serial ports, so multiple peripherals can use the same connections.

Controlling a device often requires monitoring that device too. That idea is illustrated here with a drive circuit controlling a motor and a sensor monitoring motor speed. Other useful external interfaces include a JTAG port for device programming and a CAN bus to communicate off the board.

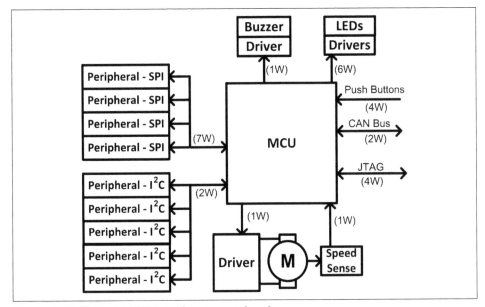

Figure 2-25. Typical MCU interface to peripherals

The examples presented here give insight into designing the global system architecture.

Determine Peripherals and Interconnects

The first strategic step to picking a DCU is to determine everything that needs to be controlled or sensed. Some things are obvious, but many items are often forgotten in an initial tally of peripheral interconnects.

Figure 2-26 shows the possible peripherals by category. The following subsections describe the peripherals.

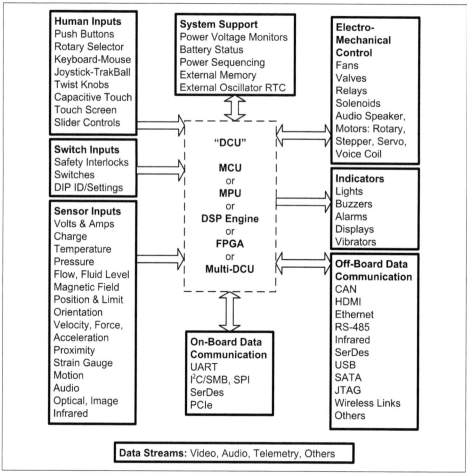

Figure 2-26. Control, sense, and data in/out categories of peripherals

Human inputs

Most human inputs are simple pushbutton switches, but there are exceptions. Most QWERTY keyboards interface through a USB port. Keypads are often implemented using a matrix switch structure, so a 4 × 5 keypad uses 9 ports instead of 20.

Slider controls, twist knobs, and rotary selectors can be implemented using either an optical encoder or a wafer switch to give binary outputs, or a potentiometer, which requires an ADC to determine position.

Determine specific human input devices and consult their specifications to determine an appropriate interconnect.

Switch inputs

Switches are not limited to people pushing buttons. For example, electromechanical systems use safety interlocks to stop active machines when their safety covers are off. Safety regulations frequently demand that a machine be rendered nonfunctional with exposed moving parts, and a safety strategy needs to be included. A safety interlock can include shutdowns for covers, latch solenoid to keep covers secured while things are moving, and motion sensors to determine when things have stopped moving.

DIP switches have been used in the past to set a binary identity or configure a control parameter. However, the availability of flash memory within the MCU allows most applications to use a memory location instead.

Minimizing the use of mechanical switches is preferred due to reliability and wear considerations.

Sensor inputs

Sensors come in many forms. A partial list is shown in Figure 2-26. In addition to what they are designed to detect, sensors can be categorized by how their outputs provide information:

Threshold sensor
    Changes state when the input exceeds a particular value. In the case of a pressure sensor, it could be designed to change state at 35 PSI as part of a dedicated module used for determining low pressure in tires.

Analog output sensor
    Changes voltage proportionally with the input. The voltage needs to be quantized to be useful to the MCU. A thermocouple is a common example. The typical thermocouple comes with a reference table linking temperature to output voltage. A suitable ADC will be needed to find the voltage, and the MCU can interpolate between the voltage and the sensed temperature.

Pulsed output sensor
    Often used for motion detection. Typically these devices emit a digital pulse for a fixed amount of linear or rotational motion. Using this information, the MCU can determine position, velocity, and acceleration.

Data output sensor
    Includes a fully integrated set of processing electronics. The sensor output is digital data from a serial port, and signal processing and an ADC are embedded within the sensor body. This approach fits well with the mostly digital approach encouraged here.

Chapter 9, "Sensing Peripheral Devices" explores this topic in detail.

On-board data communication

Communication on the PCB to peripheral devices is usually done using serial data ports. Commonly used methods at low data rates are I²C, SPI, and older UART methods. Faster methods include PCIe and SerDes serial ports. See Chapter 3, "Robust Digital Communication" for further details.

Off-board data communication

Within a multi-MCU system, CAN bus and RS-485 are often employed (FDI: Digital). Between computer systems, Ethernet is a reliable and suitable solution. Wireless methods (Bluetooth, WiFi, ZigBee, etc.) are useful options. Depending on the specific application, other options are suitable (FDI: Digital).

Indicators

Indicator types include "front panel" indicator lights, buzzers, alarms, vibrators, and displays that the user interacts with. Some displays will require a specialty port within the MCU.

In addition to front panel indicators, it is valuable to include indicators that aid debugging and troubleshooting. LEDs placed on the PCB can be used for in-the-field debugging using binary error codes or a blinking "heartbeat monitor" showing that the system software is running. Items with high failure rates often benefit from a status indicator to tell a field technician of specific problems.

Electromechanical controls

Fans, motors, solenoids, and servomechanisms all require interface driver circuitry to connect with the MCU. On/off control is generally sufficient for things like valves, relays, and solenoids. However, variable duty cycle drive capability (aka pulse width modulation or PWM) would be applicable for fans, rotary motors, and linear servos (FDI: Drive).

A speed sensor is commonly used to monitor fans and other rotary motors. A common strategy is to close a spindle speed control loop through the DCU. Linear servos and other voice coil motors require a location sensor to provide position information, and location control SW can be implemented within the DCU.

Stepper motors require a driver circuit that provides both direction and step controls. Also, stepper motors need to be supported with a "home" location sensor.

System support

The structure and control of the internal system requires support from the DCU as well. Some commonly included features are:

Battery state of charge
A voltage monitor or a charge monitoring system is often developed for the battery status or fuel gauge. This frequently needs a supporting ADC.

Power sequencing
The DCU is configured with control lines to start and stop the system power supplies. This is important for electromechanical devices to allow orderly device control and avoid destructive behavior. The DCU power is always on, thus allowing live control when any power is present. The motor power is controlled by the DCU (FDI: Power).

Power monitoring
Voltages of the various internal power supplies can be connected to the DCU through an ADC to allow the DCU to monitor the supply voltages.

External memory
A dedicated port interface may be required if large amounts of data are being processed.

External clock source
A common strategy uses an external real time clock reference at 32.768 KHz. This easily divides by 2^{15} to create an accurate 1 Hz reference.

To make all of these peripherals function reliably with the DCU, most inputs/outputs will require some form of SW signal processing, driver/receiver circuitry, ESD protection, or EMI limiting. Later chapters in this book are dedicated to these issues and the details thereof.

With these examples, a list of all peripherals and their needed connections should start to emerge.

Data streams

The processing of data streams needs careful attention because it is time restricted. Designers need to define data rates from all devices, including samples/sec as well as bits/sample to quantify throughput needs.

In addition, understanding the DSP methods to be used in processing each data stream will guide decisions on DSP HW and SW needs.

Avoid Serial Communication Bottlenecks

Serial communication (SerCom) ports (I^2C, SPI, etc.) are frequently configured with multiple peripheral devices attached. As an example, a single I^2C interface can have up to 128 devices attached. To avoid traffic bottlenecks, an estimate of the worst-case scenario for traffic through the port should be made.

Determine the time period of a data transfer for each device, and how frequently each device is active. A summation of all device times gives an indication of how "busy" the port is. No well-defined rules exist here, but staying below 25% is suggested. Even at 25%, probability is such that for every one in four attempts to use the port, it will be busy servicing a different device. A higher percentage of use is possible with only the DCU requesting data rather than multiple uncoordinated devices trying to access the port.

Two other options can reduce SerCom collisions. First, running the port at a higher data rate reduces the percentage of use. The I^2C protocol has four different data rates in the specification. Second, MCUs are available that have multiple serial ports. Generally, MCUs run at a faster clock rate than serial ports, so the MCU can service multiple ports at the same time.

Use Direct Memory Access for Data Transfer

A useful feature when moving large blocks of data in and out of the system is the capability to do direct memory access (DMA). A DMA control feature has the capability to access internal memory and the internal data bus of an MCU, without CPU involvement.

Figure 2-27 shows how a DMA device can work independently from the internal CPU and efficiently transfer data between memory blocks or memory and an external port. The DMA device can transfer data with fewer clock cycles than using the CPU to read and then write each data byte.

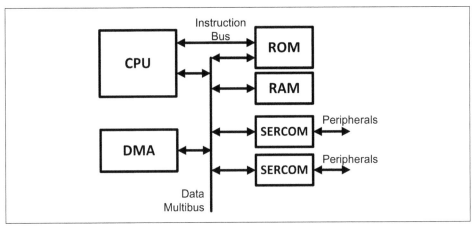

Figure 2-27. DMA methods used between memory blocks and peripherals

Determine DSP Methods

Until now, DSP has been treated as a black box. Selecting an appropriate DSP strategy requires a closer look. Either the DSP requires dedicated hardware, or a software algorithm can suffice.

Figure 2-28 depicts a DSP-implemented filter. On the left side of the filter is an HW-based implementation of a four-tap finite impulse response (FIR) filter. Three shift registers, plus the input, hold four sequential samples of the signal. The four samples are multiplied by four different values (WT1–WT4) and then summed to create the output.

On the right side of the figure is a flowchart of an SW implementation of the same FIR filter. The four sample values are held in a memory stack, and each is multiplied by WT1–WT4 and summed, providing an output. Then the oldest sample is deleted and a new sample is put on the stack, repeating the process.

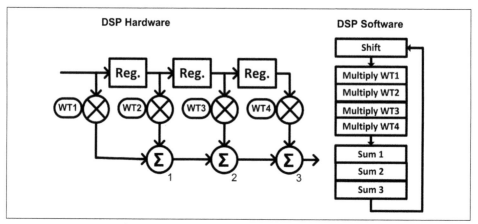

Figure 2-28. DSP FIR filter, HW versus SW methods

Both HW and SW methods are commonly used. In determining which method to use, certain things come into play:

- Sampling rate of the data stream
- DCU clock rate
- Complexity of the DSP algorithm

Essentially, selecting SW versus HW for the DSP is a question of how quickly the data comes in versus how quickly it can be processed. How "busy" the DCU is with other tasks also needs to be considered.

In addition, the delay time of the DSP may affect some systems. Figure 2-29 shows that the DSP can be considered a pipeline. With an incoming data stream and some math processing in the data pipeline, the output has a time lag from the input. That lag is referred to as *time latency* or simply *latency*. In Figure 2-29, an output ready at time B has less latency than at time C.

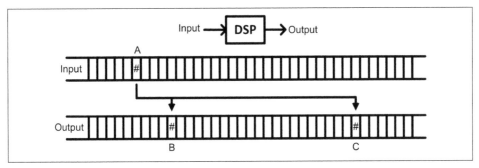

Figure 2-29. Data stream latency

In certain situations, latency can be an issue. However, in many cases it is invisible to the user. Here are two examples to consider:

Audio latency
An audio path that has latency under 100–150 ms is generally not a problem. Longer delays can be problematic in duplex communication (e.g., two people having a conversation over the phone).

Control system latency
This is latency that exists within a control feedback system. If considered from the perspective of analog feedback, latency can be an additive phase within a system. A digital system with too much additive phase in a feedback system can cause the system to become unstable (FDI: Control).

Designers need to determine how much latency is tolerable and accommodate that requirement with an appropriate DSP methodology.

Check for DSP Bottlenecks

A deeper look (Figure 2-30) at two extreme examples of data rates can be useful. At low speeds, a thermal monitor needs to transfer an 8-bit number once per minute. At high speeds, a high-resolution video data stream uses 15 Gbit/s coming from the camera sensors. Most data rate situations lie between these two extremes, but these two examples illustrate the issues. The slow data rate can easily be handled by an SW MCU system running at a low clock rate.

The fast data rate is more challenging. The 15 Gbit/s comes into the DSP already divided into red-green-blue pixel data streams at 5 Gbit/s each, which serializer/deserializer (SerDes) and PCIe interfaces can readily handle. Once serial data has been input to the DCU, the pixel data is transformed to parallel data of 10 bits/pixel for an internal data rate of 500 Mbit/s, which is well within (2023 era) digital ASIC and FPGA capability.

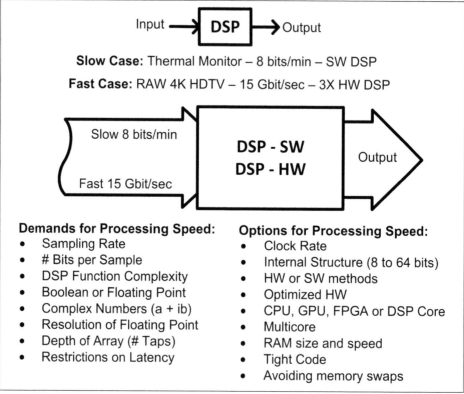

Figure 2-30. Data processing bottlenecks

Many state-of-the-art MPUs are capable of dealing with 30 bits of data at 500 MHz. Be warned that some MCU devices are designed for lower (<100 MHz) clock speeds and cannot process this data stream.

Changes to the following items will create demands for processing power and speed:

Sampling rate
How quickly are the data updates coming?

Bits per sample
A single bit is easier to process than a sample defined by 128 bits of data. Single data samples are easier to process than an array of samples, such as the data in a video image frame.

DSP complexity
A DSP algorithm can be any math function, and the processing procedure can be simple or complex.

Complex numbers
Data points that use a number value other than scalars increase processing time.

Floating-point resolution
Floating-point math increases processor time for higher-resolution results.

Size of data array
Multiple sequential samples are often used to determine a single output sample. This is illustrated with an FIR filter and how many "taps" exist in the filter structure. Image compression algorithms use multiple image frames to create each compressed frame.

Latency restrictions
Is there a time restriction from input to output?

Those are the common demands that can weigh down a DCU.

Improve DSP Speed

Multiple factors improve DSP execution time:

Clock rates
Faster clocks yield faster results. But remember that faster clocks also consume more power. If there is no need for a maximum-frequency clock, don't use it. Savings will be seen in both lower power consumption and less radiated EMI. If faster clock rates are needed, remember that most MPUs are faster than MCUs. Also, both MCU and MPU devices designed to operate using lower power supply voltage will generally be capable of faster clock rates. If devices that use 5 V or 3.3 V power are not fast enough, investigate devices designed to function at 2.5 V, 1.8 V, or lower.

Internal structure

MCUs exist with 8-, 16-, 32-, and 64-bit internal bus and CPU structures. MCUs with 8- or 16-bit structures are capable of Boolean decision-making and control, but 32-bit devices are more suitable to systems that need floating-point math. Many 8- and 16-bit devices were originally fabricated on older 5 V CMOS processes, making them slower. If you're using one of these devices, a modern functional equivalent is often available.

HW versus SW methods

A dedicated HW solution (DSP engine, FPGA, ASIC) can be designed to run faster than an SW solution.

Optimized HW

Many DSP engines are marketed with structures optimized to run math functions common to most DSP functions.

Multicore partitioning

If the function can be partitioned to run on multiple cores, a parallel processing approach can be used.

Tight code

In the era of faster clock rates and more powerful processors, improving SW efficiency is not that commonly done. Instead, use of a faster MCU is the most common technique for reducing execution time. SW structure can change execution time significantly, especially for array processing and repetitive floating-point operations.

Avoiding memory swaps

Most MCU systems will not need to deal with memory array swaps to mass storage. This issue is more common for an MPU running a multitasking OS. It takes time to move data, and avoiding transfers of big blocks of data can expedite things.

Determine DCU Internal Features

With external connections determined and a better understanding of DSP requirements, the next step is to define what internal features are necessary. The features depicted in Figure 2-31 and described in the following subsections are available on various MCU devices.

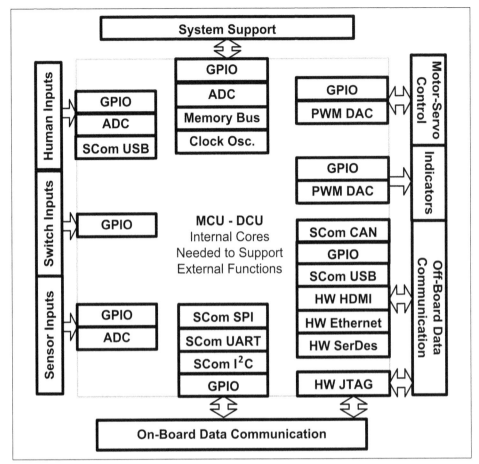

Figure 2-31. Determine needed controller cores and features

General-purpose input/output (GPIO) ports

From a list of peripheral devices, an estimate of the number of GPIO ports can be made. GPIO capabilities should also be determined, namely:

- Port enabled to service interrupt inputs
- Current sink and source capability
- High and low voltage limits using internal drivers
- V_{max} output capability using external pull-ups
- V_{max} input tolerance when driven by external devices

- Programmable port configuration capability:
  — Open drain with an external pull-up
  — Selectable internal pull-up
  — Adjustable drive strength and speed
  — Capability to tri-state the output
- For unconnected ports, how to render the pin inactive
- What internal functions can be routed through each GPIO (PWM output, comparator inputs, ADC input, etc.)

Serial data communication

Serial communication ports include CAN, RS-232, RS-485, UART, USB, I²C, SPI, Ethernet, HDMI, memory card interfaces, PCIe, SATA, JTAG, and others.

Some SerCom ports are generic and can be configured for various purposes. Some require specialty internal HW to be standards compliant. Others may require external circuitry to make them functional. All require some form of SW driver to respond properly. Check to see if the MCU vendor has a suitable SW driver (FDI: Digital).

Internal ADC capability

Many MCUs offer built-in ADCs. Check for a suitable number of bits and a suitable conversion rate. Some include programmable gain input amplifiers, which should be checked for noise, linearity, and bandwidth (BW) performance. Preferably, the ADC input is differential and not ground referenced. Differential inputs reject environmental noise better (FDI: ADC & DAC, FDI: EMI & ESD).

Internal DAC capability

Many MCUs have PWM DACs available. Some will offer sigma-delta DACs or other variants. Check for the number of bits, the conversion rate, and what external support components are needed (FDI: ADC & DAC).

Comparators

Voltage comparators are often useful in detecting a voltage transition over a trip point. Hysteresis capability or a programmable voltage reference can be offered on some devices. Check specifications for input voltage limitations, response time, and input offset parameters (FDI: Sense).

Real-time clock

An RTC is for time-based measurements. This is usually implemented using a crystal at 32.768 KHz, which provides a 1-second clock after dividing by 2^{15} times. Also, calendar functions are available that will keep track of date-time-weekday data.

Clock generation system

Most MCU devices have an internal oscillator to create a system clock. Many include a programmable clock multiplier, internally implemented with a phase locked loop (PLL), to create a higher-frequency clock. The basic clock frequency can be inaccurate when using an internal clock. Therefore, many devices also provide the internal circuits to support a crystal oscillator for a highly accurate clock frequency.

Power consumption

MCU power consumption will be determined by the MCU design and the clock frequency. Faster clocks consume more power. Understanding power consumption is especially important if the device is battery powered. Many MCUs have been optimized for low power consumption and are defined as "low-power" products. Specifics of power consumption need to be checked on a case-by-case basis.

Power supply voltage

As the CMOS transistor has shrunk, the maximum power supply voltage has also been reduced. Depending on what CMOS generation was used, the power supply may be from 5 V (older) to 0.7 V (newer). Different MCUs are fabricated on different generations of CMOS, and a broad selection of devices exist for 2.5 V, 3.3 V, and 5 V power. Specialty devices with higher-voltage interfaces also exist. Many "level shifter" chips are available to allow devices with different digital signal voltages to communicate with each other.

Programmable memory

Programmable memory (aka flash memory) is implemented with some form of EEPROM that retains data when the power is off. There are different types of EEPROM and some are limited in the number of times they can be overwritten. If many (>1,000) overwrites are expected, check this specification.

Needed memory size depends upon program complexity. Compiling the code for the device will give an estimate of needed memory size. Lacking a complete code build, a safe start is to get the largest memory option for the MCU selected. As the project finalizes, memory size can be reduced as needs become better defined.

RAM size

The majority of MCUs have RAM under 1 Mbyte. Most embedded control systems don't need huge amounts of RAM. If large arrays of data are needed, an MPU with an external RAM array may be more appropriate. Also, additional memory (both RAM and EEPROM) can be added to any MCU as a serial port connection.

Interrupt handling capability

Check to see which GPIO ports can handle interrupt inputs. Also, review how the internal interrupt controller functions and processes an interrupt. Different vendors do this in different ways.

Floating-point math capability

Floating-point math is best performed on a 32-bit (or more) processor. Although floating-point routines exist for 8-bit devices, the HW limits performance. If a need for floating-point math is expected, check that the vendor has available a set of suitable code routines for the processor and that they meet the IEEE-754 standard for floating-point arithmetic.

Operating temperature

Most MCUs will function between −10°C and 85°C. Devices can be found that will function down to −55°C and some up to 225°C.

Power on reset and brownout detection reset

MCUs need to respond to power activation in a predictable and safe manner, especially when they control power equipment and motorized devices. Many newer MCUs include appropriate control circuits to safely assist with power cycling.

Radiation hardened designs

MCUs with structures that can withstand radiation ions are a specialty item for the military, satellites, and aerospace applications. Radiation hardened (*rad hard*) devices are available from vendors that support the defense and aerospace sectors.

Direct memory access capability

As described before, DMA can expedite the moving of data without involving the CPU.

Watchdog timer

A watchdog timer (WDT) is a counter timer used to detect whether the system has stopped operating normally. If the system SW is operating properly, the WDT counter is being cleared frequently. If the system freezes, the WDT counter is not being cleared, and the WDT is activated. Depending upon the specific features, a hardware reset occurs or a special restart routine is called. WDT parameters can be configured. A WDT is a useful safety feature if the code was not configured to self-recover (FDI: Code).

Timer counter circuits

Keeping track of time by counting clock cycles is the primary function of a timer counter circuit (TCC). A TCC in an MCU can be used for:

- Event timing
- Creating waveforms and PWM duty cycle controls
- Creating delays
- Periodic execution of service routines
- Creating a slower clock

Multiple timers can be very useful. Check the features of the TCC devices to see if they have a sufficient bit range for your expected uses and capabilities to select both input and output connections.

Sleep and low-power modes

Some devices can shut down most of the chip functionality until something requests a wake-up. Other devices can be configured to sleep for a defined period, then wake up, perform a check status and action routine, and go back to sleep. These methods are unique to each vendor. Sleep and low-power modes can be a valuable power-saving feature, especially for battery-powered devices.

In sum, these features are available on various MCU devices, and they should allow you to itemize a list of needed internal features.

Physical Package Considerations

Frequently, the same DCU is offered in a multitude of package options. Newer options have gotten smaller, requiring designers to spend more time over a microscope to deal with the small form factor.

Shown in Figure 2-32 are four of the most common 48-pin packages, illustrated in a scalar size comparison. As you review the figure, note the following:

Dual inline package (DIP)
This is a through-hole device still available for some DCUs. The large package size is needed to support the wide pin spacing of 0.1 inch (2.54 mm), which

creates a large amount of connection inductance in each pin. The DIP is not for modern commercial designs due to the large size and high-frequency limitations.

Quad flat package (QFP)

This is a surface mount device with pins that are easily accessed outside the perimeter of the package. The QFP is commonly used with external clocks under 300 MHz. Presently in widespread use, it is presently available up to 256 pins.

Quad flat no lead (QFN)

This is a surface mount device without externally protruding leads. Contacts are at the perimeter of the package. Eliminating the protruding leads reduces the inductance of the connections as compared to QFP devices, and the package has better high-frequency characteristics. QFN packages are used with external clocks over 1 GHz and are available up to 400 pins.

Ball grid array (BGA)

This is a surface mount device with contacts arranged in a grid across the bottom of the package. These devices presently approach 4,000 pins and are very popular for high-frequency and small form factor designs.

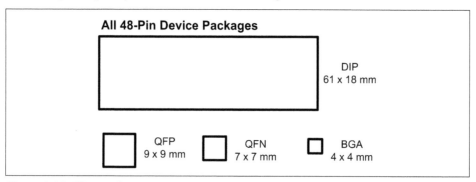

Figure 2-32. Chip packaging size comparison

Many other package options also exist. The ones listed are common options, and they illustrate the limits of size, pin count, and frequency (FDI: PCB).

Off-Chip Features and Support

In addition to internal MCU features, items beyond the chip need to be considered. As evidenced by the following list, ease of use, documentation quality, and anything that expedites development is important:

IDE

Chip vendors provide SW tools to develop code and HW to program the MCU. This usually includes a code editor, simulator, compiler-linker, interface for download, and debug system. Vendors want their chips designed into products, so they must provide a quality IDE and make it free to download and experiment with.

Documentation

Some developer's manuals for complex MPU and MCU devices are thousands of pages long. Simpler devices are less so. Make sure the material is sufficient in detail, well written, organized, and easy to understand.

Scalable chip family

Many MCUs are part of a family of chips with different numbers of GPIO ports, variations on internal features, and other options. These devices try to maintain both software and design tool compatibility across the chip family. This allows designers to change between chips without conducting a total redesign or learning new design tools.

Software libraries and application programs

Does the vendor provide properly documented and debugged support software? As an example, downloading an SW control protocol for any serial port structure is a valuable time-saver. If a vendor provides a particular port interface or internal feature, check that support code for it is also available.

Support community

Many user groups with online support and discussion forums exist for MCU product lines. Some are independent of the vendor and others are vendor supported. Having access to designers who use a family of chips and IDE tools is invaluable when you need a question answered quickly.

Evaluation boards and development kits

Vendors frequently produce evaluation boards and other development kits to support products. An evaluation board is a quick and low-cost way to start running code on an actual device. A common strategy uses an evaluation board for code development while a PCB is being created, giving a few weeks of head-start time.

Cost and source availability

MCUs are a commodity and many viable solutions exist. However, plug-in compatibility between vendors is not generally available. Consequently, a reliable vendor that can provide the needed quantities when your product ships is important.

 Device cost may be counterintuitive. Some modern 32-bit MCUs cost less than 8- or 16-bit devices using older CMOS technology. Stiff competition in a flooded vendor field has resulted in lower-cost solutions with more features. Check the cost for volume purchases, and don't assume fewer bits or features means lower cost.

Pulling It All Together

At this point, enough information has been presented here for you to:

- Understand the "mostly digital" design approach
- Define your system architecture (single MCU, multiple MCUs with or without a manager)
- Determine your peripheral devices (buttons, switches, sensors)
- Itemize your peripheral control connections (GPIO, DAC, etc.)
- Itemize your devices to sense (GPIO, ADC, etc.)
- Define which devices need both control and sense connections (motors, servos)
- Itemize devices connected to serial ports (I²C, SPI, UART)
- Define off-board communication needs (CAN, RS-485, Ethernet, etc.)
- Understand the time demands of data streams and their DSP
- Define where DSP needs enhance processing or dedicated HW
- Decide between an MCU (SW) and FPGA/CPLD/ASIC (HW) approach
- Avoid SerCom bottlenecks
- Determine what internal features are needed for the MCU

Many of these items require appropriate drive-sense circuits to interface with devices. Detailed coverage of those methods is provided in the following chapters.

Picking a DCU Configuration and Your MCU/MPU

Suggestions for how to pick a device can be broken down into several considerations, which are covered in the following subsections.

Specialized Niche Function or Feature

If an MCU with some very specialized form factor or feature is needed, your choice of vendors may be limited. Some specialty functions on the same chip with an MCU may not be available from many sources. Frequently, that specialized function can be an external peripheral to the MCU, but getting it all on the same chip saves space, money, and power.

Multi-MCU Systems

The author strongly suggests that these systems use the same MCU in all locations. Although each device may have a different function, there will be a large amount of overlap in design tools and support coding. This will save both design and development time.

An alternative approach uses the same MCU everywhere except the manager.

If the manager needs PC-type features (video interface, keyboard/mouse input, running a full-feature OS), an MPU is often a better fit. A single-board computer, with a CAN bus or RS-485 bus out to multiple MCU satellite boards, is a common solution here.

General-Use MCU Systems

For most systems, many vendors will have multiple devices that will get the job done. The first priority should be the considerations covered in "Off-Chip Features and Support" on page 66, namely:

- Does the vendor have a product line of suitable devices?
- Is the IDE easy to use and free of bugs?
- What's available for ready-to-go SW and drivers?
- Is product documentation well organized and are solutions easy to find?
- Does the online support community show promise?
- What's available for evaluation boards and development kits?

 The suggested strategy is to pick a manufacturer and its product line first, then select a specific MCU. Frequently, ease of use and a quality support infrastructure are more important than a specific device selection.

Picking a Specific MCU

After narrowing the scope to a vendor, use the lists of ports and necessary features itemized earlier to choose a suitable device. Remember: allow enough capability to support design changes and include extra ports, extra memory, and processing capability. Design details change, and desired product features invariably grow, so using an MCU with some excess capability is a good strategy for a first-generation build.

With a fully functional design, a cost-optimized design can be pursued in a second-generation build that has fewer ports and less memory. With an MCU that is part of a product line (all using the same internal CPU), this allows changes with minimal difficulty.

"Further Reading" on page 71 has a list of some of the more popular vendors of MCUs and component distributor search tools for specific devices.

Summary and Conclusions

This chapter is about "big-picture" ideas and global approaches to designing an electronic system. Following are some of the important ideas in this chapter:

- Modern design strategies try to minimize analog circuitry.
- DSP methods are versatile and widely applicable.
- HW-based control uses CPLD, FPGA, and ASIC devices.
- SW-based control uses MCU and MPU devices.
- DSP can be HW or SW based.
- Chip sets or single SoC solutions exist for most high-volume products.
- Many system architecture options exist, with single and multiple DCUs.

With that knowledge, a designer should be able to:

- Determine system architecture at the box level
- Itemize a peripheral list and needed interconnects
- Evaluate SerCom requirements and avoid bottlenecks
- Determine needed DCU features
- Evaluate an IDE and support environment
- Select a suitable DCU

In addition, designers should now have a suitable knowledge base to navigate the confusing maze of device specifications and feature descriptions of various processors.

Remember that there are probably multiple devices suitable to the design project at hand. Having more than one correct answer can make the decision easier, although confusing nonetheless.

Further Reading

Audio Latency
- Section 7.4 of International Telecommunications Union G.107 (06/2015) ITU-T Series G

Floating-Point Math in a Microprocessor
- IEEE-754 standard for floating-point arithmetic

Simulate or Build
- Jerry Twomey, "Factor Time and Money before Using a Simulator," May 3, 2014, *Electronic Design Magazine*, *https://www.electronicdesign.com/technolo gies/analog/article/21799647/factor-time-and-money-before-using-a-simulator.*

MCU/MPU Manufacturers
- Analog Devices
- Atmel
- Cypress Semiconductor
- Freescale Semiconductor
- Infineon Technologies
- Intel
- Maxim
- Microchip Technology
- Motorola
- Nuvoton Technology
- NXP Semiconductors
- Renesas Electronics
- Rohm (Lapis) Semiconductor
- Silicon Labs
- STMicroelectronics
- Texas Instruments
- Toshiba Semiconductor
- XMOS
- Zilog

Component Search Tools
- DigiKey Electronics (*https://www.digikey.com*)
- Mouser Electronics (*https://www.mouser.com*)
- Allied Electronics (*https://www.alliedelec.com*)
- Jameco Electronics (*https://www.jameco.com*)
- Arrow ECS (*https://www.arrow.com*)
- Newark (*https://www.newark.co*)

Useful Websites and Design Forums (VI = vendor independent, VS = vendor supported)
- Texas Instruments Forums (VS) (*https://e2e.ti.com*)
- Microchip Discussion Forums (VS) (*https://microchip.com/forums*)
- All About Circuits (VI) (*https://www.allaboutcircuits.com*)
- Electro Tech Online (VI) (*https://www.electro-tech-online.com*)

- Arduino (VS) (*https://www.arduino.cc*)
- Raspberry Pi (VS) (*https://www.raspberrypi.org*)
- 6502 Specific User Group (VI) (*http://6502.org*)
- Microcontroller (VI) (*https://microcontroller.com*)
- Atmel, AVR and Atmel-ARM (VS) (*https://oreil.ly/5gP5U*)
- EDAboard, electronics forum (VI) (*https://www.edaboard.com*)
- DigiKey Engineer Resources (VI) (*https://www.digikey.com/en/articles/techzone*)

All of these sites are useful to designers. The VI sites will tend to be wide ranging on topics and the VS sites tend to focus on the vendors' products. Search engines will point out many other sources, of course.

CHAPTER 3

Robust Digital Communication

Digital communication is the backbone of any system. Low-frequency, short-transit data across a printed circuit board (PCB) is usually straightforward. But when data moves off a board, goes a distance, or encounters electromagnetic interference (EMI), challenges arise. Also, ensuring reliable data at high speeds requires special attention to signaling methods, transmission lines, error handling, and clock/data phasing. Countless systems work on the bench and then fail in the field due to environmental challenges. Properly designed systems must survive real-world EMI, electrostatic discharge (ESD), variable power, and noisy grounds.

This isn't a chapter on data communications theory. The intention is to give an understanding of widely used methods so that readers can make informed decisions regarding what interface to use or where they need a transmission line. Limited coverage is also given to older methods that designers should be aware of.

Commentary here is about CMOS logic. Special considerations for other logic families (RTL, TTL, ECL, etc.) are not given. Modern methods put logic functions in controller software or single field-programmable gate arrays/complex programmable logic devices (FPGAs/CPLDs). (Nobody uses a PCB full of discrete NAND gates anymore.) Multiple high-speed data paths are now inside the chip, not across the PCB. Consequently, most PCB digital communication now falls into one of two categories: low data rate setup and configuration control of chips, or high-speed data pipelines between devices. Figures in this chapter may use single gates for illustration, but the discussion is applicable to the I/O ports of larger FPGA/CPLD devices, which are invariably CMOS. Signals off the board or out of the system are special situations that also need to be dealt with.

This chapter covers a broad swath of digital communication topics, including:

- Ground-referenced versus differential signals
- Point-to-point versus multipoint networks
- Lumped versus distributed networks
- When to use transmission lines with termination (TLT)
- Clock distribution methods and timing skews
- Parallel versus serial data interfaces
- Common serial ports and network methods
- Wireless data methods

Designers can use the information presented here to make proper choices for system communication methods.

Digital Signals, Physical Considerations, and Connections

Before looking at suitable data interfaces, some background information on how connections are made, what types of signals to use, when to use terminated transmission lines, and clock distribution methods is necessary.

Limitations of Ground-Referenced Digital Signals

Digital data is subject to signal integrity issues (Figure 3-1). Noise, amplitude loss, transmission line reflections, impedance mismatch, and other issues contribute to the problem. These issues tend to grow as connections get longer or data rates increase.

A digital signal can be corrupted by noise on power and ground (P&G) at both the transmit and receive sides of a data connection. Also, the bandwidth (BW) limitations of the connection will cause higher-frequency parts of the signal to be degenerated in amplitude and distorted in phase. Adjacent signals in close proximity contribute crosstalk noise by capacitive and electromagnetic methods. In addition, the rise/fall edges of digital signals make the impedance mismatches of a connection very visible. A connection with fast signal transients and lacking a properly terminated transmission line creates signal reflections that further corrupt the signal.

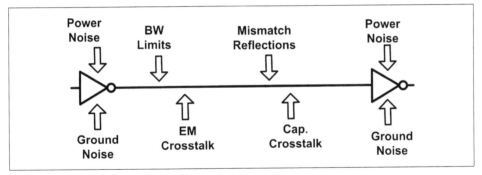

Figure 3-1. Digital signal corruption

Low-Voltage Differential Signaling

Digital drivers can be ground referenced or use differential signals (Figure 3-2). Frequently, differential signals use lower voltages than rail-to-rail ground-referenced logic, thus the low-voltage differential signaling (LVDS) nomenclature. A standard (TIA/EIA-644) has adopted the LVDS name, however this text refers to all differential signaling using non–rail-to-rail signals as LVDS.

A ground-referenced digital signal is dependent upon P&G to create and detect a signal (V_{in}). P&G noise shows up on the transmitted signal and dynamically shifts the switching point of the receiver. A differential signal is defined as the difference, ($V_{pos} - V_{neg}$), between signals and thus minimizes sensitivity to P&G noise.

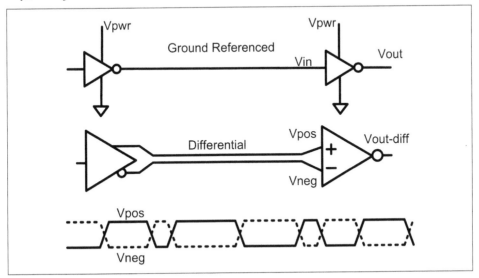

Figure 3-2. Differential versus ground-referenced signals

LVDS signals can be connected through either a differential stripline (PCB) or a twisted pair wire set (cable). Signals are equal and opposite in polarity (Figure 3-3).

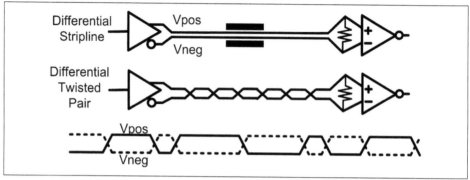

Figure 3-3. LVDS through matched differential paths

Reception of LVDS as the difference between the two signals provides numerous advantages:

Less noise sensitivity

Differential signals are routed closely together, making noise coupling common to both signals. A differential receiver rejects common mode signals.

Less sensitivity to power noise:

P&G noise is common mode to both signals, and is also rejected by the receiver.

Lower timing jitter

Accuracy of the switching point is more consistent due to both common mode noise not contributing to time spreading of the signal, and a faster dV/dT of the differential signal. Because it is differential, the effective dV/dT enjoys a 2X improvement.

Less bandwidth sensitivity

BW limitations can cause attenuation of a ground-referenced signal and create asymmetry in the received output. BW limitations of a differential signal will attenuate the signal, but LVDS receivers can detect a smaller signal without creating asymmetry.

Lower power

LVDS commonly functions with 300 mV$_{pp}$ for each signal. At high frequencies, two signals with lower voltage swing frequently require less power than one rail-to-rail ground-referenced signal.

Less radiated EMI

With equal and opposite currents in two closely placed signals, EMI is greatly reduced. From a distance, the electromagnetic fields cancel each other out. Switched current surges in ground-referenced logic act as a transmitter, with the connection creating a transmit antenna.

For these reasons, differential signaling and LVDS methods have been widely adopted for both high-speed and long-distance digital connections. Ethernet, USB, PCI Express, and CAN bus, among others, use LVDS methods.

Organizing Interconnects for Speed and Signal Integrity

With slow (<10 MB/sec) data rates, soft (>3 ns) rise/fall times, and short (<25 cm) connections, getting digital data across a PCB is generally not a problem. The Slow, Soft, Short Net (SSSN) is easy to deal with and represents a lot of the digital connections in use. A more careful approach for faster data rates will be outlined shortly. Before that, the topic of what makes a good network connection needs to be examined.

Any electrical connection without controlled impedance and with multiple connections at multiple locations is not friendly to high-bandwidth signals. Disorganized connections can be thought of as a mixed network of RLC elements and unterminated transmission lines without consistent impedance. High-frequency signals quickly exhibit problems in such connections.

Figure 3-4 shows a driver (A) that is connected to multiple loads (B, C, D, E, F) with no well-defined placement or routing. Consequently, the possibilities of transmission line reflections, impedance mismatches, and signal integrity problems abound here, especially when the data rates get pushed up. It is difficult to define what is wrong, because connection lengths and data rates are unknown. If it is an SSSN, it might work if device A can drive five loads and their associated RLC interconnects. Better definitions are needed.

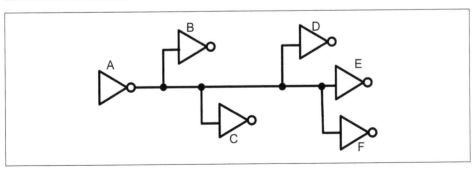

Figure 3-4. Problems with multipoint networks

The devices in Figure 3-5 are in two tightly placed target groups, so you can treat groups BC and DEF as two lumped nodes (more on this shortly). This minimizes the undefined transmission connections. However, this structure still lacks a termination resistance, and it includes a signal stub in the middle of a transmission line, which can cause impedance mismatches and local line reflections as high-frequency signals pass through. The BC group is referred to as a *stub connection* and is seen in multidrop bus connections. Stub connections should be avoided or kept as short as possible.

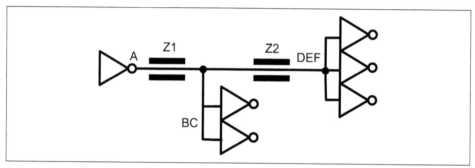

Figure 3-5. Reducing transmission reflections

Ideally, single point-to-point transmission is the preferred method.

Figure 3-6 uses a controlled impedance layout for the connection and a termination resistance at a single receiver. For high-frequency signals, this minimizes line reflections. A single point-to-point connection does not serve for all situations, but multiple drivers can provide a solution.

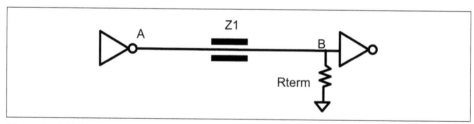

Figure 3-6. Point-to-point transmission

Figure 3-7 shows a configuration that provides a single transmission line between A and DEF as well as a single line between AA and BC. Using separate drivers for each transmission line minimizes interaction between the two. Loads DEF and BC are placed close together and can be treated as a lumped network.

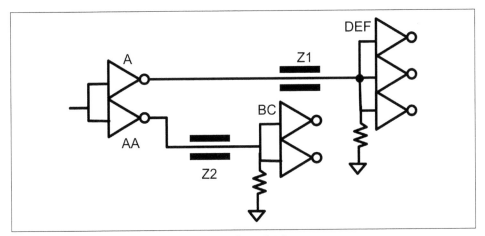

Figure 3-7. Point-to-point transmission with multiple targets

A digital connection can be treated as a lumped or distributed network. A distributed network needs transmission lines and terminations. Next, we'll decide which approach to use.

Lumped Versus Distributed Networks

It takes time for a signal to travel down a wire or across a PCB. In a *lumped network*, all connected components see the applied signal at effectively the same time. In a *distributed network*, transmission delays cause applied signals to occur at significantly different times on the attached devices.

Determining which way to treat a connection becomes a function of two items: the connection distance and the transient signal applied. For analog and linear systems, phase as a function of frequency would be used. For digital systems, the rise/fall time of the signal versus the connection length is useful.

Figure 3-8 shows an ideal signal, A, with zero rise time. Reality is closer to signal B, with finite rise and fall times. Signal B has RC time constants for both rise and fall transitions. For simplicity, these are illustrated in Figure 3-8 as ramps (C). CMOS logic is usually designed such that T_{rise} and T_{fall} are equal. Due to semiconductor variance, they will never be perfectly matched (D), but they are similar.

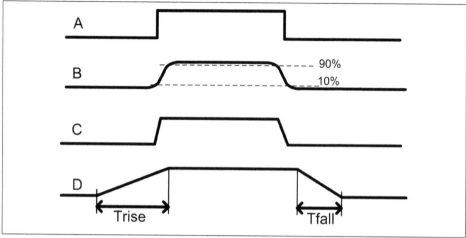

Figure 3-8. Digital signal rise and fall times

Most integrated circuit (IC) specifications define rise/fall times as the 10% to 90% time (signal B in Figure 3-8) for the rise/fall period. Most digital devices have rise/fall times ranging from 0.1 ns to 5 ns. Modern FPGA devices often allow the programming of drive strength to adjust that time period. As will be shown, rise/fall times under 1 ns can be very problematic in board designs.

Going forward, *rise time* refers to either rise or fall time. In cases where the rise and fall times are not matched, the shortest time should be used for calculations.

Since digital signals are not a linear function, an easier way to analyze this is to look at rise time as it compares to propagation delay of the digital signal. Figure 3-9 shows signal A being sent out. The time it takes for the signal to arrive at location B is called the *propagation delay* of the signal, indicated here as T_{delay}. For a typical FR4 circuit board, the propagation delay is about 60–70 picoseconds per centimeter. The relationship between T_{delay} and T_{rise} becomes important.

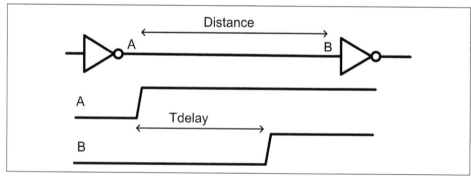

Figure 3-9. Propagation delay due to connection length

Figure 3-10 shows the relationship of T_{rise} to T_{delay} such that the propagation delay is much smaller than the rise time. Consequently, the same signal, at about the same time, is seen at both A and B. Either a short propagation delay (a physically short connection) or a longer rise time will cause this. This situation can generally be treated as a lumped connection; for this situation, A and B can be treated as the same node.

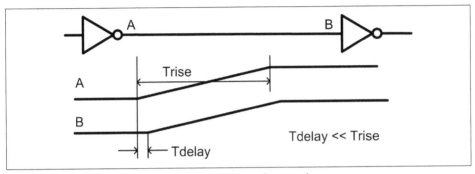

Figure 3-10. Lumped network: slow rise time or short path

A distributed connection has a shorter rise time or a longer propagation time. Consider Figure 3-11, where the rise time is shorter than the propagation delay time. The voltage at point A is different from that at point B during logic state transitions. This needs to be dealt with as a distributed network. Also, this connection may need to be turned into a controlled impedance transmission line and require termination resistance to avoid impedance mismatches and line reflections.

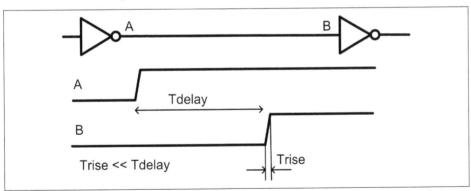

Figure 3-11. Distributed network: fast rise time or long path

Signal length, or the length of the rising edge, is the next useful measurement here. Approximating a propagation delay of 60 ps/cm allows estimating the physical length of a transient signal. Some common rise times for digital logic, from 0.1 ns to 5 ns, and their signal lengths are shown in Table 3-1.

Table 3-1. Rise/fall time and length of signal propagation

Digital rise time[a]	Length of transient edge[b]	20% of length of transient edge[c]
0.1 ns	1.7 cm	0.34 cm
0.5 ns	8.3 cm	1.7 cm
1.0 ns	17 cm	3.4 cm
2.0 ns	34 cm	6.8 cm
3.0 ns	51 cm	10 cm
4.0 ns	68 cm	14 cm
5.0 ns	85 cm	17 cm

[a] Use the fastest edge for devices with unbalanced rise/fall times.
[b] A propagation time of 60 ps/cm is used here, which is typical for FR4 PCBs.
[c] Note that 20% of the length of the transient edge is used here.
Multiple reference texts utilize values between 16% and 25%.

The connection length must be much shorter than the signal length to treat a connection as a lumped network. About 20% of the signal length is the maximum connection length before the connection must be treated as a distributed network. In Table 3-1, 20% of the length of the transient edge is used, which serves as a good metric to determine whether a more stringent analysis is needed.

Designers rarely need a detailed analysis of a digital connection. Checking device load versus drive strength, using transmission lines with termination where guidelines indicate they are needed, and some common sense in signal routing generally gets the job done. However, a brief look at an example of a distributed model (Figure 3-12) is useful to better understand high-frequency connections.

There are multiple ways to simulate and model a distributed connection. Transmission lines and electromagnetic simulators are available, or an equivalent circuit model can be created for Simulation Program with Integrated Circuit Emphasis (SPICE) simulation.

The equivalent circuit model shown in Figure 3-12 represents a 1 mm wide PCB trace that is 10 cm long. Such a device would have a distributed resistance (50 mΩ), inductance (40 nH), and capacitance (10 pF).

Breaking this up into discrete RLC segments is an approximation of behavior. A suitable number of RLC segments can be found using the transient edge length of the digital signal. Using 10 (or more) RLC sections would be suitable for the entire transient edge length.

Since a rise time of 1 ns is a signal length of 17 cm, and the connection length is 10 cm, proportionally fewer sections are suitable. A cascade of six RLC sections is shown in Figure 3-12, with a termination capacitance of 8 pF representing the capacitance of the receiving device. This can be simulated to provide some basic waveform information.

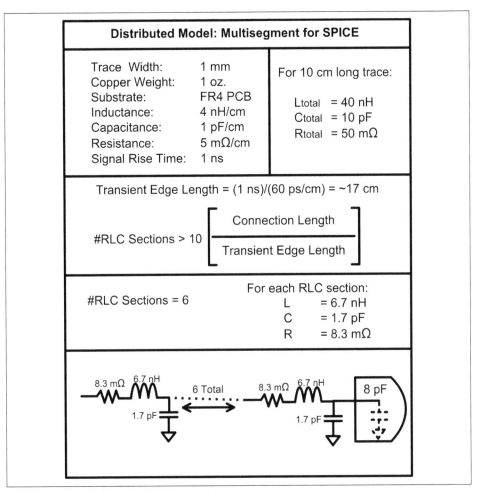

Figure 3-12. Distributed model example

When to add an impedance-defined transmission line with a matching termination is the next consideration. A TLT strategy is generally used when the T_{rise} is less than twice the T_{delay}. Figure 3-13 shows three cases: A1,B1 needs TLT, whereas A3,B3 does not; the A2,B2 situation sits on the edge. A summary of when to use transmission line and termination is provided in Table 3-2.

For simplicity, the resistor termination, R_{term}, is shown here as a single resistor. PCB circuits typically use a *Thevenin termination*, which includes a resistor to ground and a second resistor to the digital power supply. For a 50 Ω transmission line, this would become a 100 Ω resistor to ground and a second 100 Ω resistor to the power. This method balances the DC load of the circuit for both logic states, while still providing an equivalent termination resistance.

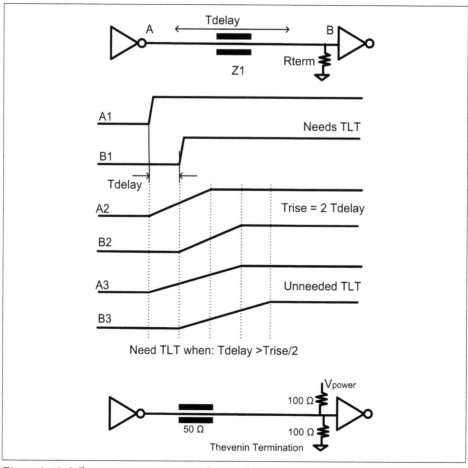

Figure 3-13. When to use transmission line and termination

Table 3-2. Transmission line and termination: rise time versus distance

Digital rise time[a]	Use TLT beyond distance[b]
0.1 ns	0.8 cm
0.5 ns	4.2 cm
1.0 ns	8.3 cm
2.0 ns	17 cm
3.0 ns	25 cm
4.0 ns	33 cm
5.0 ns	42 cm

[a] Use the fastest edge for devices with unbalanced rise/fall times.
[b] A propagation time of 60 ps/cm is used here, which is typical for FR4 PCBs.

Use transmission lines with controlled impedance layout and resistor termination, as shown in Table 3-2. As shown, devices with rise times under 1 ns are very restricted in a PCB environment. Most board-level digital devices use rise times in the 1–5 ns region.

Fast devices require special efforts in PCB layout to avoid problems with transmission line reflections and impedance mismatches. Devices with slower rise times are easier to work with. Modern FPGA/CPLD devices often provide a feature for selecting the drive strength of an output pin. Using a weaker drive strength results in slower rise times and fewer problems. If a slower device is not viable or if the ability to adjust the drive strength is not available, there are other options.

Inserting resistance at the driver end of the connection (Figure 3-14) can slow the rise time enough to create a longer connection. Short connections (<5 cm) can use the parasitic capacitance of the PCB and the IC being driven. Longer connections (>5 cm) need a capacitor closely placed with the resistor to avoid fast transients on a long inductive connection.

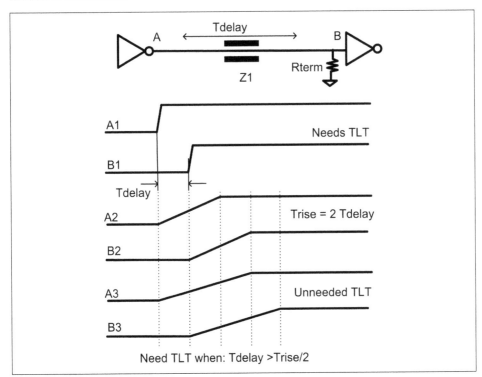

Figure 3-14. Increasing rise time to allow longer path transmission

This technique can be useful in certain situations where a longer-distance connection is needed and other methods are not readily implemented. This will skew signal timing slightly, but small amounts of resistance (10–100 Ω typically) can change the rise time enough to fix a problematic signal without redesigning the circuit.

Remember that ground-referenced digital signals have limitations, even when run through properly terminated transmission lines. LVDS methods through differential terminated transmission lines are preferred when speed, distance, and noise immunity are important.

Clock Distribution

Low-frequency (<1 MHz) clock distribution should be straightforward. As frequency goes up, connection distance increases, and as more destinations are added, the signal integrity, timing jitter, and time skews between destinations need to be analyzed. A comparison of two cases with commonly used clock frequencies illustrates the issues.

A real-time clock (RTC) runs at 32.768 KHz. This frequency is suitable for crystal fabrication and is easily divided to create a 1-second clock. Using an RTC is a common strategy for many controller systems. The RTC period is 30.5 us.

The PCIe interface standard uses a 100 MHz clock that is distributed as part of the PCIe bus. The PCIe clock period is 10 ns.

Comparing RTC to the PCIe clock illustrates the differences:

Clock buffering
A 5 ns gate propagation delay is typical for a clock buffer. For the RTC, that is 0.016% of the clock period, which is insignificant and can be ignored. For PCIe, the same buffer is 50% of the clock period and significantly affects timing.

Propagation delay
A 50 cm long PCB connection has a propagation delay of about 3 ns (using 60 ps/cm). The RTC gets a clock skew of 0.01% of the clock period and the PCIe skews 30% of the clock period.

Rise/fall time
As an arbitrary number, allocate 5% of the clock period to each rise/fall cycle. This means the RTC needs rise/fall times less than 1.5 uS, and the PCIe needs 0.5 ns. The PCIe system needs a transmission line beyond 5 cm. The RTC can have slower rise times and avoid a TLT for most situations.

 Many logic devices exhibit metastability when inputting slow transition edges. Therefore, device specifications for the slowest transition edges, or the use of devices with hysteresis, needs to be investigated. Slower rise times work, but only up to a point.

Both propagation delay and gate delay can seriously affect accuracy, or can be ignored as inconsequential, depending on clock frequency. The path to take can be determined by looking at the clock frequency and determining performance similar to the preceding example.

Too much capacitive loading can degenerate signals. Logic vendors use the concept of output driver fan out and input device standard loads to simplify the problem for designers. For CMOS devices, standard loads are a simple method of adding up the capacitance attached to a signal.

Figure 3-15 shows a clock at A as it passes through a single clock buffer; B is distributed to nine locations with long connections (Z1, Z2, Z3) and no path organization.

The first problem here is the capacitive loading presented by nine different devices. Check the capability versus loading of the clock buffer driver and the loads due to the nine devices and the interconnect. If the setup has low clock rates and slow rise times, this may be sufficient, but there is room for improvement.

The second problem here is the existence of a long stub (due to C to D and Z3) in between two other long signal paths (Z1, Z3). Connections branching off other connections are often a problem with impedance mismatches and line reflections.

Single point-to-point connections are highly preferred. Cleaning up signal paths will produce more predictable behavior.

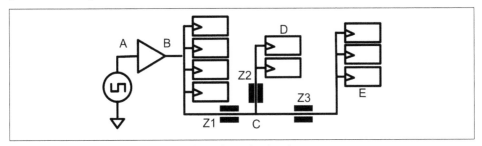

Figure 3-15. Skewed clock trees with multipath edge degeneration

Figure 3-16 uses a structure that has the same tightly grouped loads at B, D, and E. Instead of a branched connection, there are two separated transmission lines (B to E and B to D) and no stubs. The two paths have been replaced with a TLT to minimize mismatch reflections. Several items still need to be either checked or resolved:

- Examine the total driver loading with nine logic loads, transmission lines, and DC loading from termination resistors.

- Transmission lines B to E and B to D share a common node, and can interact and cause signal integrity issues.

- Timing skew errors between signals at B, D, and E need to be checked.
- Rise/fall characteristics at B, D, and E will be somewhat different due to different impedance and loading, which may or may not be significant.

Minimizing these issues by using a multi-output clock driver is often a solution.

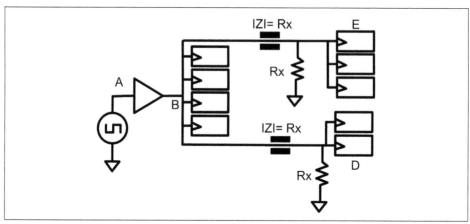

Figure 3-16. Clock spider network driving multiple lines

Figure 3-17 shows three separate drivers with TLT being used. Interactions between transmission lines are eliminated and driver loading is reduced. Signals at B, D, and E should have similar rise/fall characteristics. Clock buffers are available with timing skew errors under 100 pS (ground referenced) and under 10 ps (LVDS).

The importance of timing skew errors due to physical length mismatch of F2E, F2B, and F2D needs to be checked.

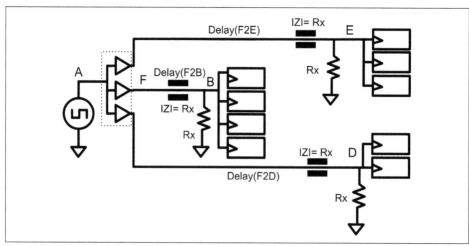

Figure 3-17. Clock distribution with buffer path/load balancing

Reminder: R_x here should be implemented as a Thevenin termination, as discussed earlier.

Do F2E, F2B, and F2D need to be length-matched to match propagation delay time? Transform the length differences into time differences and compare them to the time period of the clock to determine whether physical path balancing is a significant time skew component.

If length differences are a significant part of the clock period, serpentine interconnects in the shorter paths (Figure 3-18) can reduce time skew. Serpentine paths take up significant space and frequently are unnecessary. Important questions here are:

- Is the path length difference a significant portion of the clock period?
- Is a time skew tolerable between separate parts of the system?

Older designs, with many logic gates, frequently needed careful clock path balancing. Modern systems may have a single FPGA as the destination, and careful path balancing is not needed.

Figure 3-18. Serpentine path balancing by the numbers

Figure 3-19 shows that time skew can also be created with:

- Series resistance, which creates delay through use of RC time constants
- Gate delays in series

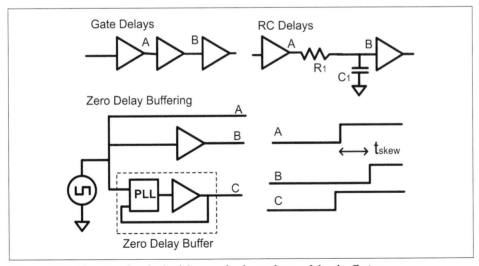

Figure 3-19. Skewing the clock, delay methods, and zero delay buffering

Gate delay can change due to semiconductor process variance and needs to be carefully examined over specified variances. The introduction of limited amounts of series resistance produces a more consistent result, but time-shifting the clock should be kept at under 10% of the clock period to allow sufficient time for the RC time constant to settle out.

As a general rule, avoid introducing intentional signal delays unless there is a compelling need for it. In the single-FPGA era, intentional signal delays are unnecessary most of the time. If shifting a signal to an earlier time is needed, zero delay buffers are also available.

The zero delay buffer output occurs at the same time as the input. To do this, a phase locked loop (PLL) is used internally to regenerate a clock output with near-zero time skew to the input signal.

High-frequency clocks within noisy environments should consider using an LVDS method (Figure 3-20). LVDS clock transfer using differential transmission lines (aka differential microstrip, FDI:PCB) and impedance-matched differential termination works well for high frequencies and noisy applications. PCIe uses this method to distribute the 100 MHz reference clock required by the standard.

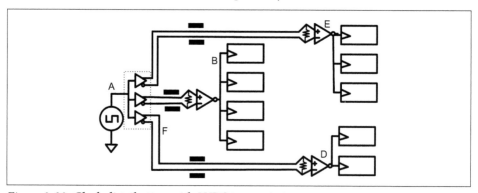

Figure 3-20. Clock distribution with LVDS transmission

Digital Communication: Parallel Versus Serial Ports

Data can be transferred between devices in a parallel or serial manner (Figure 3-21). Parallel communication requires multiple pins and connections between devices. A large amount of area inside both ICs is used to provide the external signals.

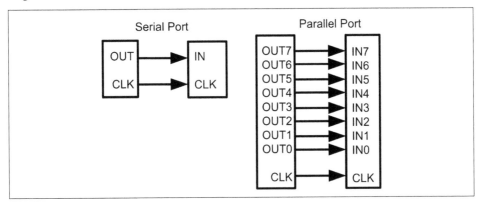

Figure 3-21. Parallel versus serial communication

Due to the cost of silicon and the complexity of interconnects, parallel ports are mostly obsolete. Serial ports pass data sequentially between devices with a reduced pin count. Modern systems use serial port methods and are the focus here.

Clocking Methods for Serial Ports

In addition to passing data, some method of providing clock synchronization must be used. Four methods are commonly used:

- Starting edge synchronization
- Parallel path clock
- Self-clocking data
- Embedded clocking

Starting Edge Synchronization

Starting edge synchronization (Figure 3-22) requires that the transmitter (TX) and receiver (RX) have a predefined data rate and access to a reference clock. In this example, when the TX goes low, the RX recognizes that as the starting edge of the signal and uses its reference clock to sample the expected middle of the data window for a short series of data bits.

Clock phase errors get progressively worse after synchronization, so this only works for slow and short data fields. Error prone and originally used in RS-232 and UART devices, this method is not encouraged for new designs.

Figure 3-22. Starting edge synchronization

Parallel Clock

A parallel clock is time-synchronized to the data (Figure 3-23). Clock-to-data synchronization needs to be properly phased for sufficient setup and hold times. The limitations of parallel clock systems necessitate a second connection and the need to maintain a good clock/data phase relationship. High-speed data systems can have problems maintaining this relationship, so embedded clocking methods are preferred.

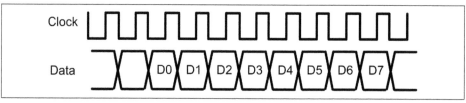

Figure 3-23. Parallel clock

Manchester Code Self-Clocking

A self-clocking data stream (Figure 3-24) is commonly created with a Manchester code. Non-return-to-zero (NRZ) data is transformed so that the data is represented as rising or falling edges. In the positive-edge version of the Manchester code, a binary 1 becomes a rising edge and a binary 0 is a falling edge. (The code set is inverted for the negative-edge version.) This set of transitions can be decoded and synchronized at the receiver without a reference clock.

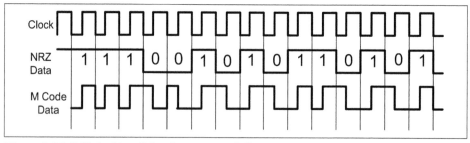

Figure 3-24. Self-clocking (Manchester encoded)

Manchester coding doubles the bandwidth of the signal, which can be problematic at high data rates.

Embedded Clock and Run Length Limited Codes

Embedded clock systems create a clock based on the data stream (Figure 3-25). This is called *clock recovery* and is done using a PLL. The PLL generates a clock that is in alignment with the transient edges of the incoming data. The received data needs frequent transient edges to keep the PLL properly aligned. This can be a problem with NRZ data, which can have long strings of the same state and lacks transient edges for timing alignment.

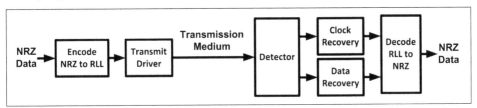

Figure 3-25. Embedded clock system

Avoiding a long string of ones or zeros requires encoding the NRZ transmission data. NRZ data is converted to run length limited (RLL) data, which guarantees a min/max limit between transient edges. This allows successful, phase-aligned clock creation.

Figure 3-25 illustrates a typical system. The RLL data controls a transmit driver, and the data passes through a transmission medium and is detected asynchronously at the receiver. This data stream is used to recover a clock and provide synchronous data to a decoder that converts the data from RLL to NRZ.

Most high-speed data communication systems use embedded clock methods. But all of these clocking methods are commonly used with different serial interfaces.

Digital Communication: Features and Definitions

Before examining the communication methods used commercially, understanding some commonly used criteria and terminology is useful:

Data rate
> Sometimes called *data transfer rate*, the *data rate* refers to how much data you can get through a connection per unit of time. This can be measured in bits per second, bytes per second, baud, or symbols per second. Vendor-specified data rates may not include supplemental time used for support communication protocols or other overhead processes.

Distance

Any communication method is distance limited. Often, data rates need to decrease as distance increases.

Addressing

The purpose of including an address is to provide a definition for the final data destination. The number of devices allowed on a connection is limited by digital addressing, capacitive loading, or other system specifics.

Error checking

Data integrity is an important part of communication. Data errors between two closely placed chips on the same PCB are generally not a problem. But as distance increases, data rates go up, or noise is introduced, error rates go up and integrity must be checked.

Protocol or protocol structure

This is the predefined procedure that devices use to communicate. Protocol features can be very different depending on the application. Common features include the definition of data sequence and size, encoding methods, error checking methods, acknowledgment and handshaking procedures, addressing structure, requests for start/stop/repeat, and others.

Data packet

This is the block of digital information that is communicated, as defined by the protocol. A generic data packet is shown in Figure 3-26. Common components of a data packet include:

Preamble

Used to prepare the communications path for what is to follow. Receivers use the preamble to make signal processing adjustments and synchronize a clock recovery PLL to the incoming data.

Addressing

This includes a destination address and sometimes a source address. Depending on the protocol, this can be a hardware-defined register address, an IP address, or a MAC address. Addressing is needed to define the destination, and in some protocols the original source of the data is needed as well.

Payload data

Also known as *data* or *data field*, this contains the actual information being conveyed. Depending on the protocol, this can be fixed or variable in size.

Error check

This information exists to quickly determine whether the data field was delivered correctly. Error checking is commonly done with cyclic redundancy check (CRC) code.

Connection configuration

This can have several forms (see Figure 3-27). A *point-to-point (P2P) configuration* is two devices communicating, with one on each end of the connection. P2P connections are high frequency and transmission line friendly. A *multidrop system* is a single driver with multiple receivers. A *multipoint system* has multiple receivers and multiple drivers. A system can have multiple receivers while still having a P2P configuration. Separate drivers for each connection are utilized to separate connections from each other.

Hot plugging

This refers to plugging and unplugging a device from a powered-up system. Many devices can't do this without damaging the electronics. Properly designed hot plugging can insert or remove a device without causing damage, won't disturb normal host functionality, and can seamlessly self-configure.

Physical medium

The connection between the transmitter and receiver has multiple methods including PCB traces, discrete wires, twisted pair wires, coaxial cable, optical fiber, infrared light, magnetic coupling, and wireless radio.

Communication channel capability

Telecommunication terminology (Figure 3-28) is used to define the capability of the data path:

Simplex systems

These communicate in only one direction. Loading data into a register is an example of a simplex process if there is no way to verify or read back the data.

Half-duplex systems

can These communicate in both directions but not at the same time. Interleaving in time is a common strategy to implement a half-duplex system.

Duplex or full-duplex systems

These can communicate in both directions at the same time. Depending on the device, the two-way communication can occur on the same path/channel/wires, or a second path can be used to communicate in the other direction. Two simplex systems can be used together to create a duplex system.

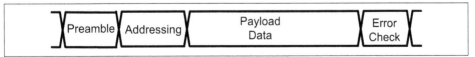

Figure 3-26. Typical data packet content

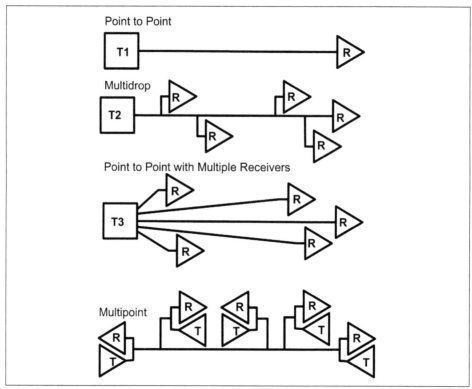

Figure 3-27. Multipoint versus point-to-point interconnect

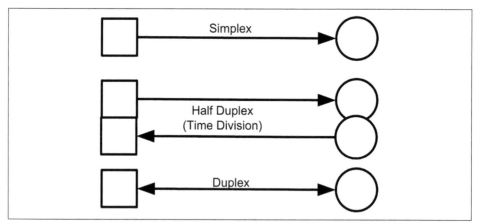

Figure 3-28. Simplex, duplex, full-duplex, and half-duplex communication

Serial Data: Shared Ground, Low Speed

Although ground-referenced logic can have signal integrity issues, it is the backbone of most digital systems. A low-impedance common ground is a must-have for such systems to function reliably. Some digital interfaces offer no detection or correction for errors, so they need to operate in a shared ground environment with high data integrity.

Universal Asynchronous Receiver Transmitter

The universal asynchronous receiver transmitter (UART) structure (Figure 3-29) uses start edge synchronization and is a simplified variant of the RS-232 interface. The UART has a receiver-to-transmitter connection for both directions and nothing else. Device clocking is performed internally using two independent time references where the clock time period (referred to as *baud rate*) has been programmed into both devices.

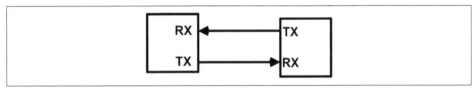

Figure 3-29. UART connections

Figure 3-30 shows the associated signals. Data transfer is started by the falling edge of either TX output. This falling edge (START) provides a zero phase reference for receiver synchronization. By knowing the baud rate of the expected data, the RX internal clock can time-sample data in the expected middle of each bit for 9 bits.

Since independent TX and RX clocks will differ in frequency, the sample point timing will drift off-center after start synchronization. This limits both the maximum data rate and the length of the data field. This method works for short runs of data only.

A parity bit (PB) is added to the end of the data to determine whether an odd number of errors occurred. An even number of errors would not be detected.

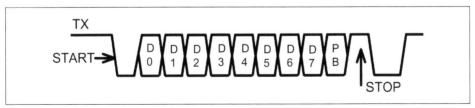

Figure 3-30. UART signaling strategy

The UART is simple but performance limited, as summarized in Table 3-3.

Table 3-3. UART performance summary

Parameter	Performance
Connection	Single point to point, full duplex
Data rate	1 Mbit/s max, unique to device
Addressing	None
Validation	No acknowledgment or parity error check
Medium	Wired, two connections
Signaling	Ground referenced, active push/pull drivers
Distance	On PCB, needs common ground
Other	Obsolete, no standard; many variants exist

The UART is limited to roughly 1 Mbit/s and a data field of 1 byte. The method has many existing legacy designs with a multitude of variants. UARTs are not suggested for new designs, but designers may have to use the methodology to interface with older devices.

Inter-Integrated Circuit and System Management Bus

The Inter-Integrated Circuit (I²C) bus structure was defined by Philips Semiconductor circa 1982. I²C is a simple, low-speed communication method for use between chips on a common PCB. Many semiconductor vendors provide an I²C interface for their products to do setup and control of internal registers.

The System Management Bus (SMB or SMBus), developed by Intel, is a variant of the I²C interface. In general, SMB and I²C devices are compatible on the same connections.

I²C consists of two signals (Figure 3-31): a clock (SCL) and serial data (SDA). Multiple devices can be connected to the SDA/SCL signals, and a pair of pull-up resistors (R$_{pu}$) are also connected. The connection pins (SDA, SCL) of any attached chip can pull down the signal, with the resistors providing a passive pull-up. Any device can initiate communication and becomes the manager, which controls communication with subordinate devices.

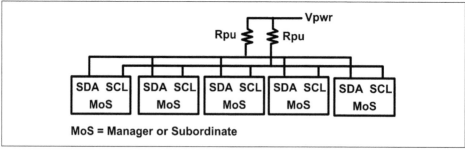

Figure 3-31. I²C connections

An example of writing data is shown in Figure 3-32. A manager device initiates a transfer by dropping the SDA, and then lowers the serial clock (SCLK). Following that, an address, read-write bit, and data field are sent. Acknowledgment (AK) is a response from the subordinate device.

The clock is specified to do a full rise and fall pulse in the middle of each bit sent. With 7 bits of addressing, the system is limited to 128 locations, but a more modern variant allows 10 bits of addressing.

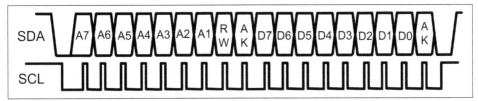

Figure 3-32. I²C signaling strategy

I²C performance is limited by its data transfer rate. Also, with no error checking, the device is limited to situations where good signal integrity is a must-have. Consequently, I²C is limited to local communication where all chips use the same low-impedance ground, as seen on a PCB. An I²C performance summary is provided in Table 3-4.

Table 3-4. I²C performance summary

Parameter	Performance
Connection	Multidrop, half duplex; multiple managers, multiple subordinates
Data rate	100 Kbit/s standard, 400 Kbit/s fast, 3.4 Mbit/s high speed
Addressing	7 bits, 128 locations, read/write
Validation	Receive acknowledge, no error check
Medium	Wired, two connections
Signaling	Ground referenced, active pull-down, resistor pull-up
Distance	On PCB, needs common ground
Other	10-bit addressing optional

This method fills an important need for easily implemented communication, and it enjoys widespread use, with many chips utilizing I²C interfaces.

Serial Peripheral Interface

The Serial Peripheral Interface (SPI) bus structure was defined by Motorola circa 1985. A single manager (Figure 3-33) with the capability to select individual subordinates for full-duplex communication is the basic structure. Many vendors have adopted the interface with some variants to suit their needs.

SPI is serial data with higher speeds than a passive resistor pull-up system can provide. With a dedicated subordinate select pin, there is no address information in the data packet. Many vendors use the SPI interface for data streaming.

SPI signals include a clock (SCLK) and two data paths: MOSI (Manager Out, Subordinate In) and MISO (Manager In, Subordinate Out). A manager device requires multiple control lines to select from multiple subordinates, and subordinate devices require an enable line known as *subordinate select* (SS). Unless a subordinate has been enabled by the manager, the subordinate maintains the shared MISO line in a high-impedance state.

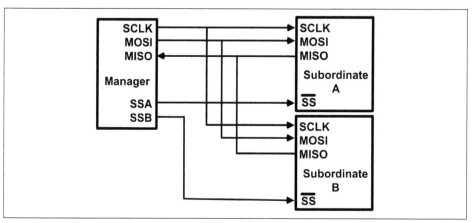

Figure 3-33. SPI connections

The signaling strategy is simple (Figure 3-34): the manager asserts a low SS connection, and SCLK is used to transfer data in both directions. Depending on the application, the data field may be longer than shown, and some subordinate chips may only have a MOSI and no MISO, thus functioning only as a receiver. Many variations on this standard exist.

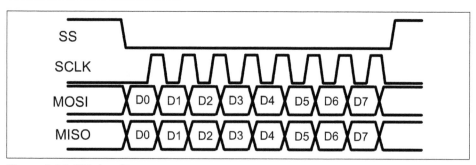

Figure 3-34. SPI signaling strategy

With actively switched up/down drivers, SPI devices can get higher data rates than resistor pull-up techniques such as I²C. However, with no error checking, the method is limited to short connections on a common ground. Table 3-5 summarizes the performance of SPI.

Table 3-5. SPI performance summary

Parameter	Performance
Connection	Single manager, multiple subordinates, full duplex
Data rate	10 Mbit/s is common and unique to the device
Addressing	None; SS pin activates device
Validation	No acknowledge, no error check
Medium	Wired, four connections
Signaling	Ground referenced, active push/pull drivers
Distance	On PCB, needs common ground
Other	No formal standard; many variants exist

Single-Wire Interfaces

Serial bus structures have been devised that use a single-wire connection. Two of these have come into common use, namely 1-Wire and UNI/O.

Dallas Semiconductor (now part of Maxim Integrated) developed the 1-Wire interface, which combines data, address, and power on a single wire, referenced to the ground connection. Generally, this may be used in peripheral devices such as sensors and identification devices. Microchip developed the UNI/O bus specification, which is designed for small form factor microcontroller buses, single-wire access to EEPROMs, and other low pin count applications. Both devices are in limited use where lowest pin count is a priority.

Serial Data: Shared Ground, High Speed

Serializer and deserializer (SerDes) interfaces are designed to transfer serial port data at high rates of speed. There are multiple devices on the market with no single standard being dominant. Many vendors provide both a serializer (transmitter) and deserializer (receiver) designed to their proprietary standard.

All SerDes interfaces share some common features (Figure 3-35). The serializer and deserializer blocks exist within separate chips, where a high-speed data port is needed. SerDes interfaces use an embedded clock method, as described earlier. The serializer input is a data bus, often internal to the chip. That data bus provides a stream of NRZ data, which is then RLL encoded.

The RLL controls the transmit driver. LVDS is used as the physical strategy for transmission. Some variants use differential current steering to create a pair of signals. Some variants may also use signal processing methods at transmission (pre-emphasis) and reception (equalization). The transmission medium can be a differential stripline on a PCB, or twisted pair wires in a cable.

At the receiver, the signals are received differentially and converted into a binary asynchronous data stream. Data and clock recovery is performed, and the RLL data is decoded to NRZ synchronous data.

All SerDes interfaces use a similar architecture. At high data rates, maintaining a separate accurate clock is not readily done, so using embedded clock methods makes sense.

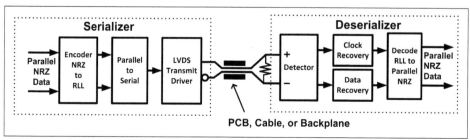

Figure 3-35. Typical SerDes block diagram

This is a typical architecture for one SerDes set. Depending on what is needed, two sets can be used for duplex communication, or multiple sets in parallel can be used to push data rates up. Table 3-6 summarizes SerDes performance.

Table 3-6. SerDes performance summary

Parameter	Performance
Connection	Point to point, simplex 2-wire
Data rate	Mostly <6 Gbit/s; many variants exist
Addressing	None
Validation	Depends on protocol used
Medium	Cable twisted pairs or differential matched lines
Signaling	LVDS or differential current steering
Distance	Depends on data rate versus distance
Other	No standards here; vendors offer chip sets to serialize/transmit and receive/deserialize

Many communications protocols use a SerDes approach to the physical layer. The specifics of a protocol to work with the SerDes interface will define things such as preamble, addressing, control data, data fields, error detection, and related items.

Data Between Boards or Between Systems: Wired Methods

Certain features are necessary to ensure reliable communication between PCBs, across a cable, or between systems. In the protocol, error checking and handshaking methods need to be defined to make sure the data arrived properly. The physical system needs to be differential to reject common mode noise and ground variance. It is difficult to maintain a good high-frequency ground connection over distance. Widely separated systems will completely DC isolate the connection through electromagnetic or optical methods.

RS-232 functions without differential signals, but it uses large (+/−5 V to +/−15 V) voltage swings to do so. Using a 30-volt difference between logic states is a brute force technique.

RS-232: Serial Data over Cable

RS-232 is the grandfather of serial communication; it was devised in 1960, when digital logic was done with discrete transistors. The original application was for teletypes with little capability to make digital decisions. Consequently, the protocol includes control signals (Figure 3-36) not needed in modern methods.

RS-232 was widely used on early personal computers to access printers. It is still widely used where modern methods are lagging, such as in older laboratory and industrial equipment.

RS-232 connects between data terminal equipment (DTE, now called a *host*) and data circuit-terminating equipment or data communication equipment (DCE, now called a *peripheral*).

The signals are:

- RD—Receive Data in
- TD—Transmit Data out
- RTS—Request To Send; comes from the host
- CTS—Clear To Send; comes from the peripheral, ready to receive data
- DTR—Data Terminal Ready; host is ready
- DSR—Data Set Ready; comes from the peripheral, tells the host it wants to send data
- DCD—Data Carrier Detect; indicates that the peripheral has an active input
- RI—Ring Indicator; provides an indication of incoming data

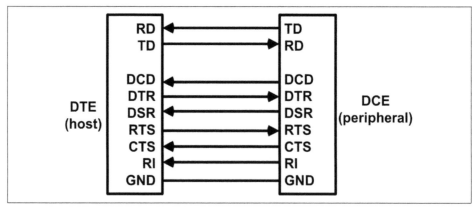

Figure 3-36. RS-232 connections

A typical data exchange is shown in Figure 3-37 and proceeds as follows:

- The host takes RTS high, telling the peripheral it is ready to send.
- The peripheral acknowledges with a CTS high.
- The host indicates it is ready with a DTR high.
- Data from the TD goes low, followed by a byte of data and a parity bit.

The first falling edge of the TD is for time synchronization. The TD connection functions the same as described in the UART exchange. Also similar to UART, both sides of the system must be set to the same data rate to communicate successfully, since timing is derived from the data rate and the phase location of the initial falling edge.

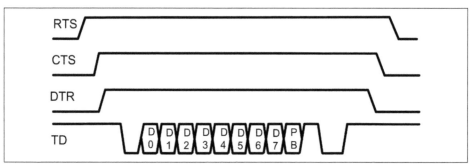

Figure 3-37. RS-232 signaling strategy

RS-232 is limited in speed and distance. Cable capacitance typically limits functionality to 15 meters. RS-232 uses ground-to-wire referenced, large-signal amplitudes in a brute force method to get a signal across. Table 3-7 summarizes RS-232 performance.

Table 3-7. RS-232 performance summary

Parameter	Performance
Connection	Single point to point, full duplex
Data rate	20 Kbit/s max
Addressing	None
Validation	No acknowledgment or parity error check
Medium	Cable wired, 9-pin and 25-pin variants exist
Signaling	Ground referenced, wide voltage swing logic
Distance	15 m max, limited by cable capacitance
Other	Obsolete; standard exists, but many "not fully compliant" variants also exist

RS-232 is obsolete but is frequently seen in older designs. The UART variant is still in use. If possible, RS-232 and UART should not be used for new designs, since there are better methods.

RS-485: Differential Serial Data over Cable

RS-485 was defined as an industry standard in 1983. At that time, the need for differential signaling and impedance-matched terminations had become evident. RS-485 is a physical layer standard where the communication protocol is not defined. Communication protocols to use the RS-485 physical layer include Modbus, Profibus, DH-485, and others. RS-485 is widely used in industrial automation and robotic control systems. It was originally specified to connect up to 32 devices, although some vendors have created interface chips that support more.

Figure 3-38 shows a twisted pair interconnect that makes the bus connection that is used between devices. Devices attached to the bus are daisy-chained in a row, avoiding transmission line stubs branching off from the bus. Termination resistance exists at both ends of the daisy chain to minimize transmission line reflections. Essentially, a transmitter in the middle of the chain is transmitting out to two transmission lines at the same time. Signal amplitudes are specified as +/−1.5 V at the transmitter and as +/−200 mV (or greater) at the receiver. This allows proper functionality, even with amplitude losses in the wiring.

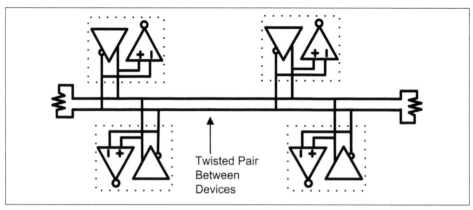

Figure 3-38. RS-485 connections

The capabilities of RS-485 are suitable to many situations where data rates are modest, multipoint connections are preferred, and differential signal noise immunity is needed. Table 3-8 provides a summary of RS-485 performance.

Table 3-8. RS-485 performance summary

Parameter	Performance
Connection	Multipoint, half duplex (2-wire) or full duplex (4-wire)
Data rate	40 Mbit/s (<10 m), 0.1 Mbit/s (1,000 m)
Addressing	Depends on protocol used; max 32 devices
Validation	Depends on protocol used
Medium	Cable wired, unshielded twisted pair
Signaling	LVDS, +/−200 mV receiver
Distance	1,200 m max, limited by cable resistance
Other	Is a "physical-only" standard; protocols include Modbus and others

Many vendors produce RS-485–compliant transceiver chips. Consequently, RS-485 is widely used and is still being implemented in new designs.

Controller Area Network

Controller Area Network (CAN bus) was developed for communication in automobiles but is also useful elsewhere. Bosch published the first version in 1983 intending to simplify automotive wiring interconnects. Production autos using CAN bus emerged after 1991. Virtually all cars manufactured after 2008 use CAN bus.

In addition to automotive, CAN bus is widely used in industrial automation, robotics, medical devices, electromechanical prosthetics, airplanes, and other industries. Systems using multiple MCUs that need communication and coordination are well suited to a CAN bus interface.

CAN bus uses a multipoint half-duplex architecture connected by a daisy-chained twisted wire pair (Figure 3-39). Signaling is differential but is not LVDS. Instead, high-low differential signaling (HL-DS) is used, where a control line can be either dominant or recessive. The advantage of differential signaling and common mode noise rejection is still retained.

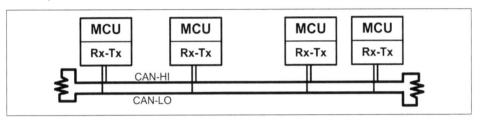

Figure 3-39. CAN bus connections

In the HL-DS method, CAN-HI and CAN-LO take either a dominant or recessive state (Figure 3-40). When recessive, both signals are roughly the same mid-scale voltage, which indicates logic 1. When dominant, CAN-HI is pulled up and CAN-LO is pulled down, which indicates logic 0. With this, any device can force a dominant 0 over another device with a recessive 1. This is valuable in bus arbitration. When a device starts to use the bus and puts out an address, another device can force an address change to a lower value. Address locations with lower values take precedence. Assigning addresses by priority gets faster response to select devices. For example, in a car, the brakes take higher priority than climate control.

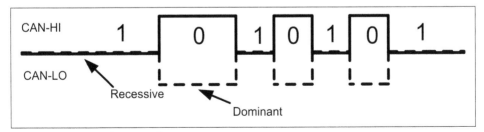

Figure 3-40. CAN bus signaling strategy

CAN bus was optimized to support a distributed multicontroller system with up to 30 devices on the bus. Many semiconductor vendors provide CAN-embedded drivers and software for their microcontrollers. Table 3-9 summarizes CAN bus performance.

Table 3-9. CAN bus performance summary

Parameter	Performance
Connection	Multipoint, half duplex (2-wire)
Data rate	1 Mbit/s
Addressing	11 or 29 bits, max 30 devices
Validation	Receive acknowledgment, CRC error check
Medium	Cable wired, one twisted pair, shielded or unshielded
Signaling	HL-DS, 3 V/1 V output, min +/−200 mV receiver
Distance	40 m
Other	Capable of address-defined priority, MCU-friendly protocols, CANOpen, DeviceNet, CAN Kingdom

CAN bus performance is limited to 1 Mbit/s and 40 meters of distance. For the target application of telemetry and coordination communication common in cars, the data rate and distance are suitable. CAN bus is widely used in aerospace, industrial robots, and factory automation.

Serial Data for Computer Systems

Some serial data methods were developed for use in the support of computer systems. Some remain associated with computer motherboards (serial ATA, PCI Express), while others are used in other applications (USB, Ethernet).

Universal Serial Bus

Universal Serial Bus (USB) was first developed in 1996 as a single-cable connection for computer peripherals. USB is now the predominant method of single-cord, plug-and-play peripheral connection for computers, laptops, tablets, and smartphones. Also, the USB cable includes power connections, leading to its adoption for battery charging.

Figure 3-41 shows that USB 1 and USB 2 are four-wire (5 V power, data+, data−, ground) connections. USB 3 adds two additional data pairs for "super speed" data cables.

USB specifications extensively define cables, power restrictions, connector design, and backward compatibility issues to ensure reliable plug-and-play device compatibility.

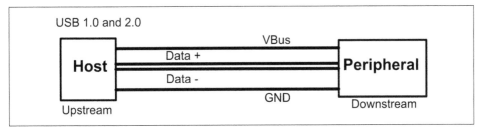

Figure 3-41. USB connections

The digital content (Figure 3-42) of a USB transaction, called a *packet*, is fairly straightforward:

Preamble
> Used for receiver synchronization.

PID
> Packet identification. The PID defines the purpose of the packet: confirmation handshaking, data exchange, configuring data rate, tokens, direction of transaction, initialization, etc.

ADDR
> Target address, for a maximum of 127 devices.

ENDP
> Endpoint. The ENDP provides supplemental information to support the PID. Depending on the PID value, the ENDP can have unique functions.

FRAME#
> Frame number used as a sequencing tool to properly order multiple packets.

DATA
> Data field. This can be from 1 to 1,024 bytes in length.

CRC
> Cyclic redundancy check; essentially error checking for the data field.

Data bytes are transmitted LSB first, MSB last.

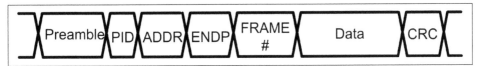

Figure 3-42. USB signaling strategy

USB devices must be user-friendly. Cable lengths are limited, but easy plug-and-play scenarios, automatic configuration, and hot plugging are well supported. Table 3-10 provides a USB performance summary.

Table 3-10. USB performance summary

Parameter	Performance
Connection	Point to point, half duplex (1.0, 2.0)
Data rate	1.5 Mbit/s, 12 Mbit/s, 480 Mbit/s, and 5 Gbit/s versions
Addressing	7 bits, 127 devices max
Validation	Handshaking protocol, CRC error check
Medium	4-wire cable, single twisted pair, power, GND
Signaling	LVDS with added series resistance
Distance	5 m (1.0, 2.0), ~3 m (3.0)
Other	USB 1, 2 are 4-wire; USB 3, 4 are 9-wire; USB 3.1–4.0 are up to 40 Gbit/s; can hot-plug

The price paid for an easy end-user experience is the complexity of the support software. USB software can be a significant part of the design effort.

Serial Advanced Technology Attachment

Serial Advanced Technology Attachment (SATA) is a specialty computer interface for mass storage devices. Magnetic hard disk drives (HDDs), solid-state drives (SSDs), and optical media (DVD, CD) use this interface when internally installed. Standalone mass storage devices have adapted to a USB interface for its easier plug-and-play setup.

SATA (Figure 3-43) has five wires: two (A, B) differential twisted pair connections and three grounds. Power is independently supplied. Although separate transmit/receive paths are available, the interface is not full duplex. Half-duplex capability is specified, due to the inability of an HDD to write and read simultaneously.

The address format reflects HDD technology using the address format of an HDD. The address information includes Head (recording head selection), Cylinder (track cylinder), and Sector (rotational "wedge" area).

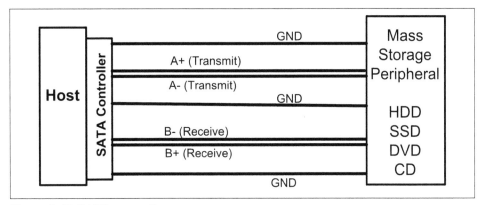

Figure 3-43. SATA connections

SATA is a specialty interface that designers will need only when dealing with a mass storage device. Table 3-11 summarizes SATA performance.

Table 3-11. SATA performance summary

Parameter	Performance
Connection	Point to point, half duplex
Data rate	150 Mbit/s, 300 Mbit/s, and 600 Mbit/s versions
Addressing	Magnetic media address: head, cylinder, sector
Validation	CRC error checking
Medium	7-wire cable, two twisted pairs, GND
Signaling	LVDS
Distance	1 m
Other	Specialty interface for mass storage devices; power on separate connection

Peripheral Component Interconnect Express

Peripheral Component Interconnect Express (PCIe) originated in 2003 as a computer bus expansion standard. As a serial bus interface, it replaces the older parallel bus approaches of PCI and PCI-X.

PCIe can be thought of as multiple SerDes channels (Figure 3-44). In PCIe, a full duplex communication "lane" uses two SerDes channels, one for transmission and one for reception. Up to 16 lanes can be included where high data rates are needed. A 100 MHz LVDS clock is provided across the bus connection. This clock serves as the reference for the receiver's clock and data recovery.

Other features include 12 V and 3.3 V power, multiple ground connections, a JTAG interface for testing, and SMBus (aka I²C) for configuring the remote device. Supplemental control connections to quickly determine how many lanes are present (X1, X4, X8, X16 variants) are provided by PRSNT#1 and PRSNT#2 pins. WAKE and CLKREQ complete the needed controls.

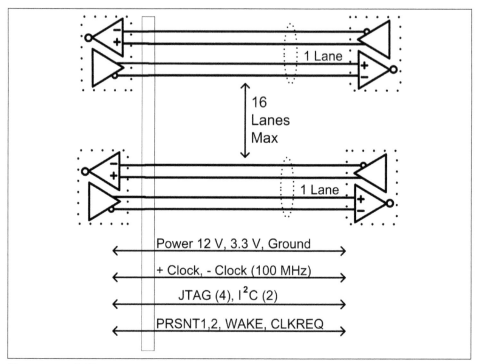

Figure 3-44. PCIe interconnect structure

PCIe performance is quite fast, but is distance limited. Version 1 (2003) is 250 MB/s per lane, and version 5 (2020) is 3,938 MB/s per lane. Using 16 lanes of version 5 provides 63 GB/s throughput.

High speed comes with a distance restriction of 50 cm. PCIe is designed to interface plug-in cards to a motherboard backplane. If high-speed data on a cable is needed, conversion to a cable-friendly format is the solution. Both Ethernet and USB are suitable, depending on the needed data rate and distance. Table 3-12 summarizes the performance of PCIe.

Table 3-12. PCIe performance summary

Parameter	Performance
Connection	Point to point, full duplex
Data rate	Up to 4 Gbit/s per lane; 16 lanes max, ~64 GB/s
Addressing	32 or 64 bits
Validation	Acknowledgment protocol, CRC error check
Medium	18- to 82-pin connector, data across PCB
Signaling	LVDS for data and clock
Distance	50 cm
Other	Fully specified standard for signals, connectors, protocol, and testing

The PCIe specification fully defines electrical, mechanical, and protocol criteria to ensure plug-in compatibility between vendors. PCIe capability is included on microprocessors, FPGAs, digital video interfaces, and other devices where high-speed data is needed

Ethernet

Ethernet is the LAN workhorse of computer systems. Early versions used coaxial cable and some modern variants use optical fiber. The bulk of Ethernet connections are done using differential signaling over twisted wire pairs. Here the focus is on the ubiquitous twisted pair Ethernet embodied in the 10Base-T, 100Base-TX, and 1000Base-T variants.

Ethernet over twisted pairs (Figure 3-45) is done with four differential signals, using eight wires. The signaling method changes with the data rate. An important part of the Ethernet connection is the magnetic coupling at each end. Using a transformer and common mode choke at both ends eliminates common mode signals. This gives up to 1,500 volts of DC isolation and immunity to voltage differences between the grounds at each end.

Ethernet uses point-to-point connections, avoiding signal corruption issues associated with multidrop systems. Connections between cables are avoided using active circuits that perform signal regeneration and retransmission, thus isolating Ethernet cables from each other.

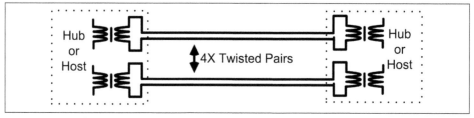

Figure 3-45. Ethernet connections

Information contained in an Ethernet message is referred to as an *Ethernet frame* (Figure 3-46). Information in the frame must support a system where source and destination are not always the two devices on each end of the cable.

The frame contains the following:

Preamble
Used for receiver synchronization

SOF
Start Of Frame; indicates the end of the preamble and the start of the address field

MACD
Media Access Control address Destination; the 48-bit address of the final destination of the frame

MACS
Media Access Control address Source; the 48-bit address of the original source of the frame

TAG
Information about the priority and protocol to be used for the frame

LEN
Length of the data field

DATA
Data field; can be 46 to 1,500 bytes long

CRC
Cyclical Redundancy Check; error checking code, also known as a frame check sequence

GAP
Required idle time before the next frame can be sent

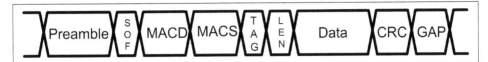

Figure 3-46. Ethernet signaling strategy

Ethernet capability covered here includes up to 1 GB/s variants. There are Ethernet variants over copper that will function at 5 GB/s and 10 GB/s. Table 3-13 summarizes Ethernet performance.

Table 3-13. Ethernet performance summary

Parameter	Performance
Connection	Point to point, full and half duplex (4- and 8-wire versions)
Data rate	10 Mbit/s, 100 Mbit/s, and 1,000 Mbit/s versions
Addressing	48 bits, MAC addressing
Validation	Receive acknowledgment, CRC error check
Medium	Cable wired, 2 or 4 twisted pairs
Signaling	DS, Manchester (10), MLT-3 (100), and PAM-5 (1,000)
Distance	100 m
Other	Other Ethernet variants exist; herein covering 10-Base-T, 100Base-TX, and 1000Base-T

Wireless Serial Interfaces

Wireless methods of making a data connection are commonplace and easy to implement. Many options exist to fit specific needs.

WiFi

WiFi came into use in 1997 and quickly became the foremost method to obtain internet access without a cable. Designing WiFi into a system is now a modular approach with complete WiFi modules requiring only a digital interface. Designers need very minimal knowledge of RF design to successfully include WiFi in a system.

Conceptually, the application of WiFi is straightforward (Figure 3-47). An internet service provider (ISP) is connected to a WiFi router. Devices with WiFi capability (laptops, tablets, smartphones) can obtain internet access through the wireless link.

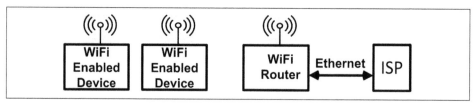

Figure 3-47. WiFi typical application scenario

WiFi operates on frequencies suitable to line-of-sight communication (ISM bands at 2.4 GHz and 5 GHz in the United States). At 100 mW of power, distance is limited to roughly 50 meters. Directional antennas and special care with obstruction-free transmission paths have yielded longer distances, but for general-purpose communications, with obstructions, interference issues, and omnidirectional antennas, 50 meters is a suitable guideline. Table 3-14 offers a WiFi performance summary.

Table 3-14. WiFi performance summary

Parameter	Performance
Connection	Multipoint wireless
Data rate	1–54 Mbit/s versions are common
Addressing	48 bits, MAC addressing
Validation	Error check and acknowledgment
Medium	900 MHz, 2.4 GHz, 5 GHz, and 5.9 GHz are common
Signaling	100 mW typical; QPSK (rev. B), 16/64 QAM (rev. A, G)
Distance	50 m, omnidirectional antenna
Other	Multiple updates with different performance of distance and data rates; those listed in this table are typical

Bluetooth

Bluetooth (BT) was developed for battery-powered, close-proximity wireless connections. BT has been widely adopted. Some example applications include:

Computers
　　Wireless keyboards, mice, game controllers, speakers

Cell phones
　　Wireless headsets, ear buds, smart watches, speakers, linking phone to car

Others
　　Medical body sensors, machines needing unrestricted motion, hearing aids, handheld remotes free of alignment sensitivity

All of these devices predominantly use BT or BTLE methods (BTLE is discussed in the next section). Figure 3-48 shows a typical application.

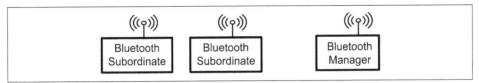

Figure 3-48. Bluetooth typical application scenario

BT works in a manager–subordinate configuration with up to seven subordinates. BT is suited to the creation of a personal area network (PAN). BT operates in the unlicensed ISM band at 2.4 GHz. Using variable power output from 0.5 mW to 100 mW BT, distance is roughly limited to 100 meters (high power) and 0.5 meters (low power). For things like cell phone headsets, data rates are sufficient to support high-quality audio and small form factor battery life with one to two days of general use. Table 3-15 offers a Bluetooth performance summary.

Table 3-15. Bluetooth performance summary

Parameter	Performance
Connection	Wireless; 1 manager, 7 subordinates max
Data rate	1 Mbit/s (1.0), 3 Mbit/s (2.0)
Addressing	48 bits, MAC addressing
Validation	Error check and acknowledgment
Medium	2.4 GHz
Signaling	Frequency hop; GFSK (1.0), DQPSK/8-DPSK (2.0)
Distance	10 cm (1 mW), 10 m (2.5 mW), 100 m (100 mW)
Other	Optimized for battery-powered wireless, with rates up to 3 Mbit/s

Many devices don't need a data rate to support audio, or they have a low amount of on time associated with transmitting data. This motivated the creation of Bluetooth Low Energy.

Bluetooth Low Energy

Bluetooth Low Energy (BTLE, also known as Bluetooth 4.0) has similar applications as BT in closely spaced networks. BTLE is optimized for items that require low average data rates and intermittent data transfers. By maintaining a low power sleep mode with brief wake times for communication, BTLE can achieve a battery life of years while BT devices are measured in days.

BTLE in version 5.1 (introduced 2017) and beyond includes the capability for mesh networks (aka Bluetooth Mesh or BT-Mesh) catering to the Internet of Things (IoT) market. Table 3-16 offers a BTLE performance summary.

Table 3-16. BTLE performance summary

Parameter	Performance
Connection	Wireless; 1 manager, with the number of subordinates dependent on application
Data rate	125 Kbit/s, 1 Mbit/s, and 2 Mbit/s versions are available
Addressing	48 bits, MAC addressing
Validation	Error check and acknowledgment
Medium	2.4 GHz
Signaling	Frequency hop GFSK
Distance	10 cm–50 m (1–10 mW)
Other	Optimized for low power for extended battery life; not suitable for duplex audio

Similar to WiFi, using BT or BTLE in a device is a black box approach requiring only a digital interface to get on the air. Designers need minimal RF knowledge to include BT/BTLE in a system.

ZigBee

ZigBee, introduced in 2005, is designed for low data rates, low-power sleep mode, brief communication bursts, and mesh network capability and offers over two years of battery life. ZigBee is optimized to work in a mesh network of many devices. Devices connect wirelessly (Figure 3-49) through multiple paths of communication. Home automation networks of distributed devices such as alarm system sensors, electronic locks, smart appliances, and other IoT devices are possible.

Every ZigBee mesh network has one coordinator (ZC) and can have multiple routers (ZR) and multiple end devices (ZED). ZR devices can pass signals through and serve to extend the network. ZED devices can connect to the mesh but not extend the network. The ZC orchestrates the system.

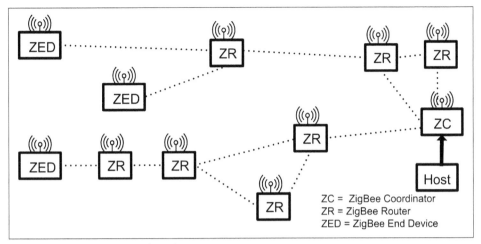

Figure 3-49. ZigBee typical application scenario

For a device to be ZigBee certified, a battery life of two years needs to be met. Table 3-17 summarizes ZigBee performance.

Table 3-17. ZigBee performance summary

Parameter	Performance
Connection	Wireless; 1 coordinator (ZC), multi-ZR and multi-ZED
Data rate	250 Kbit/s
Addressing	64 bits, MAC addressing
Validation	Error check and acknowledgment
Medium	2.4 GHz
Signaling	OQPSK
Distance	10–100 m (1–100 mW), fixed-location devices
Other	Optimized for low power; requires 2-year battery life to be certified

ZigBee devices use a 64-bit MAC address similar to a 48-bit MAC Ethernet address, allowing a unique address. The largest number of devices allowed on a mesh network is 64,000.

Z-Wave

Z-Wave is a wireless mesh network similar to ZigBee. Instead of the 2.4 GHz ISM band, it operates at 865–930 MHz, depending on country (908 MHz in the United States). Other differences include a limitation of four hops in the mesh connection and a maximum of 232 devices in the mesh.

Adaptive Network Topology

Adaptive Network Topology (ANT) started as a wireless link for sports and fitness sensors. BTLE and ANT devices compete for market share where ultra–low power consumption and low data rates are a priority.

Other Data Communication Methods

Some additional useful data communication methods don't fit the prior categories and are summarized in the following subsections.

Infrared

Infrared light is commonly used as a digital data link for wireless remote controls and communication where a device must be electrically isolated. Infrared is limited to distances under 2 meters and is directionally sensitive. Handheld remote controls for home entertainment devices are still predominantly infrared due to low cost.

Some remote controls are migrating to noninfrared methods (BTLE is common) to avoid directional sensitivity problems.

Fiber-Optic Data: Go Fast, Go Far

Data over copper has limitations in both speed and distance. As of 2023, the maximum data rate of PCIe is 64 GB/s but is limited to 50 cm of distance. Ethernet over twisted pair goes to 10 Gbit/s but is limited to 100 meters. There will be faster capability after this book is published, but to go fast and go far requires fiber-optic links.

An in-depth look at fiber is beyond the scope of this book, but a quick performance summary is useful (Table 3-18).

Table 3-18. Data over optical fiber performance summary

Parameter	Performance
Connection	Point to point, multiple channels per fiber
Data rate	50–500 Gbit/s per channel
Addressing	Depends on protocol used
Validation	Depends on protocol used
Medium	Optical fiber, multiple channels in each fiber
Signaling	QPSK, QAM, OFDM, and others are used
Distance	>100 km
Other	Multiple channels/fiber and multiple fibers/channel are done; cables with >100 Tbit/s are done

Optical fiber will go far (kilometers) and go fast (terabytes/sec). Consequently, fiber is the backbone of the internet. If you need high-performance data links, optical fiber methods need to be investigated.

JTAG: PCB Access for Test and Configuration

JTAG (Joint Test Action Group) is a digital serial access port for PCB testing. Most digital chips include a 4-pin JTAG access port for testing. JTAG ports on multiple chips are designed to connect together (Figure 3-50), allowing access to all chips with one external port connection. Controller chips, FPGAs, and EEPROMs are often programmed through the JTAG port after board assembly.

The JTAG signals are:

TDI
> Test Data In, serial port

TDO
> Test Data Out, serial port

TCK
> Test Clock, allows clocking of the serial data path

TMS
> Test Mode Select

TMS and TCK control an internal Test Access Port (TAP) controller that manages the JTAG interface. The TAP controller allows loading internal registers or pipelining data out through the TDO and into the TDI of the next device.

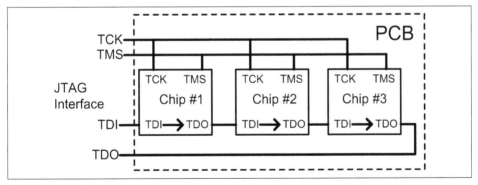

Figure 3-50. JTAG typical application scenario

In addition to internal control, JTAG is used to control boundary scan testing. This controls the digital output ports and senses the digital input ports of an IC. This allows testing of interconnects between chips on the PCB. Table 3-19 provides a JTAG performance summary.

Table 3-19. JTAG performance summary

Parameter	Performance
Connection	Wired daisy chain (2-, 4-, and 5-connection variants)
Data rate	No standard; typically under 10 MHz clock
Addressing	Serial pipeline
Validation	None
Medium	Four connections implemented in PCB traces
Signaling	Ground-referenced logic
Distance	On PCB
Other	For PCB testing, boundary scan and code downloads to digital ICs (IEEE standard 1149.1)

JTAG is commonly implemented using the four-wire variant. The speed of the JTAG chain is limited by the slowest chip in the chain. Generally, TCK is under 10 MHz.

JTAG is valuable for automated production testing and programming of on-board devices. All PCBs should have a JTAG port to make boards testable and programmable. Including a "design for test" strategy is an important part of any design.

Summary and Conclusions

The important points of this chapter are as follows:

- Digital signals may have signal integrity problems.
- Slow/soft/short networks are generally reliable.
- LVDS provides many advantages for fast or distance applications.
- Network placement and routing affects performance.
- Comparing rise time signal length to connection length is a guide to lumped versus distributed networks and terminated transmission line needs.
- The significance of time skews is guided by what percentage of the clock period the skews encompass.
- Physical length path balancing for time skews is often unnecessary. Designers need to be quantitative to determine need.
- Serial port communication is the dominant connection method.
- For chip-to-chip communication on a common PCB, use I²C (SMBus) and SPI.
- SerDes methods are useful for fast data.
- Distance communication needs common mode rejection, error detection, and handshaking validation.
- For cable communication between systems, use RS-485, CAN bus, or Ethernet.
- USB is the predominant host-to-peripheral interconnect.
- Computers interface internally with PCIe and SATA.
- Wireless interconnect options include WiFi, Bluetooth, and ZigBee, among others, and offer easy design-in methods.
- Fast and far data requires optical fiber methods.
- Include a JTAG interface in any PCB for test and programming access.

After reading this chapter, informed decisions can be made on how to get data between two points. For brevity, many specific details had to be left out. Knowing what's needed can start the implementation process. The needed details are available in many of the documents listed in "Further Reading" on page 124.

Further Reading

Digital Design Techniques

- *High Speed Digital Design: A Handbook of Black Magic* by H. Johnson and M. Graham, 1993, ISBN 978-81-317-1412-6, Pearson Prentice Hall.

- *High-Speed Digital System Design: A Handbook of Interconnect Theory and Design Practices* by S. Hall, G. Hall, and J. McCall, 2000, ISBN 0-471-36090-2, John Wiley & Sons.

- *Digital Signal Transmission* by C.C. Bissell and D.A. Chapman, 1992, ISBN 0-521-41537-3, Cambridge University Press.

- "Dealing with High-Speed Logic," Analog Devices, MT-097 Tutorial Rev 0, January 2009.

- *Signal and Power Integrity Simplified, 3rd Edition*, by E. Bogatin, 2018, ISBN-13, 978-0-13-451341-6, Pearson Education.

- F. Alicke, M. Feulner, F. Dehmelt, A. Verma, and G. Becke, "Comparing Bus Solutions," Texas Instruments Application Report SLLA067C, revised August 2017.

- *Digital Transmission Lines: Computer Modeling and Analysis* by K. Granzow, 1998, ISBN 0-19-511292-X, Oxford University Press.

LVDS

- "LVDS Owner's Manual, Including High-Speed CML and Signal Conditioning," 4th Edition, Texas Instruments, 2008.

- "LVDS Application and Data Handbook," Texas Instruments, SLLD009, November 2002.

I^2C and SMBus

- "I^2C Manual," Philips Semiconductor Application Note AN10216-01, March 2003.

- "The I^2C Bus Specification," Version 2.1, Philips Semiconductor, January 2000.

- "I^2C Bus Specification and User Manual," NXP Semiconductors, UM10204, Revision 7.0, October 2021.

CAN Bus

- "CAN Bus - ISO 11898-1:2015," *https://www.iso.org/standard/63648.html*.

- "Introduction to the Controller Area Network (CAN)," Texas Instruments Application Report SLOA101B, revised May 2016.

- Steve Corrigan, "Controller Area Network Physical Layer Requirements," Texas Instruments Application Report SLLA270, January 2008.

- Conal Watterson, "Controller Area Network (CAN) Implementation Guide," Analog Devices Application Note AN-1123 Rev. A.

RS-485
- T. Kugelstadt, "The RS-485 Design Guide," Texas Instruments Application Report SLLA272C, Rev. October 2016.
- "RS-485 & Modbus Protocol Guide," Tyco Electronics, Rev. 6, July 2002.

Ethernet
- R. Newhaus, "A Beginner's Guide to Ethernet 802.3," Analog Devices, Engineer to Engineer Note, EE-269, Rev. 1, June 2005.

SPI
- "SPI Block Guide," Motorola Inc., Version 03.06, revised February 2003.

JTAG
- "JTAG Testability Primer," IEEE Std 1149.1, Texas Instruments, 1997.

SATA
- "Serial ATA – High Speed Serialized AT Attachment," Serial ATA Workgroup, Rev. 1.0a, January 2003.

USB
- "Universal Serial Bus 2.0 Specification," Rev. 2.0, April 2000.
- "Universal Serial Bus 3.0 Specification," Rev. 3.0, November 2008.

Special Interest Groups for Protocols, Products, and Interfaces
- WiFi (*https://www.wi-fi.org*)
- Z-Wave (*https://www.z-wave.com*)
- Bluetooth (*https://www.bluetooth.com*)
- ANT (*https://www.thisisant.com*)
- ZigBee (*https://zigbeealliance.org*)
- Infrared (*https://www.irda.org*)

CHAPTER 4

Power Systems

Providing system power can be a simple matter, but for complex systems it can turn into a significant design issue. Power and ground (P&G) are common to an entire system and can be a smoky failure point, a common path for noise into the entire system, or a path where the power demand in one place affects the performance someplace else.

Safe AC power, safe failure scenarios, regulatory compliance issues, suitable noise levels on power, controlled up/down power cycles, and stable power under highly variable loads are all part of a reliable system. Understanding the options available among power converters and how to configure the power system is the goal. Designing an AC/DC converter from scratch is rarely necessary, as many off-the-shelf converters are available that are easily implemented into a system.

After initial AC/DC power conversion, additional items such as local regulation, multiple voltages for different circuits, high-frequency bypassing, selective device decoupling, power monitors, and power cycling controls are commonly needed. In addition, several source-power scenarios exist: disposable battery, battery with charging capability, external AC/DC converter, or AC/DC converter inside the system.

This chapter discusses all of these and highlights their similarities and differences.

Split Phase AC Mains Power

What does AC mains power consist of? In the United States at the residential level (Figure 4-1), the AC power grid feeds a local step-down and distribution transformer that creates a "split phase 120/240 system" output.

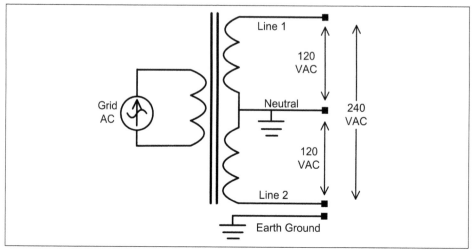

Figure 4-1. AC split phase power, step-down from commercial power grid

Figure 4-2 shows both line signals from the secondary of the step-down transformer. What reaches a typical wall outlet (120 VAC 60 Hz in the United States, and 50 Hz elsewhere) is a three-wire system comprising one line signal (aka hot), neutral, and earth ground. Neutral is the center tap from the secondary in the AC mains step-down transformer.

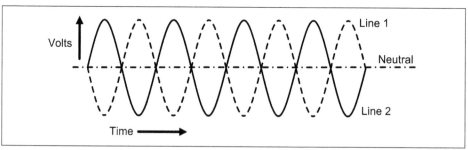

Figure 4-2. AC split phase power signals

Neutral and ground are only connected back at the AC mains distribution transformer. Prior to 1960, residential wiring was two wires only, without the ground safety, which produced unsafe devices in many failure scenarios.

AC Power Safety: Defining the Problem

When bringing AC power directly into a device, the implementation needs to be carefully considered in the design. Getting things wrong with AC power can lead to deadly devices and smoke-filled outcomes.

Most devices now use low-voltage DC power internally, provided from an external AC/DC converter, thus allowing most designers to remain blissfully unaware of high-voltage product safety requirements. Many older (prior to 1960) electrical products were inherently unsafe or could become deadly with a single fault failure. Metal enclosures were commonly used (prior to the widespread use of plastics), and lack of a ground safety connection made for a deadly combination. Thankfully, most electrical devices prior to 1960 have been retired or superseded by newer product offerings, so most of those devices are no longer in use.

High-Voltage and Low-Voltage Partitioning

Designing with power inlets lower than 50 V is classified as *extra low voltage* (ELV), and that simplifies things. Safety issue requirements are primarily placed on anything classified as high voltage (HV), namely the AC/DC converter.

Figure 4-3 shows one method that is commonly used, where both the AC/DC converter and the electronic system are enclosed in a conductive metal enclosure and a safety ground is tied back to the AC mains safety ground. This widely used approach is commonly employed in many desktop personal computers.

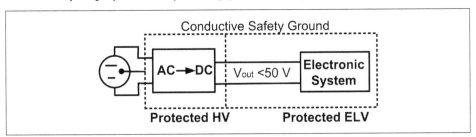

Figure 4-3. Protected HV with protected ELV DC

ELV devices are subdivided into separated ELV (no ground return path is used) and protected ELV (a ground earth safety is attached). The safety standard IEC 62368-1 (Safety Requirements for Audio/Video, Information & Communication Technology Equipment) has become the most commonly used standard for ELV safety.

Figure 4-4 shows an alternative method, where the enclosure is an insulator and is not grounded. To meet safety requirements on the HV portion, a double insulated technique is used. Double insulated devices utilize two layers of protective insulation, so if one layer fails, the device remains safe to handle.

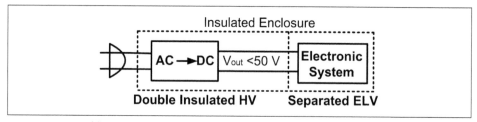

Figure 4-4. Double insulated HV with separated ELV DC

For ELV devices, safety requirements are minimal, with some form of overcurrent protection and a nonconductive enclosure generally being sufficient.

In Figure 4-5, an AC/DC power supply is in a double insulated structure and has an ELV power cord output. Many commercially available versions of this exist, with full regulatory testing already done. Using this type of AC/DC power supply, designers can create their electronic system as a separate entity that requires the least possible ELV testing. This approach is ubiquitous and is utilized in a myriad of consumer electronics devices.

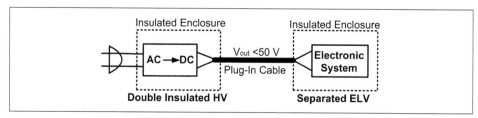

Figure 4-5. Standalone double insulated HV powering a separated ELV device

Safe Failure Methods and Single Fault Safe Scenarios

Normal operational safety is the start, but when something breaks, power systems also require a safe failure method. The *single fault safe* concept is simple: any element of the system can fail in an open circuit state or closed circuit state, or can contact any other element in the system and not cause danger to the user.

Introduction of a ground safety return path into modern power wiring allowed for a safe system in most single fault situations, where the grounded enclosure, in conjunction with a fused line connection, could keep a device safe.

The principal regulatory safety standards for power systems are:

- IEC 61140: Protection Against Electric Shock Common Aspects for Installation and Equipment

- IEC 62368-1: Audio/Video, Information and Communication Technology Equipment, Part 1 Safety Requirements

These standards are referenced in this chapter. Designers should acquire the latest versions.

According to IEC 61140, devices that use a metal chassis and enclosure structure that is attached to a ground safety connection are defined as Class I devices. Their failure safety is dependent upon the presence of a properly connected external ground. Some devices have only line and neutral connections, and they meet safety requirements using a double insulated structure. For these devices, the internal wiring and circuits are within a secondary insulating enclosure. In this manner, the device can have a single fault failure without exposing users to high-voltage contact.

Double insulated devices are considered Class II devices and the safety redundancy of the device is not dependent on an external ground. Consequently, double insulated devices are popular in electric power tools, due to remaining safe when connected to electrical power with questionable ground paths, as frequently happens in construction environments.

ELV systems are Class III devices under the IEC 61140 regulatory standard.

If the AC/DC converter is placed inside the product, the whole product needs to meet Class I or Class II safety requirements of a grounded enclosure or double insulated configuration. Using an AC/DC converter externally circumvents that restriction and is a popular approach to the problem.

Overcurrent Protection Methods and the Weakest Link

Overcurrent protection is the most basic form of system protection. There are four common ways to turn things off in an overcurrent situation:

- Fuses
- Circuit breakers
- Polymeric positive temperature coefficient (PTC or PPTC) resettable fuses
- Active circuit methods

Fuses are commonly used for their simplicity and low cost. However, fuses can be too slow, they require manual replacement to restart the system, and they are not terribly accurate. In some situations, it is desirable to control both the current and the time needed for the fuse to do its job. However, the overcurrent amount and the activation time are interactive, so don't expect fast and accurate from a fuse.

Nonetheless, the fuse is often fast and accurate enough. That, combined with low cost, simplicity, and reliability, makes fuses the go-to solution in many cases. If the fuse is expected to blow only when something major on the printed circuit board (PCB) fails, a soldered-in surface mount fuse can be used with minimal cost and excellent reliability. Fuses don't need to be replaced in many cases, and socket mounting is unnecessary.

Circuit breakers suffer many of the limitations of fuses, with speed and accuracy being "good enough" for most situations. To gain the convenience of a resettable device, the penalty is paid in cost and size.

A circuit breaker is considerably more expensive than a fuse. Also, circuit breaker reliability can suffer in hostile environments due to internal corrosion issues if they are not hermetically sealed. Generally, circuit breakers are called for when the designer expects frequent overcurrent situations due to the application.

Polymeric positive temperature coefficient devices (widely known as PTC or PPTC devices) can be considered self-resetting fuses. They warm up, due to overload current, and go into a high-resistance mode, limiting the current until they cool off, and then return to a low-resistance state.

PTC devices do have some less-than-ideal internal on/off resistance, but they serve well in situations where auto-reset is desirable. The PTC has limitations of accuracy and response time, as all thermally triggered devices do. Nonetheless, generally the PTC can be included on a PCB for a low-cost design solution.

Active circuit protection for overcurrent detection and shutdown comes in many variants. The basic concept is a set of detection circuits for current and/or voltage to determine whether the applied power is outside specified limits, which triggers suitable circuit controls to shut down or limit performance.

Active circuit systems can be fast and accurate if needed. The price paid here is in complexity and cost. Frequently, this approach is unneeded and one of the simpler methods is suitable.

Many switching AC/DC converters use active circuit methods within their design to protect circuits that can self-destruct more quickly than a fuse can respond. Seek out this feature when selecting AC/DC converters. Look for information that indicates the device is "self-protecting" or "overcurrent protected" or "current limited" in the specification.

Regardless of overcurrent protection method, the system needs to make sure the "weakest link" is the actual protection device. If the PCB traces melt at 1 A of current and the protection device opens at 2 A, something wasn't designed properly. Also, the parameters for the trip protection point need to consider both min/max sustained and startup surge currents. As a simple guideline, setting a fuse value at twice the maximum current seen in normal circuit operation is often suitable.

Systems may have high-current and low-current sections where a failure in the low-current section requires a lower value protection shutdown for that part of the system. A common design mistake is to use a single overcurrent shutdown in a situation where a low-power part of the system self-destructs since it is powered in parallel with the high-current part of the system.

Thermal monitoring of high-power devices can be useful, with protection control built into the system. For some devices, this provides another layer of safety and capability to stress devices close to their thermal limits without self-destruction. Older devices used thermocouples, or thermistors, but silicon-based thermal monitors with digital interfaces to a local controller are a more modern approach. Most modern high-performance microprocessors use thermal sensors within the IC (FDI: Sense).

AC/DC Conversion

Very few electronic devices use AC power directly. The majority of devices use DC power, and most modern electronics use 5 V and lower power sources.

Systems with DC motors or other high-power devices may need higher voltages. Due to this, the first converter in the system will be an AC/DC converter, with sufficient voltage to support the high-voltage devices.

The Classic Approach: 60 Hz Transformers

As shown in Figure 4-6, the classic approach to AC/DC conversion uses a transformer at 60 Hz (or 50 Hz, depending on the country). This seems simple: step down the AC voltage, put the reduced AC voltage through a diode bridge, and filter the rectified AC for a DC output.

While the idea is simple, certain pieces tend to be both big and expensive. The problem here is the low frequency of AC mains power. A 60 Hz transformer will be both bulky and heavy. For the same power rating, the transformer size decreases as the frequency increases. In addition, the output filter capacitor has to accurately maintain the output voltage under high-current loads between the rectified peaks of the 60 Hz waveform. This necessitates the use of a large electrolytic filter capacitor. By increasing the frequency, the size of both the transformer and the filter capacitor can be decreased. The aviation industry often uses 400 Hz AC power to take advantage of this.

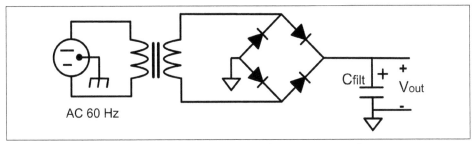

Figure 4-6. Classic 60 Hz AC-to-DC converter with iron core transformer

There are better methods than putting 60 Hz AC into a transformer.

Off-line Switchers

The limitations of the 60 Hz AC-to-DC converter are circumvented with a switched mode conversion method commonly known as an *off-line switcher* (Figure 4-7). This approach rectifies the AC line voltage, and uses C_1 to maintain a high DC voltage where ripple and accuracy are not critical. That DC voltage is rapidly pulsed across the primary of T_1 with a switch. The switching frequency is much higher than 60 Hz, with 10 KHz to 1 MHz commonly used.

Although appearing more complex, off-line switchers offer several advantages. Using a higher frequency allows for a much smaller transformer (T_1) and output filter capacitor (C_2). The high-voltage DC at C_1 can vary widely, so C_1 does not need to be very large.

This method is used in virtually all AC/DC converters today, with small form factor implementations illustrated by the numerous USB and cell phone chargers that sit readily within an AC power plug footprint.

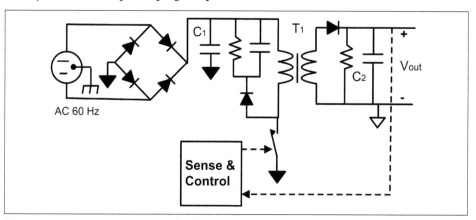

Figure 4-7. AC/DC converter with off-line switcher

A broad selection of off-line switching AC/DC converters are available for off-the-shelf purchase. The 60 Hz transformer method is much larger and more costly to implement. Due to this, the 60 Hz step-down transformer is largely obsolete in consumer electronics.

The AC/DC converter generally provides a single DC voltage and is commonly selected to be the highest DC voltage that the system needs. Other power regulators, for DC/DC conversion, are then used to complete the system power requirements.

Multi-PCB Systems: The Need for Local Power Regulation

In Figure 4-8, the impedance of power and ground wires, combined with dynamic current loads, will create both ground bounce and power sag on various circuit boards. A design can reduce this effect with shorter (less inductance) and thicker (lower resistance) wires, but will not eliminate it. Consequently, a multi-PCB design must deal with both ground and power voltage variance between PCBs.

Separate regulators on each PCB to create locally needed voltages is the widely used solution (Figure 4-8). Systems with multiple circuit boards can use a common AC/DC converter to distribute a "raw" DC voltage. This DC voltage should have "headroom," namely several volts larger than the minimum required by each of the satellite PCBs. With the added voltage headroom, the effects of ground bounce between PCBs and voltage sag can then be removed by the local voltage regulator, and each PCB will be able to tolerate P&G variance at the power input.

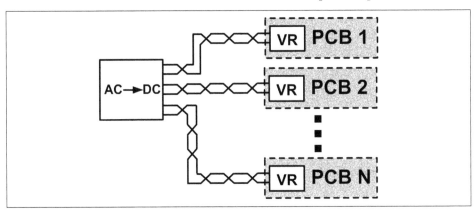

Figure 4-8. Multi-PCB systems with local voltage regulation

Many industrial systems distribute raw DC power at 36 V, 24 V, or 12 V depending on what the satellite PCBs require. Directly using a DC voltage created off the PCB without a local regulator will never be consistent or noiseless, and frequently causes failures during electromagnetic compatibility (EMC) regulatory testing. Therefore, one should regulate the power supply locally on the same PCB where it is used.

DC/DC Conversion: Linear Versus Switching

Power regulators can be divided into two groups, linear regulators and switching regulators, which use distinctly different methods to provide a consistent voltage output. Linear regulators include emitter follower regulators and low-dropout (LDO) regulators. Switching regulators include buck converters, which reduce voltage; boost converters, which increase voltage; and buck-boost converters, which can reduce or increase voltage output. Other switched-mode power supply (SMPS) configurations exist, but these are the ones most commonly used.

Linear Regulators: Conceptual

Conceptually, a linear regulator (Figure 4-9) uses adaptive voltage division. As R_{load} changes, a sense/control circuit adjusts the regulator to maintain V_{out}.

In Figure 4-9, a power transistor is serving as a variable resistor. The important thing to recognize is that this method creates an output voltage by burning energy inside the regulator. Therefore, don't expect efficient energy use when implementing a linear regulator. Also, power dissipated (P_{burned}) by the device can be quite high depending on the current (I_{out}) and the voltage drop ($V_{in} - V_{out}$) across the regulator. The power dissipated needs to be carefully considered so that the regulator does not thermally fail.

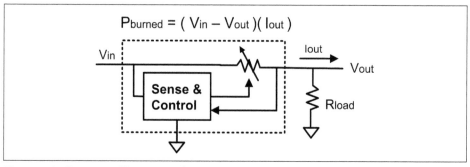

Figure 4-9. Conceptual linear regulator

The primary advantages of linear regulators are that they are electrically quiet in both radiated emissions, and they create a fairly quiet output voltage. The big disadvantages are that they are energy inefficient and often not viable for large input-to-output voltage drops.

Emitter Follower Regulators Versus LDO

Figure 4-10 shows the oldest version of the linear regulator, the emitter follower configuration. A sense and control circuit applies an appropriate control voltage to the base-emitter junction of a transistor to produce the output voltage. The transistor serves as the top half of the voltage divider and is commonly implemented with an NPN bipolar device. When the control voltage goes up, the output voltage follows along. Due to the internal circuit design, the input voltage typically needs to be 2 V greater than the output voltage to function properly.

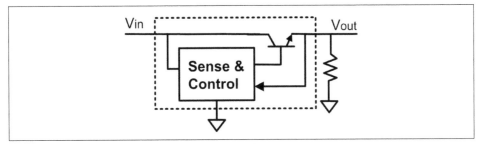

Figure 4-10. Linear voltage regulator, emitter follower

The emitter follower regulator configuration is widely used in the 7800 series (7805, 7812, etc., from multiple vendors) of voltage regulators that have been available since 1970.

Figure 4-11 shows a more modern variant of the linear regulator, the LDO regulator. The LDO regulator uses a power transistor that requires the control voltage presented to the power transistor to go down to increase the output voltage. This is done with either a PNP bipolar transistor or a PMOS MOSFET. By doing this, the necessary input voltage to maintain a proper output is greatly reduced.

LDO devices typically will work down to an input voltage that is 100 mV greater than the output. Due to that capability, the LDO is widely accepted as a more versatile linear regulator than the emitter follower configuration. The LDO still has the limitations of being an active voltage divider method and the associated inefficient use of power, however.

LDO regulators can also exhibit some feedback stability issues if not properly loaded and compensated as per vendor's recommendations for their specific devices. Carefully read the application documentation to avoid problems.

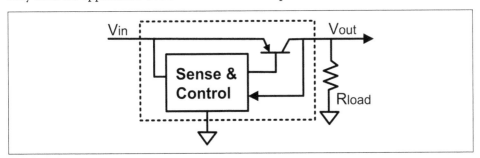

Figure 4-11. Linear voltage regulator, low dropout

Switching Step-Down (Buck) Converter

The switching step-down converter, commonly called a buck converter, uses varying duty cycle methods to reduce an input voltage (Figure 4-12). The switch is turned on and off at a high frequency (10 KHz to 1 MHz is common) while a control feedback system determines the time period during which the switch remains on.

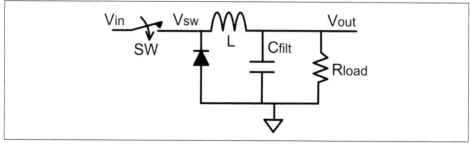

Figure 4-12. Switching step-down converter, conceptual

A higher percentage of on time yields a higher output voltage (Figure 4-13).

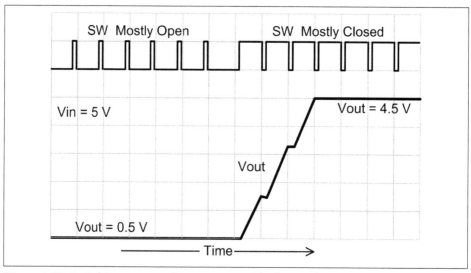

Figure 4-13. Switching constant frequency variable duty cycle

When turned on to a near 100% duty cycle, the output voltage will be close to the input voltage. A reduction in duty cycle will provide proportionally lower voltages at the output. Energy storage for a steady output voltage is maintained on the capacitor, which is fed through the inductor by either the V_{in} (switch closed) or the magnetic field collapse of the inductor (switch open).

The diode serves as a passive switch, limiting the input side of the inductor to not transition below ground when the switch is open. Higher switching frequencies allow smaller-value inductors and capacitors, thus providing small-size and low-cost implementations. The energy efficiency of the buck converter is very high, typically over 90%, with some losses due to nonideal component characteristics.

Although the overall output voltage remains constant under load, close inspection of the output voltage (Figure 4-14) shows that the switching creates some variance in the output. All switching regulators exhibit this behavior; a variant of a few mV is common in steady state situations. The noise amplitude is dependent on the switching frequency, output current load, inductor value, and capacitor value. Because of this switching noise, the use of switching regulators might not be appropriate in some noise-sensitive situations.

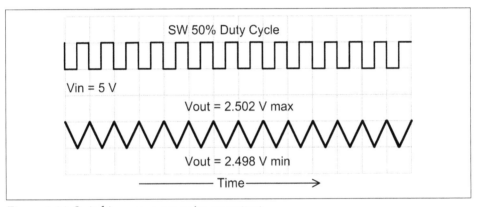

Figure 4-14. Switching converter noise on output

However, this power supply noise is still suitably low to provide acceptable power quality for most applications. Switching regulators are a high-efficiency and versatile solution for all digital power situations, and they can be used for some analog circuits. As such, the use of switching regulators to power analog circuits needs to be carefully considered on a case-by-case basis.

A typical implementation is shown in Figure 4-15, with R_1 and R_2 allowing a selectable output voltage via feedback (FB) sensing and the capability to enable (EN) the device by an external digital controller.

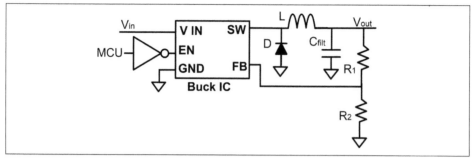

Figure 4-15. Switching step-down (buck) converter implemented

Implementations of the buck converter may include variation in switching frequency, the diode may be internal to the IC, high-power versions may use an external power transistor switch, an active switch can be used in place of the diode for better efficiency, and the feedback sense resistors may be internal to the IC for fixed-voltage versions.

Switching Step-Up (Boost) Converter

The switching step-up converter, known as a boost converter, uses the reactive characteristics of an inductor to create a voltage above the input voltage (Figure 4-16). Closing the switch causes the current to ramp up in the inductor. Opening the switch causes the $V = L \, di/dt$ to create a positive inductor voltage at the V_{sw} node, which is in series with the V_{in} voltage. This forward-biases the diode, dumping charge into the C_{filt} and resulting in a V_{out} above the input voltage V_{in}.

The switch is turned on and off at a high frequency (10 KHz to 1 MHz is common) while a control feedback system determines the switch's duty cycle. A higher percentage of the open switch dumps more current into C_{filt}, resulting in a higher output voltage.

Energy storage for a steady output voltage is maintained on the capacitor, which is fed through the diode when the switch is open. The diode also serves as a passive switch when the switch is closed, isolating the C_{filt} while the magnetic field in the inductor is replenished.

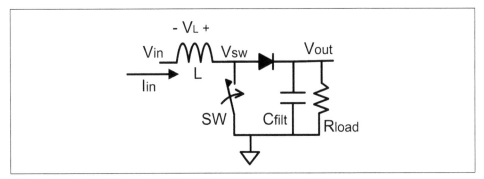

Figure 4-16. Switching step-up (boost) converter, conceptual

Similar to the buck converter, the use of high switching frequencies allows the use of smaller-value inductors and capacitors, enabling small-size and low-cost implementations. A typical implementation is shown in Figure 4-17, with R_1 and R_2 allowing a selectable output voltage via feedback (FB) sensing and the capability to enable (EN) the device by an external digital controller.

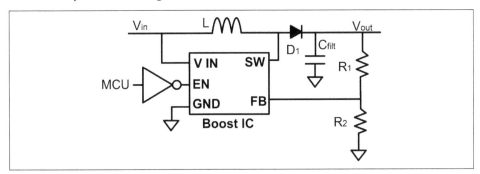

Figure 4-17. Switching step-up (boost) converter, implemented

Variations on the boost converter may include changes to switching frequency, the diode may be internal to the IC, high-power versions may use an external power transistor switch, an active switch can be used in place of the diode for better efficiency, and the feedback sense resistors may be internal to the IC in fixed-voltage variants.

Switching Buck-Boost Converter

A buck-boost converter can either raise or lower the output voltage. Referring to Figure 4-18, when SW_1 is actively controlled and SW_2 is left open, the device functions as a buck converter. With SW_1 closed and SW_2 actively controlled, the device functions as a boost converter.

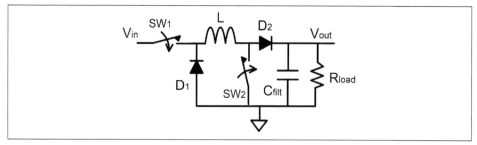

Figure 4-18. Switching buck-boost converter, conceptual

In a typical application (Figure 4-19), the internal control circuitry determines which mode is needed based upon the input voltage available and the target output voltage.

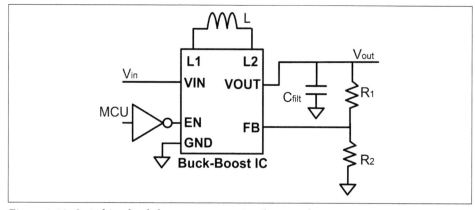

Figure 4-19. Switching buck-boost converter, implemented

The combined buck-boost converter is a useful tool when a widely variable input power supply is used, such as a battery. In these applications, the converter reduces the battery voltage to the target output voltage until the battery has discharged enough to drop below the output target, and then changes mode to a boost converter to extend the useful battery range.

Picking Regulators and Configuring a Power System

Putting a power system together is best illustrated with some examples. Figure 4-20 is the first example and has a 12 V motor driver system, a block of logic that functions from 2.5 V, a host microcontroller (MCU) that needs 3.3 V, and some analog circuitry that needs 5 V. This is a plug-in device that uses an AC/DC converter.

Configuring the power here is mostly straightforward. A 12 V output off-line switcher does the AC/DC conversion and will be used to directly power the motor driver. Motors can be high current, but a limited amount of voltage dip is tolerable during high-current starts.

Motors are also immune to most power noise, and some power variance can be compensated for by the motor controller. The current rating of the AC/DC converter needs to support the start surge current of the motor, the steady state motor current under maximum load, and the other electronics.

When selecting an AC/DC converter, in addition to voltage and current specifications, look for devices that are self-protective and current limiting, and consider the output ripple and noise specifications carefully as the application may have special requirements. If the product will go to markets outside the United States, consider devices that will function off of both 50/60 Hz and 110/220 VAC without the need to change settings or connections.

The 2.5 V logic (Figure 4-20) can be readily provided from a buck converter. The MCU needs 3.3 V and can also use a buck converter. Power for the MCU is always on, and as the central control of the system, it controls all the other power supplies. This allows selective shutdown of peripheral devices and an organized power up/down sequence that provides predictable and safe system behavior.

The analog block that needs 5 V power is not straightforward. More needs to be known, namely how much current is needed and how critical low-noise power is to the circuit. If the circuit can tolerate the power noise associated with a switching supply, it's the most energy-efficient method. But considering this is a plug-in-the-wall application, energy efficiency is not a high priority.

Using a linear regulator here gives a quiet supply but would be energy inefficient. However, the 5 V output regulator has 12 V input, and that means a 7 V drop across a linear regulator that will be dissipated as heat. In this case, if the current rises to 140 mA, the power burned in the linear regulator rises to 1 W. If the current is low and low-noise power is important, a linear regulator would work.

Check power dissipation in any linear regulator carefully. In many cases, the switching regulator with some filtering may be a viable alternative.

Finally, the system uses fuse protection of 5 A on the motor circuit, which draws a worst-case current of 3 A, and a 0.5 A fuse on the electronics power. Using two fuses is due to the high-current motor circuitry. If a single fuse were used, upon failure the low-current electronics would have to exceed 5 A and that would probably be beyond the destructive current level of the electronics. The 0.5 A fuse mitigates that.

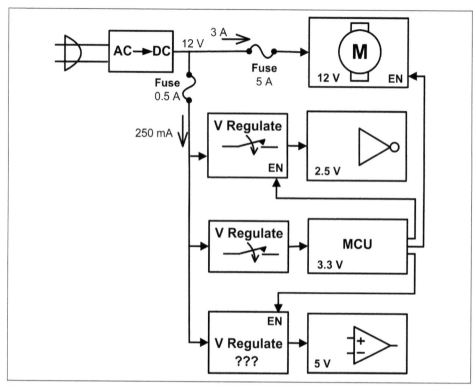

Figure 4-20. Multiple power supplies plug-in device example

Other possible options here include a PTC on the electronics for protection, and a circuit breaker on the motor circuit if the motor application is in a mechanical environment that stalls or goes into overload frequently.

Looking at the second example, shown in Figure 4-21, the power system includes a battery and charger system. This example uses a Lithium battery pack with four batteries in series and two in parallel (eight cells total, 4S-2P). When the charger is applied, the power system sees 18 V, and when the batteries are largely discharged, the power system sees 10 V (FDI: Battery).

Everything needs to function at both voltage extremes. One circuit needs 12 V, which requires a buck-boost converter that functions in buck mode until the battery pack goes below 12 V and then changes to boost mode. Examine the 12 V circuit for noise requirements, which may indicate a need for a filter at the converter output. The digital block that uses 0.9 V power is readily powered from a buck converter.

The MCU runs off 3.3 V and is always powered up from the battery system while controlling the power supplies to the rest of the system. The final block runs off 2.5 V and requires low-noise power input. This device is a good candidate for a linear regulator, using an LDO that is powered off of 3.3 V. The 0.8 V of headroom is sufficient for an LDO device to function without a large voltage drop and the associated power lost as heat.

Carefully examine the noise sensitivity of the circuitry and consider the need for filtering both before and after the LDO to provide a quiet 2.5 V. High-frequency noise can couple across voltage regulators, and a strategy of filtering input and output may be needed in very noise-sensitive situations.

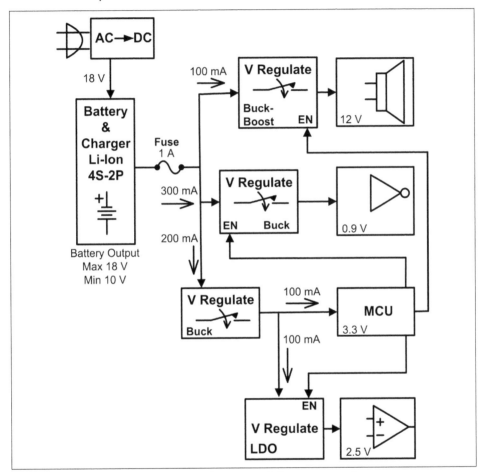

Figure 4-21. Multiple power supplies: battery and charger example

Both rechargeable battery systems and disposable batteries will have similar needs to remain functional over a wide variance in the supply voltage.

Including Power Supply Monitors

A useful addition to the power control system is the capability to monitor the voltages out of the regulators. Adding a voltage divider to an output voltage (Figure 4-22) and feeding that back to an analog-to-digital converter (ADC) port available on the host MCU can be useful in checking power sources or assisting in a sequenced power on/off cycle. The voltage divider (R_1, R_2) exists to scale V_{out} down to the midscale range of the ADC in the MCU.

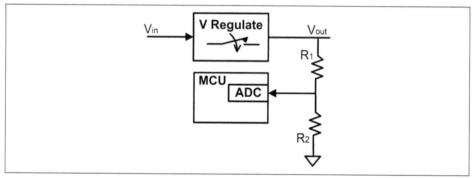

Figure 4-22. Monitoring power supplies

Power Bypass, Decoupling, and Filtering

Many of the multipurpose MCUs on the market have some limited analog capability, such as ADC ports. Some of these controllers separate the power with both digital and analog power connections. This provides a good opportunity to clarify the difference between a bypass capacitor and a decoupled power connection (Figure 4-23).

Frequently, the analog supply pins can be locally decoupled. Do this by using a ferrite bead and then both low-frequency bypass (C_{alf}) and high-frequency bypass (C_{ahf}) directly at the pin. A damped low-pass power filter can be used in place of the ferrite bead if greater rejection of power input noise is needed.

Figure 4-23 illustrates the difference between the bypassing and decoupling of power connections. The two terms are often interchanged or used incorrectly. Here, the capacitors (C_{dlf}, C_{dhf}) *bypass* the V_{power} to maintain high-frequency stability. The combination of the ferrite bead (FB-LQ) and capacitors (C_{alf}, C_{ahf}) serves as a low-pass filter (LPF) to *decouple* the analog power pin from the digital switching noise present on the V_{power} node. Decoupling involves series impedance in the power connection, and it can be useful in situations where the decoupled load does not need heavy transient currents.

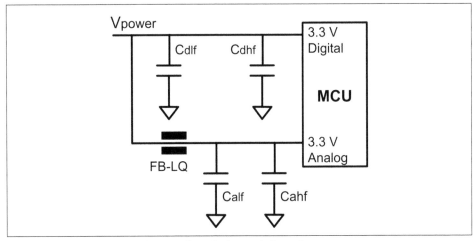

Figure 4-23. MCU power, bypass digital, decoupled analog

Radiated Noise Reduction: RC Snubbers, Ferrites, and Filters

In addition to switching regulators creating somewhat noisy power, they will also inject noise back into the power source they are fed from. Implementation of many switching converters involves components (Figure 4-24) to reduce the high-frequency noise that they pump onto the power grids, thus radiating electromagnetic interference (EMI).

In this example, both input and output have an LPF with a ferrite bead (FB$_1$, FB$_2$), and suitably sized high-frequency capacitors on both sides of the ferrite bead. If more isolation is required, an RLC network-damped low-pass network can be substituted.

In addition to the input/output filtering, an RC snubber (R$_{snub}$, C$_{snub}$) is placed at the switched node. The RC snubber provides a low-impedance grounding path for high-frequency noise generated at this switched node (FDI: EMI & ESD).

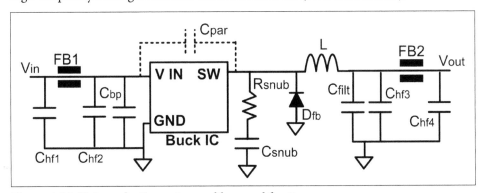

Figure 4-24. Noise reduction using snubbers and ferrites

Specific values for the RC snubber are dependent on converter frequency and internal transistor sizes. Switching converter manufacturers should provide suitable component values for their devices.

Power Output Noise Reduction: Damped LPF Networks and Cascaded Regulators

Higher-order filtering may be needed in some situations. For an in-depth design approach to damped low-pass filtering methods, Tompsett outlines a detailed design methodology; see "Further Reading" on page 157.

Other approaches here include using a switching converter, which is followed by a linear regulator that reduces noise further. This cascaded method allows a low-noise output while keeping the power losses associated with linear regulators minimized.

Power Grid Current Surges Due to Digital Logic

Figure 4-25 shows that the use of digital CMOS logic poses special challenges in keeping the DC power grid stable. The transition switching of any CMOS logic causes both NMOS and PMOS transistors to be on at the same time during the state transition. The result is a resistive short circuit between P&G on the state transition of every logic gate. This current spike, known as *shoot-through*, when replicated among the countless logic gates within any digital device creates both wideband EMI and high-frequency current demands on the power.

This is especially problematic due to I/O drivers inside a digital IC. The I/O has large transistors used in the I/O circuits of the chip, creating large shoot-through currents. The noise spectrum is distributed in frequency and is not just the fundamental clocking frequency, but includes many harmonics as well.

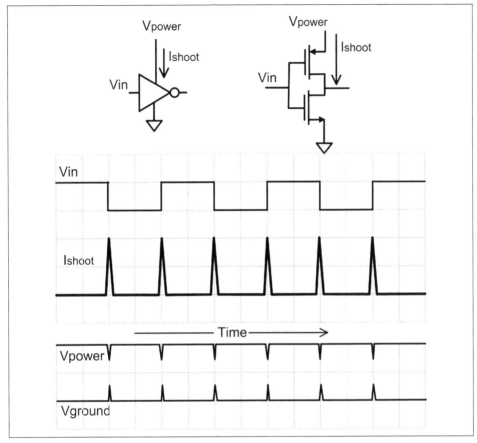

Figure 4-25. Digital current shoot-through

Due to shoot-through current, any impedance in the power or ground paths creates variance in the local P&G voltages. Low-impedance P&G planes are needed.

Low-Impedance Power and Ground Planes

Due to current surges, the impedance of P&G connections needs to be kept as low as possible. For that, dedicated P&G layers within the PCB are strongly preferred in all but the simplest of designs. Point-to-point interconnects will not suffice.

In Figure 4-26, Case A illustrates a hypothetical power interconnect with three different power values using point-to-point connections. Case B illustrates that all connections exhibit inductance and will cause the voltage across the power connection to vary dynamically as a function of any transient currents created by the devices being powered. Case C illustrates connecting the same points with a low-impedance connection. Placing this segmented power plane over a solid ground plane allows the capability to wideband-stabilize the power supply voltages with bypass capacitors.

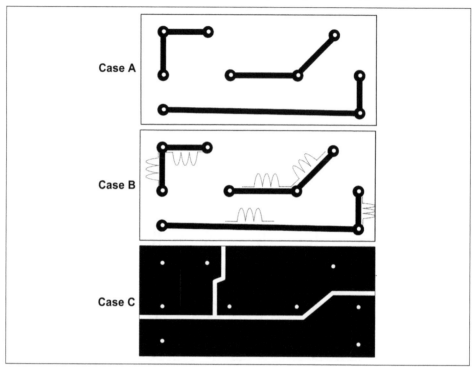

Figure 4-26. Power connections with low-inductance interconnect

Power Supply Bypass Filtering: Distributed Stabilization

In addition to the low-impedance interconnect strategy, the use of distributed bypass capacitance across the power plane (Figure 4-27) allows high-frequency stabilization of the power.

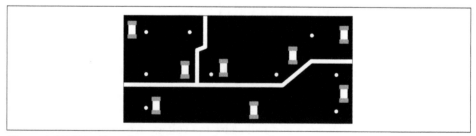

Figure 4-27. Distributed stabilization of power grid

Typically, stabilization of a power grid has three parts:

- Bypass capacitors directly at all power pins on all ICs (*at-chip bypass*)
- Bypass capacitance distributed across the power plane (*grid bypass*)
- Bypass filtering at the voltage regulator output (*regulator bypass*)

A large PCB (e.g., a computer motherboard) with multiple high-speed digital chips must have all three of these. A small low-frequency PCB (e.g., a handheld remote for the TV) would be less demanding, and the grid bypass portion might be eliminated.

Bypass Capacitors at High Frequencies

The equivalent model for a surface mount multilayer ceramic capacitor (SMT MLCC) is shown in Figure 4-28. The elements are:

- C_{main}—Principle body of the capacitor
- C_{par}—Parasitic capacitance to the underlying PCB
- ESR—Equivalent Series Resistance of the distributed device resistance
- ESL—Equivalent Series Inductance of the device
- R_{leak}—Leakage resistance between capacitance plates

The two important parameters for power bypass are ESR and ESL. The ESR limits the lowest impedance, and the ESL limits the maximum frequency where the device works effectively as a capacitor. These characteristics need to be considered in the selection of bypass capacitors.

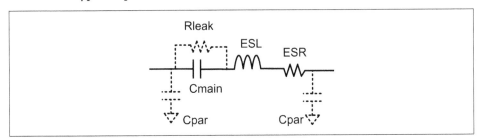

Figure 4-28. Equivalent model for high-frequency capacitors

Figure 4-29 shows the self-resonant characteristics of a typical surface mount capacitor. The low-frequency response is dominated by the C_{main} capacitance and the high-frequency response is dominated by the ESL. In between, the device goes through self-resonant frequency (SRF) and the impedance drops to a minimum, which is equivalent to the ESR. The capacitor serves as a bypass filter beyond the SRF, but the effectiveness reduces with higher frequencies.

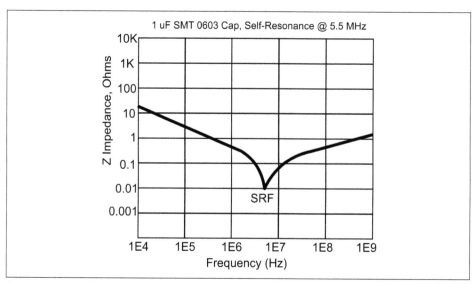

Figure 4-29. Self-resonance of capacitors

The presence of ESL limits the high-frequency capability of the device as a bypass filter for a power supply.

Table 4-1 shows a typical set of values for SRF as a function of capacitance value.

Table 4-1. Capacitor self-resonant frequency versus capacitor value

C_{value}	SRF
10 uF	2.0 MHz
1 uF	5.5 MHz
0.1 uF	19 MHz
0.01 uF	57 MHz
1000 pF	180 MHz
100 pF	560 MHz

SMT MLCC, Package: 0603 (Imperial)

Ideally, bypass filtering needs to avoid frequencies where ESL dominates the impedance. The reality is that many designs have wideband current demands that go beyond the useful range of most capacitors. There are conflicting rules of thumb and empirical studies of how to deal with this issue. See "Further Reading" on page 157.

In theory, using capacitors with different SRF points connected in parallel circumvents the SRF characteristics of singular devices. This method is a common practice, but it is often blindly implemented without looking at the SRF of the capacitors. For this particular set of values, using capacitors 100X apart in value (such as 1 uF and 0.01 uF) would space the SRF apart in a suitable manner. Check the SRF of the specific devices used, because differences in manufacturer, dielectric type, or voltage rating can result in different SRF values.

The SMT body size can also affect ESL and SRF, with an SMT 0805 body having distinctly different characteristics than an SMT 1206 body. Some vendors also offer "low ESL" capacitors as an option. These devices usually have wider contact regions and shorter bodies, or they use multiple contact points to reduce inductance. Check the specific devices to be used for the details of their parameters.

Power Bypass Capacitor Value and Distribution

Bypass capacitors are utilized to do three separate things:

- *At-chip bypass* is for the high-frequency current demands of an IC.
- *Grid bypass* stabilizes local power grid connections and reduces voltage variance due to both connection impedances and bandwidth limitations of the power source (FDI: PCB).
- *Regulator bypass* filters the power source (regulator) output and stabilizes the feedback control loop within the regulator.

Semiconductor vendors should provide their suggested devices for at-chip bypass to be used directly at their chips. If vendor information is not available, here are some suggested guidelines.

At-chip bypass capacitors:

- All power pins on all chips get dedicated bypass capacitors directly at the power pins.
- At-chip bypass capacitors go on the same side of the PCB as the chip.
- At-chip bypass capacitors should have minimal trace routing and use shortest-path connections. This is especially important with the wideband current demands of digital chips (FDI: PCB).
- PCB layout should avoid/minimize vias in connections of at-chip bypass capacitors to minimize inductance (FDI: PCB).
- Nothing smaller than 0.01 uF is suggested for at-chip bypass. Although smaller capacitance values have higher resonance points, frequently the smaller devices don't exhibit low enough impedance or are insignificant in the mixed collection of impedances of the power grid (see Archambeault, 2007).
- If possible, keep the SRF of the capacitor 10X above the most common frequency used in the IC. High-frequency chips will generally use the 0.01 uF minimum as described earlier. Some devices operating at lower frequencies may be able to use higher-valued bypass capacitors at the chip.

- Using X5R and X7R capacitors is suggested due to their limited capacitance change (+/−15%) over temperature. Y5V devices exhibit extremely wide variance (+22%, −82%) and can be unpredictable.

- NP0 and C0G capacitors exhibit little thermal and bias variance. However, these devices are limited in their capacitance values for a given volume device. Precise (<5% error) capacitance values are not needed for power bypass.

- The voltage rating of capacitors should be greater than the maximum voltage seen in the circuit. A quick rule of thumb is 2X of the power supply that they are connected to.

- Again, check with IC vendors for their preferred strategy for their chips.

Grid bypass capacitors:

- The local power plane is kept stable with distributed capacitance.

- Using many capacitors spread out is better than using a few capacitors.

- The capacitors used for grid bypass are typically larger value than the capacitors used for at-chip bypass.

- For a two-sided component-loaded PCB, put the grid bypass capacitors on the side of the PCB that is closest to the P&G layers. This minimizes inductance back to the P&G layers.

- Distributed capacitance across the PCB should still use capacitors that exhibit good high-frequency characteristics, albeit resonating at a somewhat lower frequency than at-chip devices.

- Using capacitors 100X larger in size than the at-chip bypass devices keeps the resonance points well separated. For the situation where at-chip bypass is 0.01 uF, this implies grid bypass capacitors of 1 uF.

- Grid bypass capacitors should be distributed across the power planes such that any IC connected to the power plane has grid bypass in close proximity (Figure 4-30).

- Multiple grid bypass capacitors are suggested near the perimeter of any high gate count logic IC (e.g., MPU, MCU, digital ASICs, FPGAs, memory ICs).

- Using a thin insulating layer between P&G planes provides a negligible amount of grid bypass capacitance. This idea has been touted in various places but doesn't hold up when examining the numbers (FDI: PCB, a quantitative example is examined there).

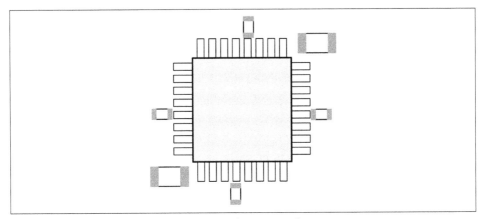

Figure 4-30. Bypass capacitance, at pin and across grid

Studies exist (see Bogatin, 2018) that deal with grid bypass capacitance as a controlled impedance versus frequency problem. Circuit simulators can analyze the interactive SRF of capacitors, but modeling the interconnection grid and everything connected to it is not quickly done. Typical design projects don't have the resources to make such a detailed study. These guidelines are an approach that can be implemented efficiently.

Regulator bypass:

- Manufacturers of any voltage regulator should be able to provide their preferred configuration for regulator bypass and decoupling.
- LDO regulators need to pay special attention to the ESR parameters of their bypass capacitors. Check the LDO specifications and application notes carefully.
- Avoid electrolytic capacitors where possible due to their poor high-frequency response and long-term reliability issues. The aluminum electrolytic capacitor (AEC) is widely used in many products, but it is most often the first component to fail. Because the AEC can put a lot of capacitance into a small area at a low cost, it is widely used in consumer electronics.
- SMT MLCC devices are preferred over AEC for their reliability and their low ESR/ESL.
- Multiple SMT MLCC devices in parallel are a viable option for larger value capacitance.
- Tantalum capacitors are more reliable than AECs and could also be used for regulator bypass.
- If AECs are unavoidable, placing parallel SMT MLCCs with the AEC will improve high-frequency performance.

Summary and Conclusions

The important points of designing a power system are as follows:

- AC power requires a strategy to avoid electric shock and ensure safety.

- When the system fails, the device must remain safe. Understand and implement methods to avoid live contacts for both normal operation and any possible failure mode.

- Overcurrent protection needs to be part of a weakest-link scenario. The fuse (or other protection) must open, not a PCB trace or an interconnect wire. This is a must-have for safety certification.

- Internal AC power needs grounding or double insulated safety methods.

- Off-line switchers for AC/DC conversion are efficient and compact DC power sources. Many low-cost options are available off the shelf, and it's usually more cost-effective to purchase one instead of designing one from scratch. Look for devices that use internal self-protection circuits.

- Regulate the power on the PCB where it is used. This minimizes problems due to connection impedance.

- Switching DC/DC converters can provide voltage reduction, voltage boosting, or both, in a compact and efficient manner.

- Linear regulators are inefficient, but they are useful where low noise on power or low EMI in the environment may be needed.

- Digital logic creates high-frequency wideband surge currents on P&G.

- P&G connections need dedicated PCB layers to create low-impedance connections.

- Impedance of P&G connections needs to be minimized wherever possible.

- High-frequency power stability requires bypass capacitors at all IC power pins.

- High-frequency power stability requires distributed capacitance across the power grid.

Techniques for PCB layout of low-impedance connections are covered in further detail in Chapter 11. A carefully thought-out power system improves signal noise levels, reduces radiated EMI, and should remain smoke free when something in the system fails.

Further Reading

- IEC 61140, "Protection Against Electric Shock," *https://www.iec.ch.*

- IEC 62368-1, "Audio/Video Information and Communication Technology Equipment – Part 1 – Safety Requirements," *https://www.iec.ch.*

- Bruce Archambeault, "Effective Power/Ground Plane Decoupling for PCB," IEEE EMC, October 2007.

- *Switching Mode Power Supply Design, 3rd Edition,* by A. Pressman, K. Billings, and T. Morey, 2009, ISBN 978-0-07-148272-5, McGraw-Hill.

- *High Speed Digital Design: A Handbook of Black Magic* by H. Johnson and M. Graham, 1993, ISBN 978-81-317-1412-6, Pearson.

- "Decoupling Techniques," Analog Devices MT-101 Tutorial, *https://www.ana log.com/media/en/training-seminars/tutorials/MT-101.pdf.*

- Paul Brokaw, "An IC Amplifier User's Guide to Decoupling, Grounding, and Making Things Go Right for a Change," Analog Devices AN-202 Application Note, *https://www.analog.com/media/en/technical-documentation/ application-notes/AN-202.pdf.*

- Kevin Tompsett, "Designing Second Stage Output Filters for Switching Power Supplies," Analog Devices, *https://www.analog.com/en/technical-articles/ designing-second-stage-output-filters-for-switching-power-supplies.html.*

- Vincent Greb, "Don't Let Rules of Thumb Set Decoupling Capacitor Value," *EDN Magazine,* September 1, 1995, *https://www.edn.com/edn-access-09-01-95-dont-let- rules-of-thumb-set-decoupling-capacitor-value.*

- Jerry Twomey, "Simple Grounding Rules Yield Huge Rewards," *Electronic Design Magazine,* April 27, 2012, *https://www.electronicdesign.com/technologies/indus trial/boards/article/21795841/simple-grounding-rules-yield-huge-rewards.*

- "Publications and Patents Authored by Jerry Twomey," Effective Electrons, *https://effectiveelectrons.com/articles-patents.*

- "Pentium III Processor Power Distribution Guidelines," Application Note, Intel, April 1999, *https://www.intel.com/design/PentiumIII/applnots/24508501.pdf.*

- *Signal and Power Integrity Simplified, 3rd Edition,* by Eric Bogatin, 2018, ISBN-13, 978-0-13-451341-6, Pearson Education.

CHAPTER 5

Battery Power

Modern electronics includes a multitude of mobile and portable devices, with the AC power cord relegated to battery charging. Battery power is a must-have for modern devices, and even plugged-in devices frequently have internal batteries to preserve critical systems when the power goes out.

Many device fires, product recalls, and explosions can be traced to improper battery system design. Also, many end users are frustrated by devices with insufficient battery capacity that can't make it through a typical day of use. Clearly, there are a lot of marginal battery systems out there in need of improvement.

Most battery textbooks focus on the chemistry and physics of batteries, with few centered on how to incorporate batteries into an electrical design. In this book, chemistry discussions are minimal and batteries are primarily treated as a black box. Readers here are trying to design better circuits and systems, not create better batteries.

In this chapter, the emphasis is on designing a battery system to fit the needs of an electronic system and on the selection criteria and support materials required to make proper design decisions. As your design progresses, make sure you obtain the latest performance specifications of the devices you plan to use. Battery technology is evolving, and specifics need to be up to date.

Battery Basics: Definitions

Figure 5-1 shows the general structure of a single-cell battery that includes three distinct parts:

Anode

Electrical connection off the negative side of the battery. The anode emits electrons when the battery is in a circuit. The anode surface that is in contact with the electrolyte serves as an electrode.

Cathode

Electrical connection off the positive side of the battery. A cathode absorbs electrons when the battery is in a circuit. The cathode surface that is in contact with the electrolyte serves as an electrode.

Electrolyte

Acts as a catalyst, resulting in ions moving between the anode and cathode. The movement of *ions*, molecules that lack electrons, within the battery causes electrons to move in the external circuit connected outside the battery. The electrolyte is often soaked in a nonconductive carrier material known as a *separator* to maintain physical distance between the anode and cathode.

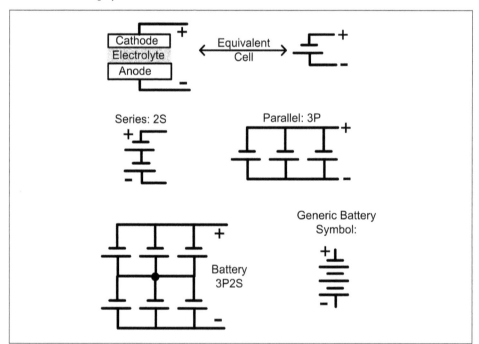

Figure 5-1. Physical structure of battery/cell

Following is some of the terminology associated with device performance:

Cell versus battery

A single *cell* is the anode-electrolyte-cathode trio shown in Figure 5-1. Individual cells are combined in series or in parallel to create a multicell battery. The terms *battery* and *cell* are often loosely interchanged. The common lead-acid 12 V automotive battery is an example of a battery created with six series (6S) cells. The common AA battery is actually a single cell. Figure 5-1 also shows how multiple cells are combined in series (2S) for higher voltage, or in parallel (3P) for higher current/capacity to create larger batteries. Combining cells in both parallel and series (3P2S) is commonly done. The battery symbol has never been rigidly standardized; the number of plates in the symbol and whether the wider plate is positive or negative varies across the literature.

Open circuit voltage

This is the voltage observed on a battery output (Figure 5-2) when the current is zero.

Internal resistance

This is the Equivalent Series Resistance (ESR) of the internal battery.

Capacity

Battery capacity is commonly measured in terms of current and time. The basic method is to measure the capability to produce a fixed current and the duration. The units of amperes × hours are normally used. The amp-hour (Ah) is used for capacity comparison and refers to how much electrical energy can be produced by the battery. To be meaningful, all batteries in a capacity comparison need to be the same voltage. In our discussion, Ah or mAh (milli amp-hour) is used where appropriate. Test methods vary among vendors, so a capacity value needs to be scrutinized for test conditions and methods. As will be shown, a single capacity value can be misleading.

C and C rate

The capacity of a battery is commonly known as C. For an ideal device, the Ah rating is the same current needed to discharge the battery in one hour. The C *rate* is a charge or discharge current that has been normalized to the capacity. If a battery has a 2 Ah capacity (C = 2 Ah), putting a 2A load on it (load current = 1C) should discharge the battery in one hour. A battery with 55 Ah capacity requires a 55A load to discharge in one hour. Any battery should ideally discharge in 10 hours if the load is 0.1C.

Load

This is the external element that discharges the battery. Battery loads are often expressed as an equivalent current (a 250 mA battery load) or as a fraction of the C (a 0.01C load).

Charging rate

This is the magnitude of the current used to charge the battery. It is not the same thing as the C rate. If a battery capacity is 10 Ah and is charged using 1A, then the charging rate is 1A or 0.1C. A battery with a capacity of 600 Ah that is being charged using 60A also has a charging rate of 0.1C.

Discharge rate

This is the magnitude of current drawn from the battery while it's in use. Frequently this is normalized to the C value. If a 10 Ah battery is discharging with 0.1A of current, the discharge rate is 0.01C.

State of charge

This is defined as the present capacity of the battery as a percentage of the fully charged capacity of the battery.

Chemistry

The battery chemistry refers to the materials used to make the cathode, anode, and electrolyte of the battery. Lead-acid, lithium-ion, carbon-zinc, nickel-cadmium (NiCd), and nickel-metal hydride are common names in use. The naming convention is not rigidly followed, however; the alkaline battery is based on zinc and manganese dioxide.

Primary battery

In the literature, a battery is considered primary if it is a single-use battery. In this text, the terms *single use* and *not rechargeable* are used.

Secondary battery

In the literature, a battery is considered secondary if it can be recharged. In this text, the term *rechargeable* is used.

Cylinder battery

This is a battery built with an external shape of a cylinder. The name doesn't indicate anything about the internal structure.

Prismatic battery

This is a battery built with an external shape that has flat surfaces, often rectangular or square.

Battery refresh

Used here, the term *battery refresh* can mean either the replacement of single-use cells or the charging of the battery.

Charge time

This is the elapsed time to recharge a battery.

Charging cycles or cycle life
This denotes the numeric estimate of how many times a battery can be discharged and charged before failure or degenerated capacity occurs.

Use time
This is the elapsed operating time between battery refresh events.

Shelf life
This refers to how long an unused, properly stored battery remains in good functional condition.

Service life
This text uses the terms *use time* and *shelf life*. Depending on the information source, the term *service life* has multiple meanings.

Energy density
Energy density serves as a comparison metric for how much energy is stored in a given volume or weight. Energy density is typically measured in (watt)(hours)/(kilogram) or (watt)(hours)/(liter). Depending on whether a system is constrained by weight or volume issues, either can be considered.

Power density
Power density serves as a comparison metric for how much output power can be created within a given volume or weight. Power density is typically measured in (watts)/(kilogram) or (watts)/(liter). Devices with high power density can briefly provide large amounts of energy to serve a particular application. As an example, an internal combustion engine uses a battery optimized for high power density to provide high amounts of "cranking current" to start the engine.

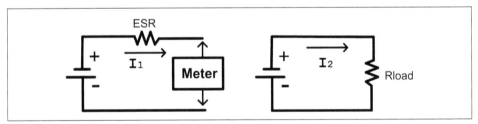

Figure 5-2. Electrical criteria of a battery

Decision Guidelines for Rechargeable or Single-Use Batteries

With some products, deciding between single-use batteries and rechargeable batteries is self-evident. With others, the decision is not straightforward. Following are certain things that should be considered when making that decision:

Price

Product cost goes up when a recharge system is included. The internal electronics are more expensive—adding the cost of an AC/DC source for the charger, the higher cost of rechargeable batteries, and possible problems with the air-shipping of restricted (Li-ion) materials. Using snap-in holders for single-use batteries and adding "Batteries Not Included" on the package is a low-cost solution.

Low-power use

Devices that use low power and require infrequent battery refresh are friendly to single-use batteries. Things like smoke detectors, remote controls, calculators, electronic keys, and wall clocks are all obvious single-use candidates.

Sporadic device use

Devices used infrequently should be configured for single-use batteries due to their better shelf life and slower self-discharge characteristics. This includes emergency devices (flashlights and radios), nonhospital defibrillators (AEDs), and similar.

Swappable battery packs

Easily swapped battery packs with rechargeable batteries fit a scenario where lightweight portability is necessary but heavy power consumption is still needed. The construction industry has embraced power tools supported by multiple battery packs, avoiding the mobility limitations of power cords.

Internally installed rechargeable batteries

These work well when power demands require frequent battery refresh and the typical use scenario for the device allows charger access. Tablets, laptops, and cell phones are common examples.

Power outage systems

Devices that activate during AC power outages are best served by rechargeable batteries. This allows an automatic reset and battery recharge when mains power has been restored. Both uninterruptible power supplies and emergency exit lights fall into this category.

After deciding whether you should use single-use or rechargeable batteries, the next step is to determine the power budget.

Defining Power Requirements

Electronic systems can have power consumption extremes depending on the application. Supercomputers run on megawatts and modern electronic watches function on 2–6 microwatts. Determining exactly where a design sits in this broad energy spectrum is an important part of defining whether batteries can be used and what the battery strategy should be.

Several things need to be determined:

- Use time
- Peak currents during use
- Long-term average current (LTAC) over all use scenarios
- Functional voltage range of the system and min/max voltages needed

Use time is straightforward. Here are some examples:

- Smoke detector: >12 months
- Cell phone: >20 hours
- ZigBee devices: >2 years mandated for certification

One variation on use time is the "number of clicks" concept. For things like a wireless mouse or a handheld remote control, specifying the number of clicks between battery refresh events is a possible approach. Devices need to consume negligible current between click events for this to be valid. The amp-hours per click is determined and then multiplied by the defined minimum clicks to determine needed battery capacity.

Determining peak current and LTAC requires examining the transient current of the device (Figure 5-3). This can be done in the lab with a test fixture, or an accounting of all active circuits can be manually estimated.

Depending on the system design, many different profiles of current consumption are possible. Case A in Figure 5-3 is the simplest scenario, where a steady state current is the same as the LTAC. For Case A, the LTAC and peak current are the same. Case B is common for many single-purpose systems, where brief bursts of high-current activity are separated by long, low-current sleep periods. The LTAC will be highly dependent on the time periods of active versus sleep and their respective currents. The peak current occurs during the active period. Case C and Case D are typical of many multifunctional systems with mixed current use profiles. In both situations, various things turn on/off for intermittent periods.

Prior to building hardware, a first-pass estimate of LTAC can be based on the high-current devices and their expected amount of active time. Electromechanical devices (motors, servos, solenoids, etc.) tend to be the elephant on the list. Peak current surges are commonly seen at the startup of a motor or servo device. Display lights, high-performance field-programmable gate arrays (FPGAs), and multicore microprocessors can also dominate a power budget.

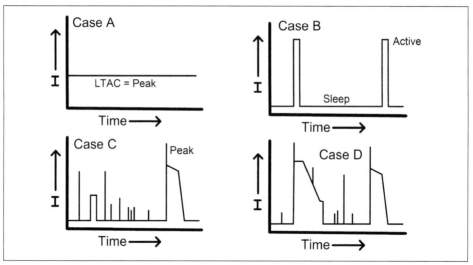

Figure 5-3. Power budget current profiles

The min/max operating voltages of the system electronics depend on the power system design (FDI: Power). If battery power feeds directly to active circuits, operating voltage is limited by those circuits. Many power systems use some form of voltage regulator up front, allowing a wider functional voltage range.

With power requirements defined, a battery can be fitted to the electronics.

Battery Discharge Versus Functional Voltage Range

A typical battery discharge profile with a fixed-current load is shown in Figure 5-4. The voltage output of the battery slowly descends in a nonlinear manner as the battery discharges.

The discharge profile starts from a fully charged device (A) where the battery is "topped off" by the charger, or illustrates a fresh single-use battery. Discharge quickly reduces the voltage to a region where a more linear discharge profile occurs (B, C, D), and discharge is considered complete at the knee in the curve (E). Discharge beyond E can sometimes damage the battery, depending on the chemistry. Single-use batteries are often used down to F, but many rechargeable battery management systems protectively cut off at E to avoid damage.

As the battery discharges, a progressively lower voltage is presented to the system. If the system can function down to E–F, it is making full use of the battery's energy supply. If the system stops functioning back at C–D, the full capacity of the battery is not being used. To utilize the full range of the battery, either a boost circuit can be used as part of the power system (FDI: Power), low-dropout (LDO) regulators may be used in the power system to increase functioning range, or the battery pack can use batteries connected in series to increase the output voltage. All are viable.

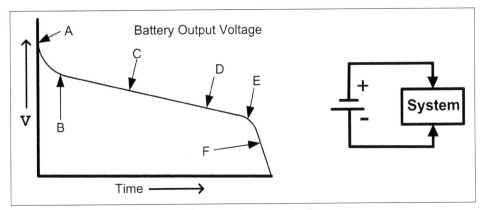

Figure 5-4. Battery discharge profile versus functional voltage range

Battery Types by Chemistry

Different types of batteries have preferred features, with the division between battery types drawn along two lines: chemistry and package characteristics (see Table 5-1). Considerations when picking a battery include:

- Price and vendor availability
- Shelf life (capacity deterioration, self-discharge, physical leakage)
- Energy density (by volume or weight)
- Recharge durability (number of recharge cycles versus capacity deterioration)
- Maintenance issues (water replenish, top-off charging)
- Capacity (amp-hour)
- Peak current capability and power density
- Safety restrictions for use and transport
- Capacity reduction due to temperature
- Life reduction due to temperature
- Disposal restrictions and toxicity
- Charging time

Table 5-1. Comparison of battery types by chemistry

Common name	One use[a]	Shelf life (years)	Self-discharge rate	Normal voltage per cell	Minimum voltage before turning off	Features and comments
Zinc carbon (ZC)	S	4	10%/year	1.5	0.8	AAA, AA, C, D, and 9 V variants available, low cost, low energy density, prone to leakage, performance very thermally dependent
Alkaline	S	7	3%/year	1.5	0.8	Many variants available, 2X cost of ZC, moderate energy density, low internal resistance, performance thermally dependent
Li-manganese dioxide	S	12	1%/year	3.0	2.0	Commonly used in coin cell batteries
Li-iron disulfide	S	12	1%/year	1.5	0.9	AAA and AA variants available, 10X cost of ZC, good energy density, low internal resistance
Lead-acid (SLA AGM)	R	5	5%/month	2.1	1.75	Many 6 V and 12 V variants, low cost, low internal resistance, good for bulk energy storage where volume/weight are not important
Nickel-cadmium	R	3	20%/month	1.2	0.9	Flat discharge, fussy charge/discharge, low internal R, fast recharge, toxic disposal issues, replaced by Li-ion in new designs
Nickel metal hydride	R	4	25%/month	1.2	1.0	Flat discharge, fussy charge/discharge, low internal R, suitable for high-surge currents, lower cost than Li-ion, good energy density
Lithium-ion	R	10	5%/month	3.6	3.0	Needs circuit protection and charge balancing, low internal R, suitable for high-surge currents, very good energy density, safety issues
Lithium-ion polymer	R	Not available	7%/month	3.7	3.0	Similar performance to Li-ion, useful for noncylindrical batteries
Lithium-iron phosphate	R	3	3%/month	3.2	2.4	Safer than Li-ion with no thermal runaway, commercial use expanding into EVs and other products, promising but still evolving

[a] S = single use, R = rechargeable

Following are some additional comments on devices by chemistry:

Zinc-carbon
> This is an older technology that largely survives due to its low cost.

Alkaline (aka alkaline-manganese)
> Alkaline batteries are the high-volume workhorse of single-use devices. They are very popular for general use and a cost-effective improvement over their zinc-carbon predecessor.

Lithium manganese dioxide (aka lithium metal)
> These battery types are widely manufactured in coin cell packages and are not readily available in other package variants.

Lithium iron disulfide (aka lithium metal)
> This is a high-cost technology and is not encouraged for new designs due to limited availability.

Lithium thionyl chloride (LTC)
> Characteristics of this technology include single use, very flat discharge curve, wide temperature range, good shelf life with low self-discharge, and high internal resistance. LTC devices are a good fit for low-current and wide-temperature applications. As of 2023, LTC devices are starting to be made by more vendors. There are restrictions and safety warnings on LTC battery use. Check local regulations and restrictions before using LTC in a design.

Lead-acid–flooded cells
> This technology is commonly used in automobiles, trucks, golf carts, and similar applications. A deep cycle variant changes the internal plate structures to be optimal for deep discharge cycling. A cranking variant uses an internal plate structure, which is optimal for short-period, high-current surges. This technology is not suggested for electronic systems due to slow charging time, maintenance issues, messy spill problems, and a need to be kept upright.

Lead-acid–sealed–absorbent glass mat (SLA-AGM)
> These are also known as valve-regulated lead-acid, absorbent glass mat batteries. The SLA-AGM battery holds the acid electrolyte in fiberglass between the cathode-anode plates; because it is sealed, no maintenance is required. Also, SLA-AGM does not need to be held upright, making it popular in airplanes. SLA-AGM batteries come in many capacity sizes, with 6 V (3 cell) and 12 V (6 cell) being the most popular voltages.

Lead-acid–sealed–gel (SLA-gel)
> The gel battery mixes the sulfuric acid electrolyte with a silica compound that results in stiffening of the mixture. No maintenance is required and the batteries

can be used in any orientation. SLA-gel and SLA-AGM have many of the same characteristics, with SLA-gel having somewhat better thermal dissipation.

Nickel cadmium (aka NiCd or NiCad)

NiCd batteries used to be a popular solution for power tools and other high surge current applications. NiCd devices have low internal resistance suitable for powering high-current devices. More modern alternatives are now using Li-ion and NiMH methods. Due to toxic cadmium waste, some countries have banned NiCd batteries. Therefore, NiCd is not suggested for new designs, and manufacturers are phasing them out.

Nickel-metal hydride (NiMH)

NiMH has low internal resistance and performs well with high surge current applications. NiMH suffers from large amounts of self-discharge. With a poor shelf life, they need to be kept on a charger when not in use.

Lithium ion (Li ion)

Li-ion batteries now occupy the bulk of the rechargeable battery market, with an established presence in everything from small electronic handhelds to large automotive power cell arrays.

Lithium ion polymer (LiPo)

LiPo has similar electrical characteristics to Li-ion devices. LiPo devices use a laminated sheet method for the electrodes, separator, and electrolyte, which allows devices to be made into many different shapes.

Lithium iron phosphate (LiFePO4 or LFP)

These are rechargeable, and are safer and less prone to combustion and thermal runaway than Li-ion. Cycle life is claimed to be much longer than in Li-ion devices. Energy density is slightly lower than Li-ion. Presently, device costs are higher than similar capacity devices, but they are expected to drop as more vendors develop manufacturing. LFP is starting to be used as the principal power in EVs.

These are the widely used chemistry types. This area is being heavily researched and improved methods are frequently emerging.

Discharging Behavior of Batteries

To better select a battery setup, it is useful to look a bit deeper at the characteristics of a battery discharging while being used. Simple amp-hour ratings and nominal voltages can be very misleading.

First, it is necessary to determine the best approach to get quantitative answers on battery performance. Does a math model for a battery exist, and can it be used in a simulator to get answers?

As shown in Figure 5-5, a basic battery model would include some internal RLC impedance, a voltage source, and the interconnect inductance. For a DC-only model, just $R_{internal}$ and $V_{battery}$ are needed. These are not fixed values, however. $R_{internal}$ changes for charging and discharging operations, and with aging of the battery. $V_{battery}$ changes with the state of the charge, temperature of the battery, aging due to storage, and aging due to recharge cycling. Simulation models can be developed, but they require a lot of test data and curve fitting to create a meaningful model. Generally, it's not worth the effort.

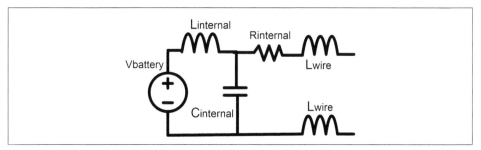

Figure 5-5. Battery equivalent circuit

Trying to simulate a battery doesn't expedite the design of a suitable battery pack. The battery discharge profiles can provide a much easier way to get a solution. For this, a typical AA alkaline battery is used as an example in Figure 5-6. As shown, this curve set shows the discharge profile of a common device for five different currents. Doing the math, the curves display Ah ratings between 0.9 Ah and 1.9 Ah. Higher discharge currents generally produce a lower Ah rating.

In addition, the vendor-stated battery capacity can be very misleading. In this case, 2.7 Ah was the stated capacity of the device. Vendors often state the Ah rating at a very low discharge current to get a better capacity number. Consequently, it is important to seek further details regarding capacity testing. A higher current means a lower Ah rating. Determining a discharge curve where the discharge current value is similar to the LTAC of the system is advised.

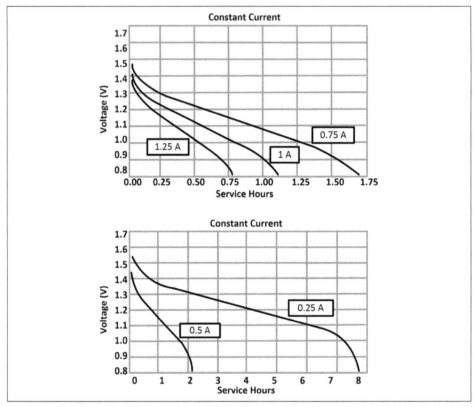

Figure 5-6. Battery discharge current versus use time (AA alkaline)

Most battery performance is tied to both discharge current and temperature. As an electrochemical reaction, battery performance is very temperature dependent. Figure 5-7 shows a performance reduction from more than seven hours to less than one hour over temperature extremes. The thermal performance of the battery and the temperature of the application environment are important considerations in the design. Many vendors test with warm batteries (25°C–30°C) to produce better capacity numbers.

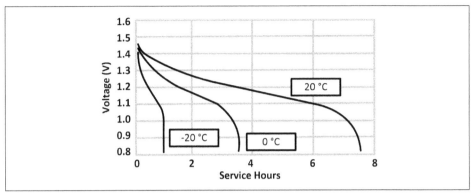

Figure 5-7. Battery discharge curves: use time versus temperature

It is important to get the performance details and check the LTAC versus use time over temperature. Using higher-capacity devices or using batteries in parallel is the most common strategy to improve use time.

Battery capacity is not a single rigid number; margin also needs to be included. Peak current must be checked as well, to avoid battery damage. Combining batteries in parallel or using a higher capacity battery are valid methods to achieve a sufficient use time. The number of batteries in series depends on the chemistry and the minimum voltage needs of the system.

Table 5-2 illustrates the expected useful voltage range from a single-cell battery, broken down by chemistry. The common alkaline battery gets run down as low as 0.8 V per cell, well over the knee of the discharge curve. Since the devices are single use, there is no concern with causing internal battery damage.

Table 5-2. Expected voltage range by battery chemistry

Type	Discharge min voltage[a]	Nominal voltage[b]	Max voltage
Alkaline[c]	N(0.8)	N(1.25)	N(1.5)[d]
Lead–acid	N(1.75)	N(2.0)	N(2.4)[e]
Li-ion	N(3.0)	N(3.6)	N(4.2)[e]
NiMH	N(1.0)	N(1.2)	N(1.6)[e]

N = Number of cells in series cells within the battery

[a] Voltage limit defined by battery chemistry
[b] Dependent on nominal load and temperature; check specific battery documentation
[c] Common alkaline devices (AAA, AA, C, and D, but not 9 V)
[d] Voltage determined by use of new battery
[e] Voltage created by presence of battery charger

For example, a 9 V transistor radio battery is a variant of the alkaline battery and uses six cells within its structure. So, $6(0.8) = 4.8$ V is the discharged minimum voltage that should be designed for. As another example, the common automotive 12 V lead-acid battery contains six cells, so $6(1.75) = 10.5$ V should be expected when fully discharged, and a maximum of $6(2.4) = 14.4$ V would be expected when a charger is attached.

A comparison of the minimum voltage for Li-ion (3 V) and NiMH (1 V) is useful. Three NiMH cells are needed to get the same voltage as one Li-ion device. These two devices are often compared for use, and a good comparison needs to include similar capacity and appropriate cell count. Alternative methods include designing voltage boosting capability in the power system.

In all cases, the system electronics need to function at both min/max voltages to make full use of the battery capacity. If the system electronics shut down before hitting V_{min}, consider either series batteries to bring V_{min} up, or a boosting voltage converter circuit in the power system (FDI: Power), thus allowing electronics functionality at lower battery voltages.

Designing a Battery Set: Single Use and Multiple Cells

Table 5-3 shows the many single-use batteries available. Single-source batteries should be avoided whenever possible. Devices summarized here are all widely available. There are cases of products being halted because the battery vendor stopped making a specialty single-source battery.

All of the devices in Table 5-3 are alkaline, unless stated otherwise. One interesting item in this table is the comparison of AA in lithium-metal to alkaline. Although the lithium metal has ~20% better capacity, vendor prices are ~5X higher for lithium over alkaline. If higher capacity is needed, two alkaline batteries in parallel are more cost-effective. The 9 V alkaline/lithium devices have a similar capacity versus cost difference.

Table 5-3. Single-use batteries (AAAA through D, 9 V)

Common name	Volt	Diameter (mm)	Height (mm)	Capacity (mAh)
AAAA	1.5	8	42	670 @ 10 mA
AAA	1.5	10	44	1,300 @ 25 mA
AA	1.5	14	50	2,900 @ 25 mA
AA (Li)	1.5	14	50	3,500 @ 25 mA
C	1.5	25	50	8,000 @ 25 mA
D	1.5	33	60	16,000 @ 25 mA
9 V	9	17×26	48	600 @ 25 mA
9 V (Li)	9	17×26	48	750 @ 25 mA

Remember that battery capacity reduces at higher currents. Capacity numbers are often published using low currents and warm temperatures to produce higher numbers. Table 5-4 shows how capacity varies with load current.

Table 5-4. Battery capacity versus discharge current: single-use alkaline (AAA through D)

Load I:	@ 25 mA	@ 100 mA	@ 250 mA	@ 500 mA
AAA	1,300	950	600	300
AA	2,900	2,500	2,000	1,500
C	8,000	7,000	5,000	4,000
D	16,000	15,000	10,000	7,000
		Capacity in mAh		

Single-use standard batteries for small applications are mostly coin cell batteries. These are also known as watch batteries, hearing aid batteries, or button cells. The IEC 60086-3 standard defines 65 different versions of coin cell batteries. Table 5-5 shows the most commonly used coin cells.

There is no standard for labeling and numbering of coin cells. Some vendors use labeling that includes diameter and thickness. But, vendor-specific names do not all follow the same standard. A CR2032 battery, for example, is 20 mm in diameter and 3.2 mm thick. Compare physical dimensions and voltage to determine compatibility. Also, make sure multiple vendors source the device before committing to use it.

Table 5-5. Single-use coin cells

Common name	Volt	Diameter (mm)	Height (mm)	Capacity (mAh)
392	1.5	7.8	3.6	44
357	1.5	11.6	5.4	150
CR1616	3.0	16	1.6	55
CR1632	3.0	16	3.2	130
CR2016	3.0	20	1.6	90
CR2025	3.0	20	2.5	163
CR2032	3.0	20	3.2	240
CR2450	3.0	24	5.0	620

The chemistry used in coin cells varies. Alkaline and various lithium-metal devices exist, and silver oxide and zinc-air devices are also offered. Coin cells that used mercury and other hazardous materials have been phased out.

When designing a battery set using single-use batteries, take the following into account:

- Long-term average current (LTAC) or burst use (Ah per click)
- Peak current
- Use time, or total clicks before battery refresh
- Min/max voltage of the system versus the battery
- Temperature of the environment

A low-power wireless thermostat is a good example for single-use batteries. As summarized in Table 5-6, the ZigBee standard requires that devices have a battery life longer than two years. In this example, a thermostat wirelessly reports over the network every 60 seconds. The device goes live for 250 ms, using 9.3 mA, and then sleeps, using 4.2 uA, for 60 seconds. The LTAC is 43 uA, providing a use time of over seven years for alkaline AA batteries and over three years for AAA batteries.

Since the operational V_{min} = 1.9 V, using two batteries in series is a suitable solution. In addition, since the minimum battery voltage at full discharge is 2(0.8) = 1.6 V, the batteries are not fully discharged when the system stops functioning at 1.9 V (0.95 V per cell), but there's ample margin on the two-year requirement if using AA batteries. If desired, a more careful analysis can be done using 1.6 V as the bottom of the discharge cycle.

Table 5-6. Example: single-use batteries, ZigBee wireless thermostat

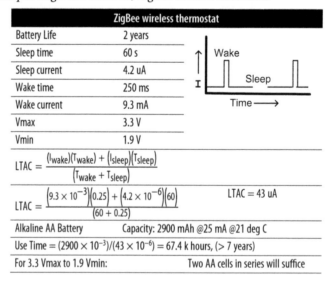

ZigBee wireless thermostat	
Battery Life	2 years
Sleep time	60 s
Sleep current	4.2 uA
Wake time	250 ms
Wake current	9.3 mA
Vmax	3.3 V
Vmin	1.9 V
$LTAC = \dfrac{(I_{wake})(T_{wake}) + (I_{sleep})(T_{sleep})}{(T_{wake} + T_{sleep})}$	
$LTAC = \dfrac{(9.3 \times 10^{-3})(0.25) + (4.2 \times 10^{-6})(60)}{(60 + 0.25)}$	LTAC = 43 uA
Alkaline AA Battery	Capacity: 2900 mAh @25 mA @21 deg C
Use Time = $(2900 \times 10^{-3})/(43 \times 10^{-6})$ = 67.4 k hours, (> 7 years)	
For 3.3 Vmax to 1.9 Vmin:	Two AA cells in series will suffice

Since this ZigBee thermostat spends most of the time in a micropower (4.2 uA) sleep mode, using a buck-boost converter to use the full energy of the battery is not suggested. The buck-boost converter would need to run all the time, and would greatly increase the current in sleep mode.

Designing a Rechargeable Custom Battery Pack

Picking the chemistry is the first step in designing a custom rechargeable battery pack. Presently there are three contenders:

- Sealed lead acid (including SLA-AGM and SLA-Gel)
- NiMH
- Li-ion family (including LiPo, LiFePO4, Prismatic, and cylinder variants)

 NiCd is prohibited in some countries, and is being phased out by manufacturers due to toxic waste issues. Consequently, NiCd batteries are discouraged for new designs.

By both mass and volume, SLA has the worst energy density, Li-ion has the best energy density, and NiMH sits in between. If low weight and small volume are your top priorities, Li-ion is the path to take. If weight and volume are not a concern, SLA can be a low-cost and reliable solution. LiFePO4 is rapidly becoming a contender in areas where higher capacity is needed. Check vendor capability and unit costs; as of 2023, this is rapidly changing.

The high surge current capability and higher energy density of NiMH was the motivation for it replacing NiCd in many applications. Li-ion can be configured for high surge current scenarios but may need specially designed safety shutoff circuits. If a specially shaped battery in a limited space is needed, a LiPo device is generally the best fit.

Table 5-7 shows the most common offerings of SLA-AGM batteries. These batteries are offered in many variants, but the most widely available are three-cell (6 V) and six-cell (12 V) devices. Since there is no labeling standard for the devices, you should verify compatibility by dimensions, voltage, capacity, and connector location. A wide range of capacities are available, and SLA devices combine easily in parallel and in series to increase capacity and voltage.

Table 5-7. Lead-acid (SLA-AGM), volt/size/capacity, multisourced

Device[a]	Volt (V)	Height (mm)	Width (mm)	Length (mm)	Capacity (Ah)	Weight (Kg)
PS-612	6	51	24	97	1.3	0.3
PS-640	6	100	47	70	4.3	0.8
PS-670	6	94	34	151	6.7	1.1
PS-6100	6	94	51	151	12.0	1.9
PS-1212	12	52	43	96	1.4	0.5
PS-1220	12	60	35	178	2.4	0.9
PS-1230	12	60	67	133	3.4	1.3
PS-1250	12	101	70	90	5.0	1.6
PS-1270	12	94	65	151	7.0	2.2
PS-12120	12	93	98	151	12.0	3.6
PS-12180	12	167	76	181	17.0	5.7
PS-12260	12	125	177	167	24.0	7.7

[a] Battery vendor (PowerStream) device numbers are given here as typical examples. Each vendor uses different numbering. Compatibility needs to be individually determined.

When combining batteries in parallel, keeping the battery currents balanced for both high current discharge and charging requires some attention to detail. Figure 5-8 shows that when combining batteries in parallel, the IR drops in the wiring need to be kept balanced so that the batteries provide equal currents to the load. The interconnect resistance and where it is placed can affect battery load in high surge current scenarios.

Case 1 (bad) illustrates a common way that two high-current batteries are placed in parallel to increase capacity and surge current capability. When a high-current surge is demanded by the load, the battery currents I_a and I_b will not be equal because extra resistance is present from R_{wire} on both sides of the left battery. Because the batteries are connected in parallel, the charge state of the two devices will balance each other out when the load current is low. However, the battery on the right will provide a larger part of the surge current and therefore will tend to degrade in performance more quickly. In all cases, R_{wire} should be kept identical. In high-current situations, small amounts of resistance can be significant.

Case 2 (good) fixes the surge current imbalance by making sure each battery has a single R_{wire} in series. Case 3 (bad) is an extension of the problem to three batteries, where surge currents will be different from all three devices. Case 4 (good) rebalances the resistive paths to all three batteries by using tie-together bus bars. In addition, it avoids stacking connections on battery connections that can be mechanically problematic. Case 5 (good) also balances the resistive paths to all batteries, without needing a bus bar tie point. This configuration can be used for multiple batteries in parallel.

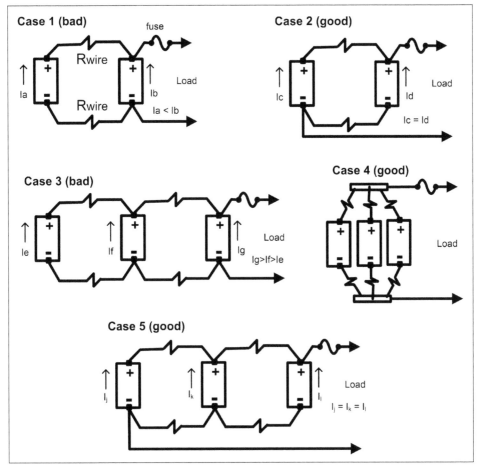

Figure 5-8. High-current battery connections

In addition, when combining devices, keep the batteries matched. Using the same age, same type/model, and same vendor gets the best balanced charge/discharge performance.

SLA devices will be individually purchased and manually configured into a connected array where needed. Battery packs using Li-ion and NiMH cells are generally assembled by a specialized manufacturer.

As shown in Table 5-8, NiMH cells are available in a multitude of sizes. Some of the more common ones are listed. Always determine long-term availability and specifics of performance before including NiMH in a design.

Table 5-8. NiMH for custom battery packs

Name	Voltage range max–min (V)	Diameter (mm)	Height (mm)	Capacity (mAh)
AAA	1.4–1.0	10.0	44.5	700 @ 140 mA
AA	1.4–1.0	14.5	50.5	1,500 @ 300 mA
A	1.4–1.0	17.0	50.5	2,450 @ 490 mA
C	1.4–1.0	25.0	50.0	3,100 @ 600 mA

Availability of Li-ion batteries is different than other devices. Custom shapes and sizes are available in prismatic variants (seen in cell phones, tablets, and other small devices). As shown in Table 5-9, the predominant Li-ion cylinder battery is the 18650 size. The 18650 is used in arrays from a single cell to large (>8,000 cells) arrays in electric cars. The 18650 was widely manufactured for laptop battery packs, and as laptops slimmed down (requiring prismatic batteries) the devices became available for other applications at attractive prices.

Larger Li-ion cylinder batteries are now being built in quantity. Due to demand, this market is evolving rapidly.

Table 5-9. Li-ion for custom battery packs

Name[a]	Voltage range max–min (V)	Diameter (mm)	Height (mm)	Capacity (mAh)
18650	3.7–3.0	18	65	2,000–3,550 DOV[b]
21700	3.7–3.0	21	70	~4,500
20700	3.7–3.0	20	70	~4,200
Prismatic	Custom, voltage range: N(3.7–3 V), capacity as desired[c]			
Pouch cell	Custom, voltage range: N(3.7–3), capacity as desired[c]			

[a] Labeling: <XX><YY><Z> XX = Diameter, YY = Height, Z = 0 = Cylinder
[b] DOV = Depends on vendor
[c] N = Integer number of series cells

Due to safety issues with Li-ion, many manufacturers are reluctant to offer unprotected batteries on the open market. Instead, a common approach is to sell unprotected batteries only to recognized manufacturers of battery packs so that the devices are properly outfitted with safety circuits, charge balancing, and support electronics.

Many battery pack vendors offer a multitude of preassembled battery array modules with battery management and safety electronics built in. Some of these come with safety regulatory certifications and can be a cost and time saver.

Conceptually, the safety cutoff device (Figure 5-9) is simple: disconnect the battery if the charger tries to exceed a maximum voltage (~4.3 V/cell) or the system tries to discharge the battery below a minimum voltage (~2.4 V/cell). For multiple batteries in series, the cutoff voltages scale up proportionally.

The safety cutoff is a last line of defense for the battery before damage or fire occurs. Charger systems, when properly configured, should stop charging at about 4.2 V, and power management should stop draining the battery at around 3.0 V per cell.

The voltage limits mentioned here (4.3, 4.2, 3.0, 2.4) are typical, and most vendors will have similar values. If planning to multisource batteries, a set of limits to accommodate all vendors can be worked out.

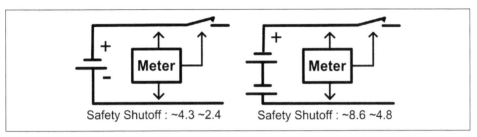

Figure 5-9. Li-ion safety cutoff

As emphasized throughout this book, nothing is ideal, including batteries. As illustrated in Figure 5-10, when a battery with multiple series cells is charged, voltages are slightly different across each cell. As the output voltage rises during charging, the charging of multiple series cells can be different. Charge balancing is needed to both maximize charge and avoid individual cells from going into overvoltage.

On the right in Figure 5-10, a controlled bypass circuit connects around each cell. While charging, this keeps all cells at the same voltage, and all cell voltages rise together. Charge-balancing control chips are readily available for cells in series from 2S to 32S.

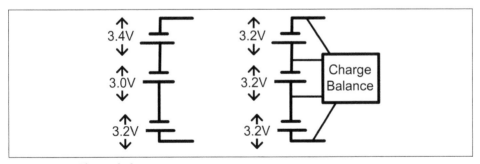

Figure 5-10. Charge balancing

Typically a charge-balancing circuit and a safety cutoff circuit are integrated into the battery pack.

Designing a battery set using rechargeable batteries needs to consider:

- Long-term average current (LTAC) or burst use (Ah per click)
- Peak current
- Use time, or burst use (total clicks) before battery refresh
- V_{min} and V_{max} needs of the system versus V_{min} and V_{max} capability of the battery
- Temperature of environment and resulting capacity changes of the battery
- Reduction in capacity due to aging
- Safety cutoff circuits
- Charge balancing circuits

Table 5-10 is an example of specifying a custom rechargeable battery for a smartphone. This example determines needed battery capacity by testing the various modes of the phone for current consumption. After determining the current used in each mode, the time for each mode is estimated for the use time. With a smartphone, several activities can be occurring simultaneously. The total Ah are summed and prorated for 20% loss of capacity due to aging. The results are typical for batteries used in smartphones presently on the market.

Table 5-10. Example: capacity calculations for rechargeable battery

Smartphone battery use: custom Prismatic LiPo	
	Power used while:
Use time: 24 hours	Suspend mode = 70 mW= 21 mA
Vmax: 4.3 V	Lit display = 300 mW = 88 mA
Vmin: 2.7 V	Web browsing = 430 mW = 126 mA
Vnom: 3.4 V	Active calling = 1,050 mW = 309 mA
Temp: 21 C	Texting = 300 mW = 88 mA
	Idle = 270 mW = 79 mA
Activity per 24 hours (h):	
Suspend mode: 8h	8(21) mAh = 168 mAh
Idle: 10h	10(79) mAh = 790 mAh
Active calling: 2h	2(309) mAh = 618 mAh
Web browsing: 3h	3(126) mAh = 378 mAh
Texting: 1h	1(88) mAh = 88 mAh
Lit display: 6h	6(88) mAh = 528 mAh
	Total: 2,570 mAh
Rechargeable battery; specify a usable system with up to 20% loss of battery capacity as the battery ages:	2,570 mAh = (0.8) capacity Capacity = 3,200 mAh @ 400 mA (current of: active call + lit display)

Charging Batteries

Battery charging methods are unique to chemistry. Each manufacturer will probably suggest a slightly different set of charging parameters to get the most performance out of their devices. In addition, the number of cells in series and in parallel scales the parameters of the charging process.

There are many custom integrated circuits (ICs) designed to control battery charging. Most are defined by the chemistry of the battery they are designed to charge. The parameters of charging are either digitally selected or controlled using external components, to set voltages and currents needed for a specific battery pack. Multiple battery-charging ICs are available for Li-ion, lead-acid, and NiMH devices.

Designers have implemented software-defined charging by a host controller. However, high-voltage controls are needed for the power transistors, voltage and current sensing are needed, and the dedicated charging IC is often a more cost- and time-effective solution.

Two examples of charging profiles are discussed here: one for lead-acid and one for Li-ion devices. The information is for one cell. As devices are added in series, the voltage will scale. As devices are added in parallel, the current will scale.

Most batteries use constant current charging, where a fixed current is injected into the battery. The magnitude of the current is determined by the capacity and the chemistry of the battery.

A lead-acid constant current charging strategy is shown in Figure 5-11. The constant current used, I_{cc}, is typically set at 10–30% of C for the battery. Higher currents would cause more energy to be lost as heat while charging. This current limitation implies that a lead-acid battery takes four to 10 hours to fully recharge. Vendors will specify the details of their charging profile and the highest currents that can be used without doing damage.

A fully depleted cell is 1.75 V or lower. When the cell rises to 2.4 V, the current is tapered off in a segment referred to as *top-off charging*. After top-off, the charger holds the cell at ~2.26 V in float or maintenance mode until removed from the charger. For a typical six-cell automotive battery, depleted is 10.5 V or lower, charging top-off is 14.2 V, and float is 13.6 V.

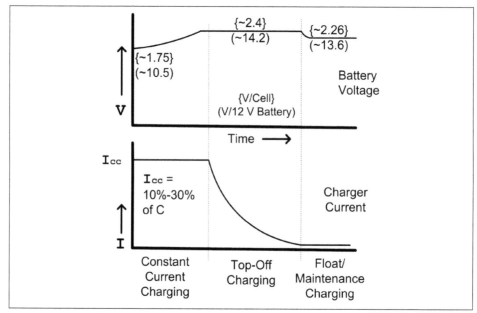

Figure 5-11. Lead-acid charging profile

Figure 5-12 shows that charging an Li-ion battery has some differences. Li-ion starts charging in a constant current mode. Most Li-ion devices use a charging current of 50–80% of C, so a full recharge needs one to three hours. Again, the numbers are vendor specific.

A fully depleted Li-ion cell is 3 V or lower, and when the voltage of the cell rises to 4.2 V, the charging current is tapered down (saturation charging) and then turned off.

A maintenance voltage (float voltage) is not utilized for an Li-ion battery. Rather, the output is monitored until the device reduces to a defined point (typically 3.7 V) where the saturation charging is reenabled to top off the cell. Multiple cells in parallel scale up the C of the battery pack and multiple cells in series scale up the voltage.

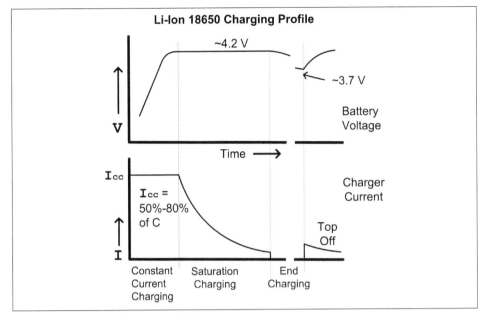

Figure 5-12. Li-ion charging profile

Smart Batteries

An integrated battery pack has features beyond just the battery cells. A battery pack with added features is often referred to as a *smart battery*. Consider the full-feature example in Figure 5-13.

The smart battery pack shown has a 4P-4S configuration. Cells are connected in parallel groups of four before the groups are connected in series to create the set. All five battery nodes are connected to a battery management system (BMS) that facilitates charge balancing and monitors/controls charging.

Thermal monitoring (T_{sense}) is often included, especially within large arrays. Current limiting (I_{limit}) can be done by using a thermally resettable fuse (aka PPTC) or by controlling the cutoff switch through the BMS.

Current sensing (I_{sense}) allows host feedback on both charging and discharging rates. A safety cutoff (cutoff) prevents overvoltage (4.3 V/cell) and undervoltage (2.4–2.7 V/cell) from happening. Voltage sensing is also included for the whole battery pack.

The voltage sensing and current sensing can be used to provide the system's charge monitoring state. Time, voltage, and current information can be monitored by the host process to create a coulomb counting system, or a fuel gauge status can be estimated by the overall voltage of the battery. Both are commonly used.

Digital serial interface control is available back to the host process commonly through an I²C/SMB interface. (Future devices should migrate to a CAN bus for better data integrity and error checking.) Of course, the smart battery still needs an external battery charger to provide the needed charging profile to support a 4P-4S battery pack.

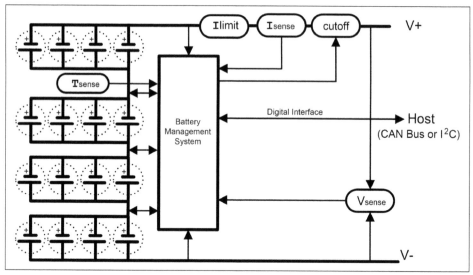

Figure 5-13. Full-feature smart battery system

Figure 5-14 shows a typical charging and battery system that would be found in a laptop computer. The system power can be provided from the AC/DC converter or the smart battery, with selection controlled by the presence of the AC/DC supply. A charger and controller provide the needed charging profile for the smart battery. All items within the dashed lines are available as a power-charger management system on a chip (SoC). The two switches in the power path are usually external power MOSFETs driven by the SoC.

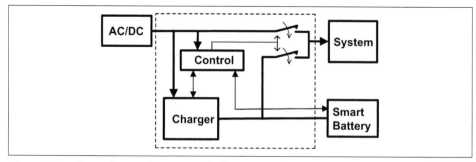

Figure 5-14. Typical smart battery with charger configuration

Regulations and Safety for Batteries

To sell and ship devices with Li-ion batteries, a suite of tests and regulatory certifications are needed, depending on the countries involved. Expect to see testing requirements based upon the following:

- UN/DOT 38.3: Transportation Testing for Lithium-Ion Batteries
- IEC 62281: Safety of Primary and Secondary Lithium Cells and Batteries During Transport
- IEC 62619: Secondary Cells and Batteries Containing Alkaline or Other Non-Acid Electrolytes – Safety Requirements for Secondary Lithium Cells and Batteries, for Use in Industrial Applications
- IEC 62133-2 and UL 62133-2: Secondary Cells and Batteries Containing Alkaline or Other Non-Acid Electrolytes – Safety Requirements for Portable Sealed Secondary Cells, and for Batteries Made from Them, for Use in Portable Applications – Part 2: Lithium systems
- UL 2054: Household and Commercial Batteries
- UL 1642: UL Standard for Safety, Lithium Batteries
- IEC 62155: Secondary Cells and Batteries Containing Alkaline or Other Non-Acid Electrolytes – Safety Requirements for Portable Sealed Secondary Lithium Cells, and for Batteries Made from Them, for Use in Portable Applications

The safety certifications for Li-ion battery packs in a consumer product can be costly to obtain, and for low-volume products they may be prohibitive. Fortunately, several alternatives are available. Battery pack vendors offer multiple standard configurations of cells that are already tested and safety certified. Also, an NiMH battery pack has fewer regulatory hurdles than Li-ion to clear and may be a possibility.

All batteries have safety concerns that need to be addressed, and those are often unique to the chemistry. Operational temperatures, limitations on charging/discharging rates, room for physical expansion within the final application, and storage and shipping temperatures are all questions that must be asked specific to the cells being used.

The most common industry strategy is to specify a battery pack configuration and engage a specialist manufacturer of battery packs to provide an assembled product. These organizations can provide printed circuit board (PCB) circuits with safety cutoffs, charge balancing, and other smart-battery features within the battery pack. As well, most of these organizations can get their products through the safety certification maze.

Safety testing is important. A large number of cell phone battery fires were traced to the mechanical enclosure of the battery not allowing enough room for battery swelling. Careful electromechanical design review and safety compliance testing should keep your product off the evening news.

Other Energy Storage and Access Methods

This book is dedicated to practical design solutions. Consequently, battery storage is presented as the foremost energy storage solution for portable electronics. Other methods are often proposed, and a brief look at these methods can be useful.

Supercapacitors

Supercapacitors are often mentioned as a battery replacement, but technology versus price point becomes the issue.

Comparing a battery versus capacitor in Figure 5-15, a typical alkaline AA battery costs under $1 USD and provides 250 mA for 7.5 hours, or a total charge transfer of ~7,000 coulombs.

Present (2023) technology supercapacitors can be found with 4,000 farads of capacitance and a maximum voltage of 2.5 V. Charging the capacitor to 1.75 V will provide the 7,000 coulombs, if discharging to 0 V. Up to 10,000 coulombs can be stored at the maximum capacitor voltage of 2.5 V. Long-term reliability and how quickly the device self-discharges are also issues.

The capacitor price is ~$300 USD, and the device is physically much bigger than the AA battery. Generally, the lowest-cost path to a good solution prevails. Here the supercapacitor loses on size, cost, and reliability issues.

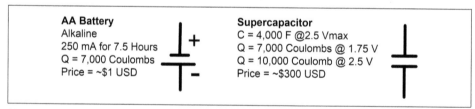

Figure 5-15. Capacitor storage versus AA battery

Hydrogen Fuel Cells

The idea of using a miniature hydrogen fuel cell has also been suggested. Shown in Figure 5-16, a hydrogen fuel cell is an electrochemical system that uses a catalyst to create hydrogen ions, which combine with oxygen to create a waste byproduct of water. The hydrogen ions travel across a proton exchange membrane, while the stripped electrons cause current to flow.

While oxygen can be taken from the environment, to have a functional system some tank or cartridge with hydrogen is needed. Large-scale systems exist; telecom systems have embraced them as backup power systems, and a limited number of cars and buses use the technology. Furthermore, being able to rapidly refill a tank instead of taking the time to charge a battery is a big plus for long-distance vehicles.

Small-scale electronic devices running off of fuel cells have been considered, but as of 2023 small form factor devices are available from very few vendors. Experimental and demonstration devices do exist in small form factors, but not yet as a commercial product. Watch this space.

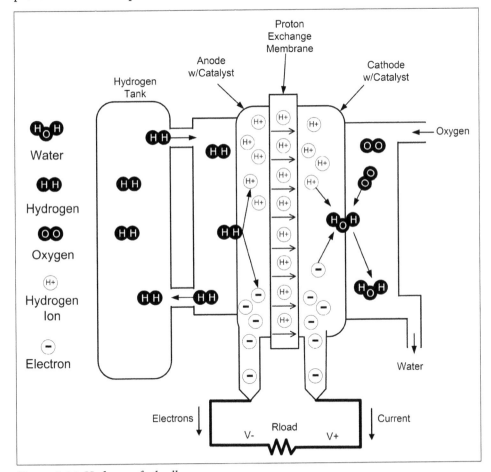

Figure 5-16. Hydrogen fuel cell

Flow Batteries

Flow batteries (Figure 5-17) are seen as a useful device in large-scale installations, but not small form factor electronics. The flow battery uses two electrodes, like a conventional battery, but relies on a flowing fluid electrolyte that can be pumped between the electrodes. Depending on the strategy and chemistry used, a flow battery can have a single electrolyte or two electrolytes (anolyte and catholyte) that are unique to each electrode.

The amount of energy stored is linked to electrolyte volume, and the maximum current is linked to electrode size. Flow batteries have long life projections, with expected life of 20 to 30 years, and the ability to scale up to mega-watt-hour capacity. This makes them suitable to utility grid and enterprise-scale applications. Industry has also shown a strong interest in flow batteries as energy storage for solar and wind energy systems.

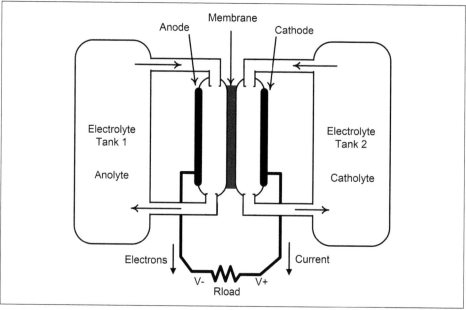

Figure 5-17. Flow battery

Since they do involve tanks, pumps, plumbing, and cooling systems, flow batteries are large-scale systems that are not applicable to small form factor electronics.

Wireless Power

Wireless power (WiPow) is most commonly defined as transferring electromagnetic energy between devices without wires (Figure 5-18). WiPow can be broadly divided into several overlapping considerations:

- Close proximity wireless battery charging
- Energy transmission over distance
- Wireless power without batteries

Close proximity wireless charging of batteries has been widely adopted in low-power (under 15 W) handheld electronics with good success. Most smartphones now include wireless charging capability. High-power WiPow charging for bigger devices, such as electric cars (3–7 KW), mobile medical devices (300 W), and robotics, has also been done, with efficiency above 80%. Widespread adoption of close proximity wireless charging is hindered by cost. The power electronics and coils for a 300 W system cost $100 to $300 to implement, and a system for quickly recharging an electric car sells for over $5,000.

Power cords are less elegant but a lot less expensive. Also, docking stations with hard electrical contacts can usually be implemented for these applications. The docking station method costs less and is simpler to realize.

Energy transmission over distance has problems due to transfer efficiency. It is important to recognize here that energy, and not information, is being transferred.

A wireless signal can easily transfer data, while losing 60 dB of the signal amplitude between the transmitter and receiver. Even with that signal loss, the data can be recovered with some signal processing.

However, a 60 dB signal loss in power transmission implies transmitting one million watts to recover 1 watt at the receiver. Devices close together or using highly directional antennas improves efficiency, but that limits device mobility.

Nicola Tesla was never able to demonstrate wireless energy transmission without huge amounts of energy loss.

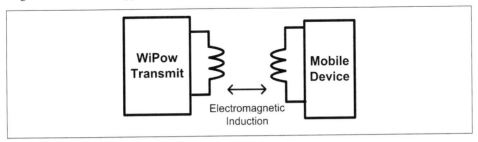

Figure 5-18. Wireless power (WiPow)

Wireless power without batteries only works if energy coupling is reliable and not interrupted. Controlled situations such as within a machine, or with devices tightly linked, have been successful.

Solid State Batteries

Solid state batteries (Figure 5-19) hold promise for improvements in energy density. The concept of a solid state battery (SSB) can involve several things:

Solid electrolyte
> The electrolyte is not a gel, paste, or liquid.

Lithography fabrication methods
> This involves using manufacturing methods that allow better control over creation of the anode, cathode, and electrolyte. Chemical vapor deposition and other lithography techniques of the semiconductor fabrication world are starting to be used.

Pure lithium anode
> This involves the creation a pure lithium metal anode instead of the lithium compounds presently in use.

All of these items create a higher energy density battery. SSBs are presently limited to small low-power devices and are both expensive and only available in limited quantities.

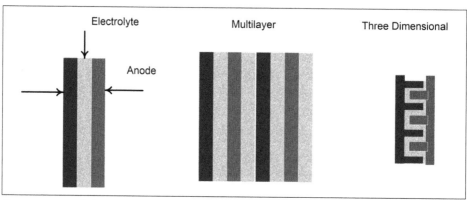

Figure 5-19. Solid state battery

Many SSB research efforts are ongoing in the industry. At the moment, issues remain in terms of processes and methods, reducing internal battery resistance, poor low-temperature performance, scaling production to volume, and making the devices cost-competitive with existing technology.

Summary and Conclusions

The important ideas and items in this chapter are as follows:

- Battery specifications are not standardized and are inconsistent between vendors. Consequently, designers need to carefully read the specification details and test conditions to determine meaningful battery capacity numbers.
- Deciding between single-use and rechargeable devices comes down to price, how often battery refresh is needed, how long devices are stored between use, and the specific requirements of the application environment.
- System power requirements can be estimated or measured by determining the long-term average current (LTAC).
- By fitting the battery discharge voltage profile to the power requirements of the system, full use of the battery capacity can be achieved.
- Battery performance by chemistry can differ widely, and new battery technology is still being created and brought to market. Always check what's available now, but be wary of using newly introduced devices that are not available from multiple sources.
- Battery capacity changes with load and is not a fixed value. Higher current use results in lower-capacity performance.
- Battery capacity changes with temperature, and typically, cold batteries have lower capacity.
- Load balancing of high-current parallel battery connections requires keeping wiring resistance balanced.
- Safety cutoff circuits and charge balancing are required for many battery packs.
- Charging profiles are unique to the device's chemistry.
- Alternative energy methods exist, but nothing has yet replaced the battery for energy storage in small electronic systems.

After reading this chapter, the reader has the knowledge to:

- Pick the battery chemistry
- Determine the needed capacity
- Pick suitable devices
- Define the serial/parallel configuration of a battery pack
- Define a charging strategy
- Define smart-battery features where needed

Manufacturers of battery packs can provide the most recent specification for cells they have available and build the battery pack specified. Semiconductor manufacturers can provide charging controllers that are suitable for a charging circuit that is fitted to the parameters of the battery pack and the device chemistry.

Further Reading

- *The TAB Battery Book: An In-Depth Guide to Construction, Design and Use* by Michael Root, 2011, ISBN 978-0-07-173990-0, McGraw Hill.
- *Batteries in a Portable World, 3rd Edition*, by Isidor Buchmann, 2011, ISBN 0-9682118-2-8, Cadex Electronics.
- S.C. Hageman, "Simple PSpice Models Let You Simulate Common Battery Types," *Electronic Design News*, vol. 38, 117–129, 1993.
- M. Reisch, "Solid-State Batteries Inch Their Way Toward Commercialization," *Chemical and Engineering News*, November 20, 2017, *https://cen.acs.org/articles/95/i46/Solid-state-batteries-inch-way.html*.
- C. Fang, X. Wang, and Y.S. Meng, "Key Issues Hindering a Practical Lithium-Metal Anode," *Trends in Chemistry*, March 22, 2019, 152–158, *https://www.cell.com/trends/chemistry/fulltext/S2589-5974(19)30028-0*.
- A. Dementyev, S. Hodges, S. Taylor, and J. Smith, "Power Consumption Analysis of Bluetooth Low Energy, ZigBee and ANT Sensor Nodes in a Cyclic Sleep Scenario," Department of Electrical Engineering, University of Washington, Seattle, *https://semiwiki.com/wp-content/uploads/2016/08/IWS20201320wireless20power20consumption.pdf*.
- A. Carroll and G. Heiser, "An Analysis of Power Consumption in a Smart Phone," University of New South Wales, *https://www.usenix.org/legacy/event/atc10/tech/full_papers/Carroll.pdf*.
- T. Scroggin-Wicker and K. McInerney, "Flow Batteries: Energy Storage Option for a Variety of Uses," *Power Magazine*, March 1, 2020, *https://www.powermag.com/flow-batteries-energy-storage-option-for-a-variety-of-uses*.
- "Battery University," Cadex Electronics, *https://batteryuniversity.com*.
- "Power and Battery Technical Resources for Design Engineers – PowerStream," *https://www.powerstream.com/tech.html*.
- "Engineering Guidelines for Designing Battery Packs," PowerStream, April 18, 2023, *https://www.powerstream.com/BPD.htm*.
- "Custom Battery Packs," Technical Engineering Guide, Epec Engineered Technologies, *https://www.epectec.com/guides*.
- "Battery Management Deep Dive On-Demand Technical Training," Texas Instruments, Video, *https://training.ti.com/battery-management-deep-dive-technical-training*.

CHAPTER 6

Electromagnetic Interference and Electrostatic Discharge

Electromagnetic interference (EMI) deals with unwanted signal coupling and other secondary effects within or between electronic systems. The terms *electromagnetic compatibility* (EMC) and *signal integrity* are often used as generalized terms to cover most of these topics. EMI and radio frequency interference (RFI) are often used interchangeably, and herein are considered the same thing. Historically, RFI has been in use since the early era of analog radios. In this book, the terms *EMI*, *RFI*, *crosstalk*, and *noise* are used interchangeably.

Electrostatic discharge (ESD) is included in this chapter as a special topic under the umbrella of "unwanted" signals. Many circuits discussed here affect both EMI and ESD performance.

Any digital electronic system creates EMI, with the noise magnitude determined by the clocking rates, number of gates, and presence of any other actively switching devices. Designers should recognize that the environment is awash in electronic noise due to active electronics, radio devices, solar noise, grid power systems, and countless others. Designers need to limit the magnitude of the EMI created by their products, and protect their electronics from the noise of the outside environment.

Within an electronic system, internal noise generated can also be a detriment inside the same system. Sometimes internal noise can reduce accuracy, degrade performance, or make a device nonfunctional.

ESD events also need to be dealt with as part of system protection. For consumer products, ESD protection must be sufficient to avoid functional destruction. For high-reliability designs, such as for medical devices, the protection must maintain proper functionality even in the presence of ESD events.

Preliminary Ideas

Many ideas presented here can be implemented with little or no cost. Some require the addition of low-cost items like ferrite beads (FB), resistors, and capacitors. Including EMI reduction measures in a design is important for gaining regulatory clearance. Many consumer products have an FB molded onto the power cord, due to failed emissions testing. That FB was added to the power cord as a last-minute fix. That could have been avoided.

Generally, systems that require comprehensive EMI suppression are circuits that switch currents at high speed. Switching power supplies are noise generators. All digital devices generate EMI. Anything with a clock over 100 MHz commonly needs noise reduction techniques in place. Small things like handheld battery-powered devices, with clocks under 1 MHz, generally don't need noise suppression. That leaves a large gray zone in the middle where noise efforts may or may not be needed.

Including many of the low/zero-cost EMI reduction techniques discussed in this chapter will help keep devices within regulatory limits. Anything with conductive connections or contacts outside an enclosure requires ESD protection.

From outside the box, Figure 6-1 shows that EMI gets evaluated in four areas:

Radiated EMI
> Measured from the perspective of externally observed noise. Does the system under test generate so much EMI that it affects the performance of other devices outside the system?

Susceptibility to radiated EMI
> The opposite side of the problem. For defined amounts of EMI external to the system, does it affect system performance?

Conducted EMI
> Deals with noise, created by the system, that is injected onto the AC power lines. This manifests itself as current transients, or high-frequency voltage transients, that the AC power mains can't suppress.

Susceptibility to conducted EMI
> Deals with a system's capability to properly function while attached to noisy, nonideal AC power mains.

Depending on the product, industry standards or regulatory compliance requirements have been defined to test these EMI scenarios. Medical, aerospace-aviation, and military devices have well-defined mandatory testing for product qualification. Other industries and government regulatory bodies have developed requirements unique to their sectors.

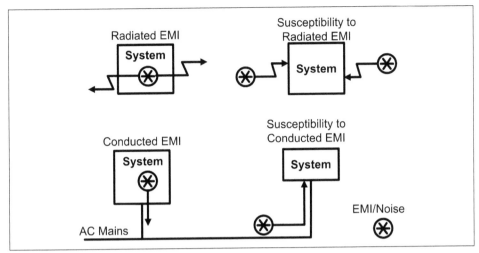

Figure 6-1. EMI: four avenues at the top level

Internal to a system, dealing with EMI encompasses several areas: radiated noise created, noise coupling within the system, and immunity to external noise.

Intrinsic Noise

Electronic noise can come from sources other than EMI. Inherent/intrinsic noise is due to the characteristics of system components and semiconductors (Figure 6-2).

Flicker noise (aka 1/F noise) is a low-frequency component coming primarily from CMOS transistors and their nonideal characteristics. The magnitude of the noise is proportional to the size of the transistor, with larger devices creating less noise.

Thermal noise (aka Johnson-Nyquist noise) is wideband noise (aka white noise) with no spectral weighting (flat over frequency) due to thermal effects within resistors. Higher temperatures or higher resistance values create more noise. Frequently an amplifier will exhibit a combination of thermal and flicker noise. A low-frequency ramp-down to a flat horizontal noise floor is commonly seen.

Shot noise (aka Poisson noise) and *burst noise* (aka popcorn noise) are due to the transition of discrete charges (single electrons or small numbers of electrons). Burst noise is semiconductor related and caused by the trapping and releasing of electrons. Shot noise is associated with current in a conductor and is seen at extremely low currents where singular electrons are a significant proportion of the current.

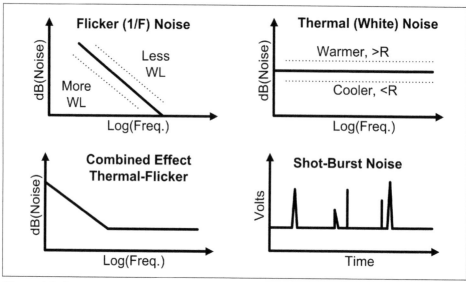

Figure 6-2. Intrinsic noise sources

Intrinsic noise sources are issues of concern within analog front end (AFE) devices and analog integrated circuit (IC) design. System-level designers will generally not need to deal with them but should be aware of their existence. The focus here is EMI, both in/out of systems and how to mitigate it.

General Strategy Dealing with EMI

The approach used here builds an awareness of noise issues that will be encountered and illustrates design methods to manage those issues.

System noise needs to be limited to remain functional and accurate. Noise radiating from the system needs to be restricted before the device can be sold. It is quicker and less expensive to deal with noise issues as part of the design process than attempting to fix problems in a finalized design.

To properly mitigate noise issues, designs need low-impedance grounding, properly protected ports, and minimization of radiated noise. In addition, a functional tolerance for both internal and externally applied EMI is necessary. In many cases, EMI-limiting enclosures (Faraday cages) are also needed to meet regulatory restrictions.

Ignoring EMI/ESD issues invariably results in one thing: devices failing regulatory testing. That quickly leads to a haphazard attempt to duct-tape in filters, ferrites, and ESD protection as the product readies for production. That approach costs more than implementing EMI/ESD measures within the design. The approach outlined here is one of implementing best practices to optimize or improve the EMI and ESD resilience of the design.

Quantitative analysis of radiated noise, coupled with an analysis of noise susceptibility, is theoretically possible but neither practical nor expedient. In a complex system, it is not one noise source and one victim of noise, but hundreds of sources and victims. To do a quantitative analysis is extremely time consuming.

Instead, an efficient approach is to recognize and resolve problematic areas while including methods within the entire design to improve noise performance. The outcome is both robust and reliable.

Regulations and Requirements

What regulatory requirements apply depends upon the device's application and country of use. All regulations and requirements consider the device being tested as a black box, either inflicting EMI or measuring EMI, both radiated and conducted, while observing defined functionality.

The international standards group that defines EMI requirements is CISPR (Comité International Spécial des Perturbations Radioélectriques, loosely translated as International Special Committee on Electrical Radio Interference). CISPR is part of the International Electrotechnical Commission (IEC). Most countries provide legal requirements that mandate CISPR standards or a variation thereon.

Within the United States, the Federal Communications Commission (FCC) regulates radiated emissions of devices. The Food and Drug Administration (FDA) regulates medical devices, and the Department of Defense (DoD) issues requirements for military electronics. Depending on the product, other agencies may issue requirements for their sector.

The FCC restriction on radiated emissions can be found in the Code of Federal Regulations Title 47 (Telecommunications), Volume 1, Chapter 1 (FCC), Subchapter A, Part 15, Subpart B (Unintentional Radiators) paragraph 109 (Radiated Emissions Limits). Other countries have similar requirements.

After digging through the details, this FCC requirement can be simplified to a "stay under the line" compliance plot, as shown in Figure 6-3. In the figure, the horizontal bars indicate the maximum allowed radiated noise over frequency. The example shown has a violation at 200 MHz but is otherwise compliant.

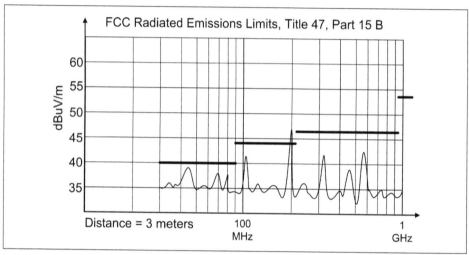

Figure 6-3. FCC Part 15B emissions restrictions

Two common ways to test for radiated emissions are to use an Open Area Test Site (OATS) or to conduct the test in an Absorber Lined Shielded Enclosure (ALSE) screen room. The ALSE setup is similar to an OATS setup except it is contained within a Faraday cage screen room. Both yield suitable results when calibrated and properly run. An OATS setup is shown in Figure 6-4.

A calibrated antenna is used to observe radiated emissions at a fixed distance (3 m or 10 m as per FCC requirements) with a spectrum analyzer attached. Since emissions from the device under test (DUT) can be directional, the DUT is rotated 360 degrees while scans are run.

All commercial devices in the United States need to meet either FCC Part 15A or 15B for radiated noise. The FCC (Part 68) also covers conducted emissions put back onto power lines, and other regulations are applicable. Depending on the product sector, applicable requirements vary. A full determination of what regulatory requirements need to be met should be made early in the design process.

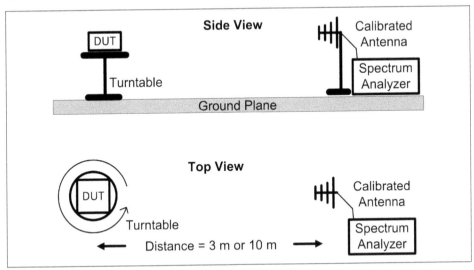

Figure 6-4. OATS testing setup

Devices purchased by the DoD must meet Military Standards 461. Items in the European Union need to meet CISPR 22 standards for EMC. Japan has its own standards body in the VCCI Council (see *https://www.vcci.jp*).

Visualizations of Noise Coupling

Some lab testing will be part of any noise analysis. This is not a book on proper test and measurement (T&M) techniques, so readers are cautioned to educate themselves on proper T&M methods to get reliable data. Much of the data mentioned here can be generated by a certified test laboratory.

One of the clearest illustrations of noise scenarios is the talker-listener model, with a transmission medium in between. Frequently this is called the source–medium–victim model, and it is shown in Figure 6-5.

The following noise mitigation methods can be fitted to this source–medium–victim model:

Silence the source
Techniques used to reduce the magnitude of noise generated by any given source.

Contain the source
Techniques that don't silence the source, but contain and localize the noise generated and keep it at or near the source.

Protect the victim
Methods such as shielding and enclosures that don't allow noise near the victim. Filtering and bandwidth-limiting connection signals are also included here.

Deafen the victim
Techniques that make the victim nonresponsive when noise gets to the victim. Circuits using common mode noise rejection or bandwidth limiting to make the victim less responsive are often useful here.

Separate the source and victim
Physically separating with distance is the most common strategy applied here. However, separation in frequency can be used in some cases, and time interleaving (aka separation in time) of noisy events and noise-sensitive events can also be done.

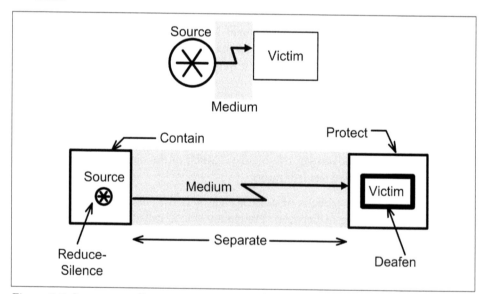

Figure 6-5. Source – medium – victim noise model

Frequency Domain Analysis of EMI

Getting useful information about noise sources and their cause is best determined in the frequency domain.

The top plot in Figure 6-6 depicts a voltage versus time domain, and there's not a lot of readily apparent information there. Instead, looking at the signal spectrum quickly tells a clear story. In the frequency domain there are spikes at 15 MHz, 45 MHz, and 75 MHz, which are harmonic multiples of 15 MHz.

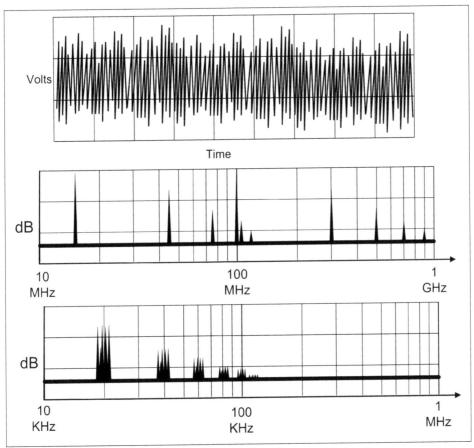

Figure 6-6. Noise in time versus frequency domain

In this example, the system has a free-running clock at 15 MHz that is radiating outside the device. In addition, this example shows signals at 100, 300, 500, 700, and 900 MHz. Again, another clock in the system is at 100 MHz, and its harmonics are part of the noise problem.

At lower frequencies, noise is seen at 20 KHz, and harmonics thereof. This would be typical noise due to a switching power supply operating near 20 KHz. When the frequency content is examined, there is often a very clear message of what is creating the noise that is observed.

When dealing with noise issues, the analogy of "peeling the noise onion" is frequently used. The author prefers the analogy of an archeology dig; as you dig deeper, more noise problems are discovered and resolved and another layer of noise issues become apparent.

As an example, Figure 6-7 shows a noise profile before and after some noise suppression was added. The upper plot (before) shows that the biggest noise contributor, D1, peaks slightly above 100 MHz. The lower plot (after) shows that the D1/D2 signal has been reduced and is no longer the dominant noise problem in the system. Most of the high-frequency noise problems above 100 MHz (D1, E1, F1, G1) have been reduced, and the most significant noise problem is now at B2 (~35 MHz).

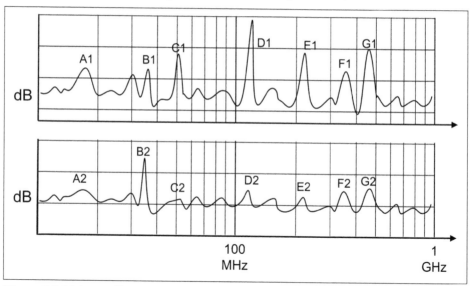

Figure 6-7. Noise profiles as sources are eliminated

These are typical results. As noise sources are eliminated or reduced, other noise sources become apparent. Most EMI is created from transient switching, primarily in digital devices and power supplies, both of which create large amounts of noise. Figure 6-8 shows how noise propagates from the power and ground (P&G) to system signals.

P&G connections are common to all circuits on a printed circuit board (PCB). Noise on either of the P&G connections gets to all circuits. Consequently, a low-impedance common ground and a power supply network with high-frequency stability are both important. However, some noise will exist on both P&G due to less-than-perfect power systems, connection impedance, and multiple switching transients throughout the system.

For digital devices, P&G noise transfers directly to logic outputs, primarily through internal transistor connections and parasitic capacitance. For synchronous digital devices, some noise is tolerable. With devices that are not synchronized, noise transfers into time domain edge jitter. P&G noise present on rising or falling digital signals transfers to time variance.

For analog and mixed signal circuits, P&G noise will transfer over to output signals or corrupt internal circuit functionality. Depending on the internal circuits, the noise can be directly transferred or be actively amplified by internal circuits. Analog and digital devices directly connected to the same power supply can be especially problematic.

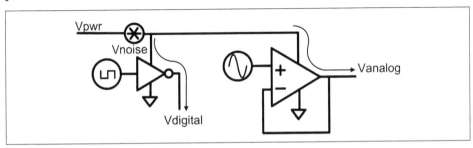

Figure 6-8. Noise coupling through P&G paths

Noise coupling also occurs through parasitic capacitance between connections and magnetic induction methods, as shown in Figure 6-9. In the figure, View A shows a circuit schematic with a digital signal traversing a PCB in close proximity to a sensor connected to an ADC. View B adds probable signal coupling paths, showing parasitic capacitance (C_{par}) between connections and electromagnetic fields from transient currents traversing A to B, which results in signal coupling between the two connections. View C completes the story, illustrating the physical layout of the connections that are "talking" to each other.

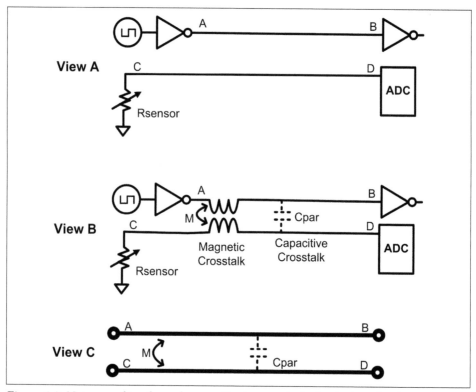

Figure 6-9. Noise coupling through capacitive/magnetic (inductive) methods

Creating unintentional antennas that radiate EMI is illustrated in Figure 6-10. In that figure, View A shows a signal creating a current along a connection with length L1. Depending on the frequency used, L1 can function as a transmitting antenna and L2 can function as a receiving antenna. The connection length versus the signal frequency is important.

View B shows the unintentional transmitter-receiver pair. The wavelength is a function of the frequency and the velocity of the medium. A good-quality antenna is one-quarter of the wavelength ($\lambda/4$), and connections shorter than $\lambda/20$ make a poor antenna. Obviously, the typical electrical device can create many possible antennas.

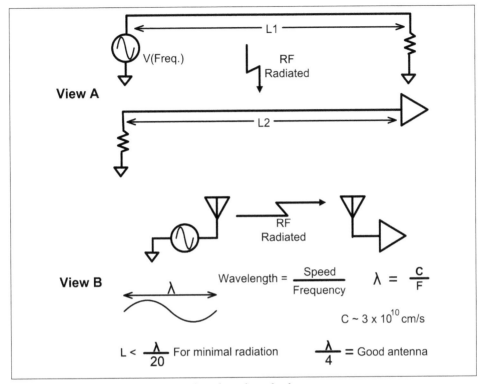

Figure 6-10. Noise coupling through radiated methods

In Figure 6-10, a first-order estimate for electromagnetic wave velocity is the speed of light. Although approximate, it is good enough to determine possible noise-radiating antennas. As an example, a 100 MHz square-wave clock includes 7th harmonic content. That 700 MHz sinusoid signal has a wavelength of 43 cm, with 11 cm wires making excellent antennas, and any connections longer than 2 cm radiating EMI.

Radiated EMI is especially problematic outside the system, where a system creates noise seen by other devices. Power cords turn into transmitter antennas when common mode noise is on all the power cord wires.

The capability to minimize transient currents in a single conductor can help with radiated emissions. Currents always need a return path, however, as shown in Figure 6-11. In the figure, View A shows the magnetic field (M1) created by a transient current from a digital signal. The I-loop current that goes out in the connection has to return somehow, and it does so by the ground, creating a current loop.

View B explicitly illustrates the loop's return current. The magnetic fields, M2 and M3, are spaced apart, showing the current loop antenna that has been created. View C illustrates that closely placed out/return currents cause magnetic fields that cancel each other out. At a distance, the EMI radiated is much less than that radiated in

View B. Closely spaced complementary currents are widely used in power and cabling strategies.

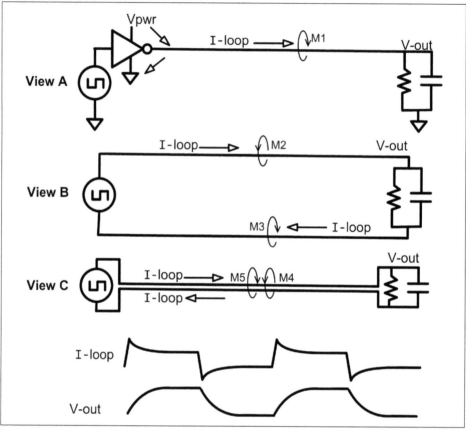

Figure 6-11. Current loops and return current paths

Generally, EMI problems increase with frequency. Some examples readily illustrate the reasons, as depicted in Figure 6-12:

Case A
    Capacitive coupling increases with frequency, due to the impedance of any parasitic capacitance reducing, and thus, signal cross-coupling increases.

Case B
    Magnetic crosstalk also increases with frequency. This is due to Faraday's Law of Induction, where the induced voltage increases with frequency.

Case C
    Radiated crosstalk increases with higher frequencies. Higher frequencies have shorter wavelengths, and more conductors become suitable antennas.

Case D

Power supply filtering becomes less effective at higher frequencies due to noni-deal capacitors and their connection inductance. Power filter capacitors (C_{srf}) are not ideal, and exhibit self-resonance characteristics due to their internal induc-tance (ESL). Above the self-resonance frequency, their function as a capacitor degrades as the impedance of the ESL increases (FDI: Power).

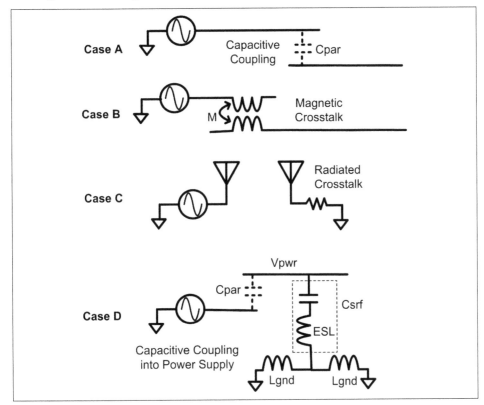

Figure 6-12. EMI coupling as a function of frequency

Grounding

A good low-impedance grounding strategy is important. Any ground has some impedance and can vary dynamically due to current surges through that impedance network. Poor grounding can affect performance and become more problematic with more complex systems, higher frequencies, large transient currents, or the stresses of ESD events.

Establishing a low-impedance ground becomes a foundation to close current loops and serves as a stable current return path. A low-impedance ground also becomes the common reference point for EMI reduction.

Depending on system size, there are several common approaches, as illustrated in Figure 6-13.

Case A shows a common scenario for small, low-power devices. The system ground is the PCB ground plane. The AC/DC supply is a wall plug-in, and the connection between uses a low EMI twisted wire pair to avoid an open current loop radiator. To be viable, the PCB needs to be low noise, due to no enclosure shielding.

Case B shows another popular approach where the internal PCB generates more EMI and needs shielding. The internal PCB still provides the system ground. A conductive shield encloses the PCB, connects to the PCB system ground, and serves as a Faraday cage to reduce emissions. A ferrite bead around the power wires reduces the common mode noise antenna characteristics of the power wires, while transient current EMI is minimized using a twisted pair. This example is typical of devices like laptop computers and tablets.

 A suggested strategy is to create a design without shielding that easily converts to the shielded version. If regulatory testing fails radiated emissions, the shielded version is ready to go as a substitute. Conductive coatings for the inside of plastic enclosures are readily available for this purpose.

Case C illustrates the most common grounding strategy used, and is commonly utilized for larger electronic systems. All devices exist within a conductive chassis that serves as both a common safety ground and a Faraday cage. Due to multiple connection impedances, there can be some dynamic variance between all the grounds that are connected together.

Case C (ground connections) illustrates the multiple inductances in the ground paths. The AC safety ground (L_{safe}) connects to a chassis structure that can have small amounts of chassis inductance (L_{chs}) in it. Ground connections in the mounting (L_{mt}) of each PCB can also have some effect.

Multiple grounding mounts for each PCB to the chassis are suggested as a best low-impedance practice. A solid conductive sheet ground underneath the circuit boards is greatly preferred. Air vents and external connections can be on the other surfaces of the box. A chassis ground configured in this manner is generally "good enough" to serve as a solid reference foundation.

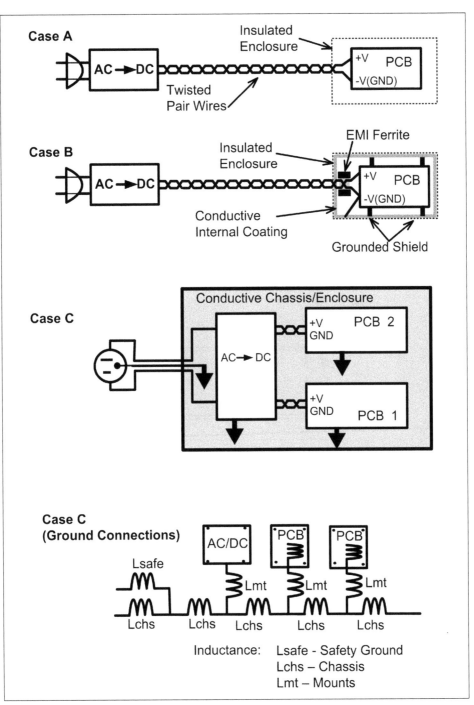

Figure 6-13. Grounds within a system

Even with a good-quality grounding strategy, some dynamic variance will exist across a ground grid. Seeing 10–100 mV transients from corner to corner of a metal chassis is not uncommon. Poorly designed devices can have much larger voltage transients across the metal chassis. Consequently, some variance in ground needs to be accounted for, as shown in Figure 6-14.

Case A in Figure 6-14 shows a sensor connected to a differential ADC input. Due to ground noise, the voltage at G1 is not the same as G2, and measurement errors will occur. Case B uses the same ADC and illustrates the idea of sensing a remote ground to remove ground noise errors. Case C illustrates a sensor that does not need to be grounded. This offers the best option to avoid ground noise and should offer the lowest-noise result. As with all analog signals, keeping the connection short is always important.

Placing the ADC in close proximity to the sensor (FDI: Sense) and then transferring a digital signal across the PCB is preferred, but that option is not always available.

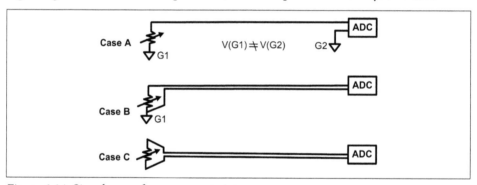

Figure 6-14. Signal grounds near sense point

After creating a low-impedance chassis ground, a strategy to do so within each PCB is also necessary. Modern PCB designs use a solid ground layer within the PCB. The most commonly employed method is four layers (Figure 6-15), where the two outer layers are used for signal interconnect: one internal layer for ground and one internal layer for power supply distribution. More complex boards may add additional layers (FDI: PCB). Early generations of PCBs did not use a dedicated ground plane and invariably suffered from reliability and ground noise problems.

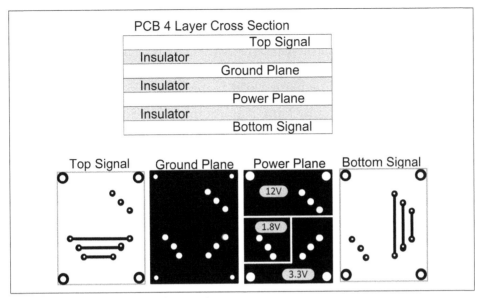

Figure 6-15. PCB ground and power layers

Splitting of grounds is not suggested for general use. The only exception is when a portion of the electronics needs to be fully isolated. This comes into use in medical devices, where patient-attached electronics must be unattached from any DC current path.

If grounds need to be split (Figure 6-16), maintain each ground plane as a solid entity as much as possible. Currents should traverse the ground with minimal impedance.

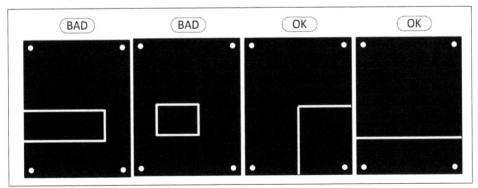

Figure 6-16. Ground splitting

Isolating grounds require special support circuitry, shown conceptually in Figure 6-17. For isolated grounds, power needs to be transferred without a DC return path. This is commonly done with a DC/DC converter. Power is transferred by magnetic methods. Data communication also needs to be done with no DC return path, commonly done with an opto-isolator or other data link that does not share a physical connection. The details of this method are discussed in "Medical Devices" on page 509.

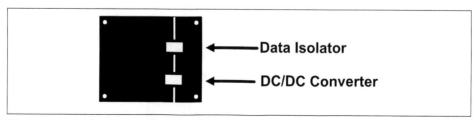

Figure 6-17. Isolated grounds, typical structure

Again, splitting grounds on a single PCB is not suggested for general use (FDI: PCB).

Reducing Conducted Emissions to AC Power Mains

Most designers don't develop AC/DC power converters, because many low-cost options are already available. One part of the AC/DC converter may need attention, however; namely conducted emissions placed on the AC mains input. This is largely done by filtering methods.

A typical filter is shown in Figure 6-18. This commercially sold filter is generally available within a chassis mount power connector. Some filters include a common mode choke (right side as shown) and others do not. If conducted emissions become a design issue, a power plug connection can be purchased that includes the filter.

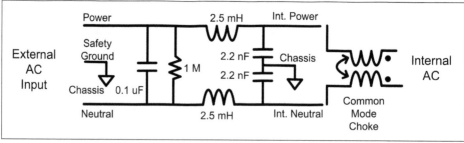

Figure 6-18. AC connection suppression-filter configuration

Cable Interconnect Strategies

Cabling interconnects can have different strategies depending upon the application. Figure 6-19 explores the more common options.

Case A is the open current loop discussed earlier. Open current loops should be avoided because they create radiated EMI and they lack balanced and equal paths on both sides of the signal.

Case B introduces twisted pair wires that minimize radiated EMI through magnetic field cancellation. With closely spaced, equal and opposite currents, the net field generated is minimized. Additionally, most EMI coupled into the twisted pair will be common to both wires, creating common mode noise. This allows a differential receiver to improve EMI rejection by subtracting the signals from each other. For power connections and differential signals, this approach works well.

Case C deals with common mode noise introduced to a cable. This can be an issue at both the sending (CM-Noise 1) and receiving (CM-Noise 2) ends. Noise introduced to all wires in a cable bundle creates an antenna to radiate noise. The use of an FB at each end effectively separates the noise source from the antenna. This is the preferred method for any cable connection going outside a shielded enclosure.

Case D looks at a special case where a cable inside a system is a noisy radiator. Using a braided shield over the conducting wires and connecting both ends of the shield to the common chassis ground will reduce radiated EMI from the cable. Currents in the ground shield should be avoided, so if there is any voltage variance between the grounds at both ends, grounding only one end should give similar results.

Case E is suggested when connecting to a sensor or other low-amplitude signal source on a cable beyond the system enclosure. A shield over the cable, connected to the ground at the receiver (ADC here), minimizes noise seen at the receiver. A ferrite on the cable minimizes the cable acting as an antenna.

Case F is common for low-voltage differential signaling (LVDS) data communication on long cables. DC isolation is created with transformers at both ends of the cable. This eliminates the need for a shared ground between the devices at each end. Common mode rejection chokes are also included, rejecting the noise pickup of any lengthy cabling scenario. This cabling structure exists within all Ethernet cable connections.

Twisted pair wires do more than just hold two wires together. With opposing currents, they reduce EMI creation and avoid open current loops. With tightly twisted wires and differential signals, they reject common mode noise when using a differential receiver.

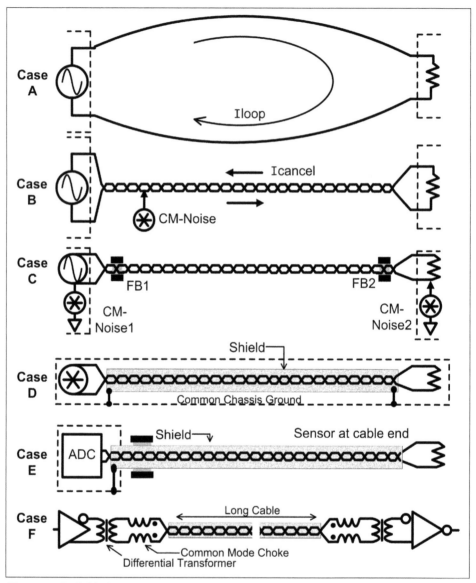

Figure 6-19. Cable interconnect strategies

The cable shield is typically grounded at the receiving end (Case E in Figure 6-19). Protection from noise is a priority at the receiver, and so the shield should be tied to the ground associated with the receiver. Current flowing in the shield should be avoided, and shield connections at both ends can create an open current loop. Case D is an exception, since the intent there is to silence a noisy cable, and both ends are on the same chassis ground.

The cable strategy for a sensor that requires both power and signal becomes two twisted pairs, one for signal and the other for P&G. Cable structures with multiple twisted pairs are commonly used.

Reducing Noise Generation at the Source

Noise reduction starts at the source. Selection of circuit methods, clock frequencies, signal distribution, and other techniques will result in significant noise reduction in any system.

The predominant EMI sources in most systems are devices that switch current or cause high-frequency transients. All digital devices, clock tree drivers, switching mode power supplies, and high-current switching amplifiers are common EMI sources. As frequency and currents increase, so does the EMI.

Slower Clocks and Softer Transitions

The first EMI source easily improved is the transition time of digital devices. As illustrated in Figure 6-20, fast transition edges and the resulting current spikes produce more EMI than slower transitions. With slower rise/fall times, harmonic amplitudes are reduced.

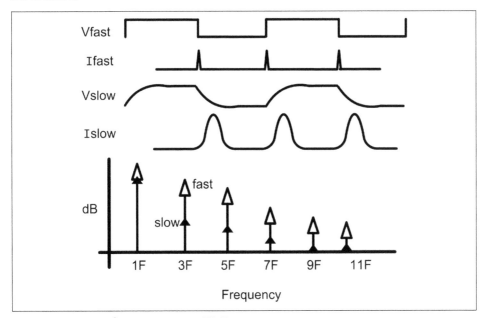

Figure 6-20. Digital rise time versus EMI

Changing the I/O driver strength is the technique available to control transition speed. Digital application-specific integrated circuits (ASICs), field-programmable gate arrays (FPGAs), and complex programmable logic devices (CPLDs) often have the capability to set driver strength for each output. Some microprocessor (MPU) and microcontroller (MCU) devices also provide this feature.

Look for a configuration option to set drive strength when programming the device. Slower data ports can use a weaker drive strength setting that reduces EMI and avoids the need for transmission lines.

Many embedded controller systems are designed using needlessly fast clocks and quick data transients to control slowly executed physical tasks. This is common with mechanical devices under electronic control. Slower clocks and "softer" logic edges result in both lower EMI and lower power consumption. Reducing the drive strength settings on an FPGA has been used to reduce EMI and clear regulatory testing.

LVDS for Digital Data to Reduce EMI

Using LVDS methods (Figure 6-21) for fast and distant digital data also has EMI advantages. LVDS uses complementary currents, similar to the twisted pair wires described earlier. This generates less EMI than rail-to-rail logic.

Figure 6-21 shows a typical CMOS (signal A) that uses 3.3 V high and 0 V low. LVDS (signals B, C) uses smaller amplitudes, 1.4 V high and 1.0 V low. The differential signals of LVDS are largely immune to P&G noise and reject common mode EMI. Consequently, smaller amplitudes work reliably and create less noise.

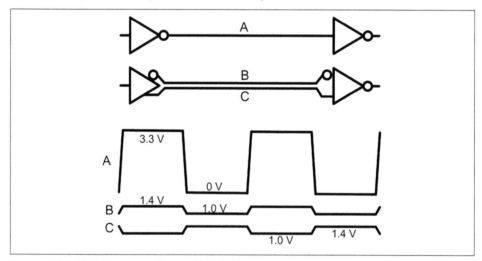

Figure 6-21. LVDS versus rail-to-rail digital

Spread Spectrum Clocks to Reduce EMI

Synchronous digital devices create large amounts of EMI at harmonic multiples of the clock frequency. Clock EMI can be reduced through spectral spreading, shown in Figure 6-22.

Also known as clock dithering or a spread spectrum clock generator, spectral spreading works by changing the clock period slightly on a cycle-to-cycle basis. The change is small and does not violate the setup/hold timing of the devices. Examining the clock in the frequency domain shows a significant change. The clock signal was a narrow, single-frequency spike and has become a wider and lower amplitude "spread" output. Improvements greater than 20 dB are common.

Many devices have failed FCC emissions testing with a conventional clock and then passed with a spread spectrum clock.

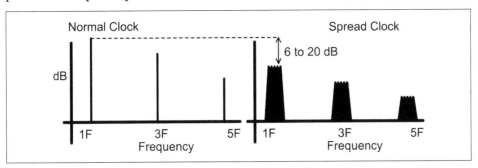

Figure 6-22. Spectral spreading of clocks

Any digital device still needs low-impedance P&G, aggressive power decoupling, and other EMI and signal integrity issues applied.

EMI Reduction for Switched-Mode Power Supplies

Spread spectrum methods are sometimes applied to switched-mode power supplies (SMPS) to reduce emissions from the switching power supply. But spread spectrum methods are usually not needed for small system SMPS. The most effective methods for SMPS noise reduction focus on snubbers and local isolation, as shown in Figure 6-23.

SMPS should always include PCB locations for a few simple noise abatement techniques. Shown in Figure 6-23 is one example of a step-down SMPS, where the node V_{sw} is the largest noise source. Due to constant on/off switching of the NFET, the V_{sw} node sees heavy transient currents that create EMI.

Attaching a low-pass filter (LPF) (R_{snub}, C_{snub}) to the V_{sw} node, often called an RC snubber, reduces EMI significantly by creating a low-impedance path to ground for high frequencies. The RC time constant is selected such that the value is one-tenth the minimum functional switched pulse time period. This suppresses high-frequency transients without changing basic functionality.

As an example, an SMPS uses a controller that has a minimum switched period of 10^{-6} seconds. That implies a snubber with an RC time constant of 10^{-7} seconds. For $R_{snub} = 10\ \Omega$, the appropriate $C_{snub} = 0.01\ uF$. Generally, vendors of SMPS controller chips can provide application notes on RC component selection suitable to their specific ICs.

In addition to the snubber, several other items are suggested. The output capacitor (C-LF) is not ideal and often exhibits poor high-frequency characteristics due to large internal inductance. Adding a supplemental parallel device (C-HF) with better HF characteristics will reduce HF emissions.

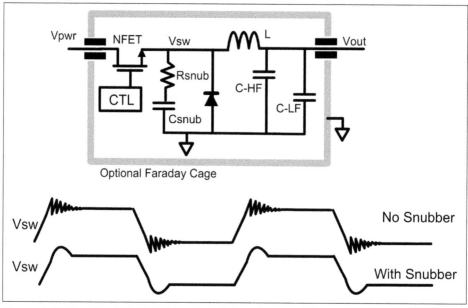

Figure 6-23. Localized EMI choking in SMPS and RC snubbers

Placing FBs (or other RF chokes) at the input and output of the SMPS reduces noise that travels out to the rest of the system. Use of a Faraday cage around the SMPS may be required in some cases, especially as the power exceeds 50 W.

All SMPS designs should include an RC snubber, C-HF and C-LF, and the FBs. If needed, a Faraday cage and better RF chokes can be considered.

Unintentional EMI Antennas

Some items can become unintentional antennas. A common antenna seen is created by a floating heat sink on a power transistor or a heat sink on an MPU, as shown in Figure 6-24. If the heat sink is not grounded, it can create an antenna. Consequently, grounding the heat sink needs to be investigated.

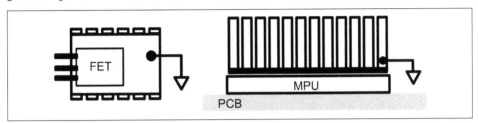

Figure 6-24. Heat sinks as an EMI antenna

EMI Suppression on Motors

Using a filter on motors can also help, as shown in Figure 6-25. For a brushed DC motor, using an RC snubber will reduce EMI from both arcing motor brushes, and the voltage transients created by motor inductance.

Pulse width modulation (PWM) is commonly used to control motor speed, and an RC snubber can be selected that does not interfere with PWM functionality. Using FBs at both ends of the twisted pair driver cable also helps keep the cable from acting as an antenna. Additionally, placing the flyback diode (FDI: Drive) at the motor, instead of on the PCB, avoids flyback current transitions through the cable, further reducing EMI.

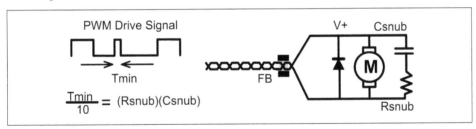

Figure 6-25. Motor drivers and RC snubbers for EMI

Reducing Noise Coupling Between On-Board Devices

After reducing noise at all sources, the next step is to minimize noise between devices. A quick review of coupling mechanisms between devices is shown in Figure 6-26.

Digital signals can tolerate limited amounts of noise, and priority should be given to protecting analog signals in the system. Incorporating that idea is a good starting strategy.

Capacitive coupling is often the most common path of noise on a PCB. Noise transmission by parasitic capacitance can be reduced by selectively spacing sensitive connections, placing grounds between signals, and managing connection impedance.

Magnetic induction noise can be avoided by using digital signals over distance, using differential signals, and avoiding long/close parallel paths to signals that have high-current transients.

Radiated noise is less problematic within a PCB due to fewer suitable antennas versus the long wires associated with going off the PCB. Radiated noise from any PCB is easily contained with a Faraday cage.

P&G connections between components provide an easy noise path, thus the need for a low-impedance ground plane under all circuits and aggressive use of power supply decoupling throughout a PCB.

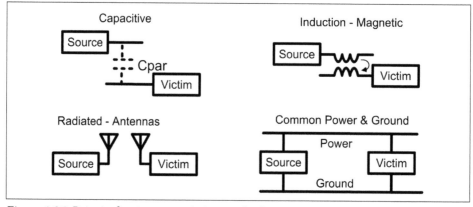

Figure 6-26. Principal noise transmission methods

On a PCB, capacitive coupling and P&G noise tend to be the most common problems.

Identifying the Big Talkers and Sensitive Listeners

Floor planning, component locations, and signal path routes all strongly affect noise performance. To properly deal with this, first understand which components and signals are noisemakers (sources) and which components and signals are sensitive (victims). Also, determine which signals need to travel long paths on the PCB and which signals are noisy with lots of transition switching. Digital control lines that have minimal activity can be recognized as a quieter class than high-speed data pipelines.

Classifying connections from the most sensitive (victim) to the loudest (source) would look like this:

- Small-amplitude analog ground-referenced signals (very sensitive)
- Small-amplitude analog differential signals
- Other analog signals
- Low activity (static) digital control lines
- LVDS digital data streams
- Ground-referenced digital data streams
- Ground-referenced digital data streams going off the PCB (most noise)

The general rule is to keep sensitive items (top of the list) away from loud noise generators (bottom of the list).

In a similar manner, the noisiest chips are switching devices, namely digital chips and SMPS devices. The higher the clocking frequency, generally the more of a noise problem it presents.

Power supply connections are not part of this. Generally, power is on its own connection layer and SMPS circuitry should be tightly grouped, avoiding lengthy connections. Keeping the analog and digital power supplies separated is suggested. Digital devices create large amounts of transient noise directly on the P&G.

Analog, mixed signal, and any AFE circuits should not share a power supply with digital devices (Figure 6-27). Using separated power supplies, created with independent regulators and filters, is the preferred approach. At a minimum, separating analog and digital power (AVDD, DVDD) with an LPF is suggested.

Some MCU vendors suggest a ferrite bead and capacitors to keep AVDD quiet, but frequently that's not enough. Analog circuits are sensitive to power supply noise and perform best with independently regulated and filtered power.

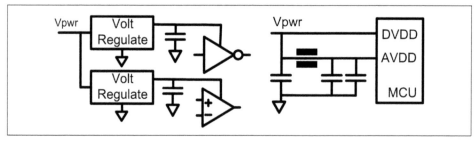

Figure 6-27. Independent power for analog-digital

Floor-Planning the PCB for Noise

With a good understanding of what's noisy and what's sensitive, a strategy for floor planning, component placement, and routing can be created. The first priority in this should be noise.

Conceptually, Figure 6-28 is a good generic floor plan for many PCBs. To the far left side are sensors and devices with analog, small-amplitude, or noise-sensitive issues. Protective shielding and grounding strategies are used to isolate these devices. To the far right are off-board digital data streams with noisy surge currents due to driving external loads and cables. Also on the right is an HF MCU with switching noise.

The power system is placed in the middle. Power can be noisy, but it is usually less problematic than digital switching noise. Additional power filtering entering the AFE region may be needed, and signal buffering between the AFE and MCU helps keep noise out of the AFE.

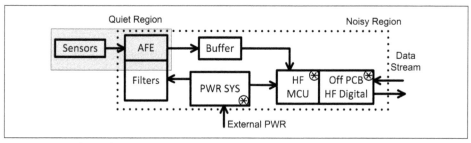

Figure 6-28. Floor-planning zones

When placing components on both sides of the PCB, a floor plan with a noisy side (digital) and a quiet side (noise sensitive) is often used (Figure 6-29). An internal ground plane serves as a shield between the two sides. This is a common strategy used in cell phones.

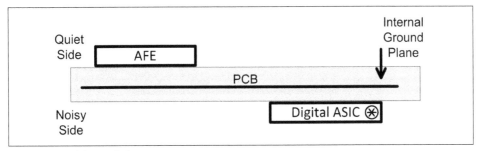

Figure 6-29. Two-sided strategy

Floor planning to keep sources and victims apart helps greatly with reducing noise coupling into victims. Selective placement of grounds and shields also reduces noise coupling. Signal routing strategies are easily implemented. Figure 6-30 illustrates some pertinent ideas.

Case 1 shows five signals coming into the pins on a chip. The connections are made but there's no thought to signal placement, and long, tightly spaced connections often have capacitance and magnetic coupling issues. Case 2 spreads the signals apart, which should reduce the parasitic coupling between signals.

Case 3 introduces grounded conductors between the signals so that most parasitic capacitance between connections is to ground and not to another signal. Selective use of grounds between signals is a useful isolation technique. Vias are used to connect the grounding strips to the internal ground plane of the PCB.

Cases 1, 2, and 3 illustrate a blind application of separation and grounding ideas. Achieving optimal results requires understanding the functional use of the connections first.

Case 4 determines connection functionality before routing. For this example, the top three signals (A) are low-activity digital lines and the bottom two (B) are analog inputs of a differential ADC. The A signals are not critical as either source or victim. The B signals need to be routed as a differential pair for common mode purposes and be protected from possible noise sources.

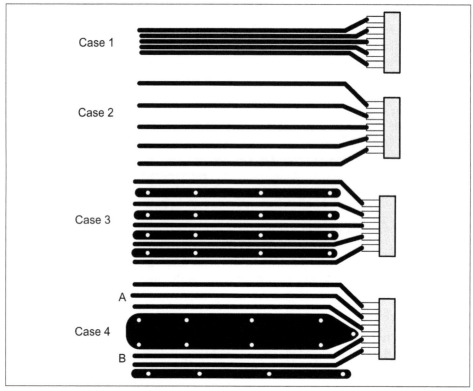

Figure 6-30. Reducing coupling, separation, and signal splitting with grounds

A routing strategy that looks into the specific needs of signals can optimize placement and noise abatement methods as needed (FDI: PCB).

Faraday Cage Methods to Contain or Protect from EMI

A Faraday cage (FC) is most often used for noise containment of entire systems. However, FCs can also be used for localized containment of noise sources and selective protection of sensitive circuits within a system. Figure 6-31 provides an example showing the use of a metal shield can that solders onto the PCB protecting an AFE. To perform properly, the PCB includes a solid metal ground plane underneath to fully enclose the device. Multiple shield structures like this are common in many cell phones. Shielding around both sources and victims is required in the tight proximity of a cell phone.

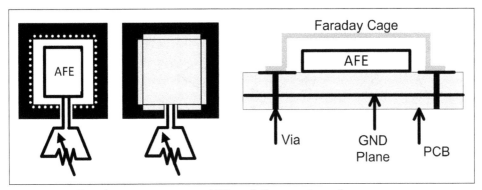

Figure 6-31. Faraday cages to protect AFE and silence noisemakers

Making Circuits Less Noise Sensitive

Inevitably, some noise couples to circuits. In all cases, it is wise to take steps that make any devices less responsive to undesired signals.

Noise-Sensitive High-Impedance Nodes

Connection impedance can directly lead to undesired coupling. In the example shown in Figure 6-32, a voltage divider defines V_{in}. For this example, a parasitic capacitor, $C_{par} = 0.01$ pF, is assumed, and is fixed as a function of the physical layout. As shown, the noise source is a digital signal with 3 V_{pp} amplitude and 1 ns rise/fall times. The transient edge injects ~30 uA of current into the resistors. Therefore:

- If $R_1 = R_2 = 100$ K Ω that 30 uA would result in a ~3 V noise transient.
- If $R_1 = R_2 = 100$ Ω the transient would be ~3 mV.

With large resistors, noise can become a significant part of V_{in}. Adding a filter capacitor (C_{filt}) makes a drastic change. Using a 0.1 uF filter capacitor creates voltage division between the two capacitors and the noise transient is reduced to microvolts.

A practical solution is shown on the right side of Figure 6-32, with R_1, R_2, and C_{filt} suitably sized. Adding C_{filt} offers one additional improvement: noise on V_{pwr} is filtered for a more stable and quieter reference at V_{in}. Additionally, most IC pins have 3–10 pF of internal capacitance due to ESD and I/O circuits inside the chip.

It is important to locally filter any DC voltage. Small parasitic capacitors are common, and this illustrates the importance of the issue.

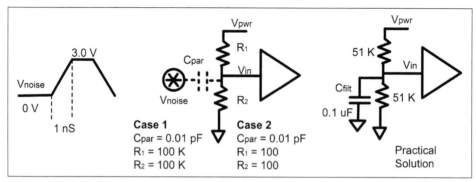

Figure 6-32. Weakness of high-impedance nodes

Most RF frontends are created using an impedance near 50 Ω, keeping thermal noise low and suitable for transmission lines. That low impedance does consume current. For low-frequency (<1 MHz) designs, staying between 100K and 10K is usually a good compromise on noise performance and power consumption.

Noise Immunity of Differential Signals

Differential signals have advantages for both analog and digital. LVDS for high-speed digital has been discussed, but more information for analog signals is useful (Figure 6-33).

As with digital LVDS, differential analog signals are defined by the difference between two signals $(+V_{sig}, -V_{sig})$ and are ideally independent from ground. The analog differential device is commonly called either a *differential output instrumentation amplifier* or a *fully differential amplifier*. In audio designs, these are also referred to as *balanced amplifiers*. All of these can be thought of as op-amps that are differential at both input and output. Differential amplifiers mostly reject common mode signals.

Consider the connections shown in Figure 6-33. P&G at both ends of this system will have noise. Noise at the sender (N1, N2) may create common mode noise on the sender output but is mostly rejected by the receiver. Any noise coupled into the closely spaced interconnect (N5) is common mode and also rejected. Noise present at the receiver P&G (N3, N4) is also rejected by the receiver circuits as common mode noise. With better noise performance, the amplitude of differential signals can be smaller while still meeting desired signal-to-noise requirements.

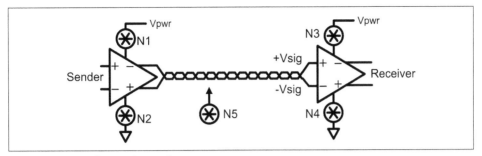

Figure 6-33. Differential signals reject noise

Noise Immunity Through Bandwidth Limiting

Response time and bandwidth limiting are additional useful methods to create circuits that only respond to the signal of interest. Figure 6-34 illustrates this idea. If the amplifier in use has excess bandwidth (BW) beyond the signal of interest, then limiting the input BW should be considered. Here, R_1, C_1 helps exclude noise beyond the BW of V_{sig}, and R_2, R_3, C_2 establish a near DC BW for the reference voltage (V_n) applied to the comparator.

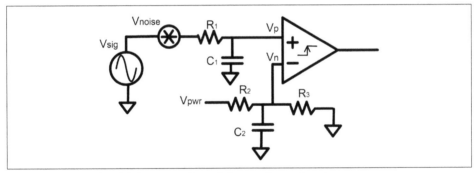

Figure 6-34. Bandwidth-limiting of amplifiers and signal paths

The response time shown in Figure 6-35 illustrates a fast versus slow response to a noisy input signal. Response time and input filtering are commonly used tools to minimize false positives due to noise.

Figure 6-35 shows a slow transient signal with imposed noise. The bandwidth (or response time) needed is defined by the signal. As an example, if V_{sig} is a tank-level monitor and the tank needs 10 minutes to empty, there is no need for a circuit that responds in nanoseconds. Consequently, a slow response comparator and/or input BW limiting can avoid erroneous responses to high-frequency noise.

Circuit response needs "just enough" speed for the signal of interest. Needlessly fast system electronics can generate false positives due to noise.

Digital methods are also available through digital averaging or time window methods to determine whether a stable state is present before recognizing a state change (FDI: Code).

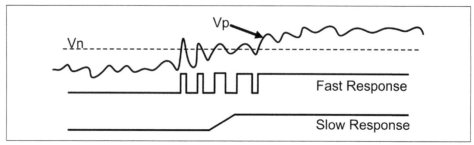

Figure 6-35. Comparator response time

Properly configured comparators help create good noise immunity. Hysteresis (Figure 6-36) is another useful technique. Hysteresis can shift the comparator switching point as it changes state. This is done internally using a Schmitt trigger circuit, or it is done with feedback around the comparator that slightly shifts the reference voltage.

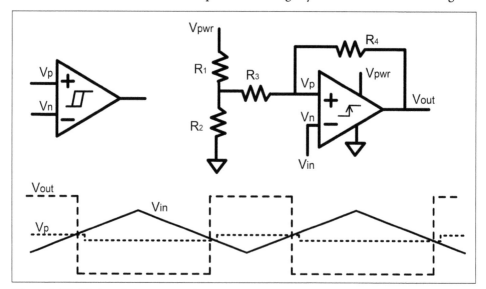

Figure 6-36. Hysteresis in comparators

Nonideal comparator characteristics (Figure 6-37) can also contribute to noise sensitivity. Consider the following:

- Small amplitude signals (V_{sig}) can be lost in the noise of the environment.
- Input offset (V_{off}) affects the switching point. Offsets up to 10 mV are common, so the signal applied to the comparator needs to be much larger than the offset.
- Comparators have a fixed gain (60–100 dB), and small amplitude signals may not create the full rail-to-rail digital output desired.

For signals less than 100 mVp-p, a gain amplifier might be needed between the signal and the comparator. The parameters of V_{sig} need to be compared to the performance specifications of the comparator to determine suitability.

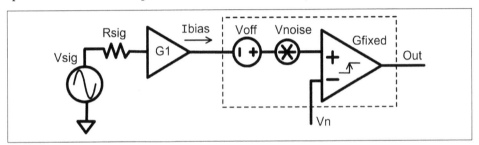

Figure 6-37. Nonideal comparator

Suppressing Noise into and Out of the System: Faraday Cage Techniques

Using an FC to partition off noise has already been introduced. Additional details are needed to implement an FC for an entire system (Figure 6-38).

An FC can be used to protect circuitry from outside noise, or to contain a noise source in some systems. Multiple FCs can be used to keep different parts of the internal system from interfering with each other. Additionally, an FC for an entire system is frequently used to contain EMI and meet radiated emissions requirements.

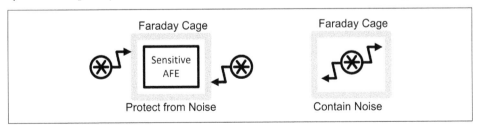

Figure 6-38. FC to protect or retain noise

When designing a system it can be unclear whether an FC is needed. Consequently, a good strategy is a design that adds FC capability with minimal effort.

The physical aspects of the FC need to be considered (Figure 6-39). The four common options for enclosure materials are:

- Plastic
- Plastic with an internal conductive coating
- Aluminum
- Steel

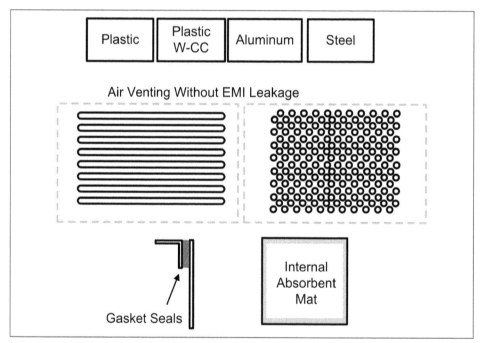

Figure 6-39. FC: physical considerations

Steel does best when containing or shielding magnetic fields. It is rarely a mandatory material, however. Both aluminum and steel will contain radiated EMI when properly implemented. Plastic enclosures are low cost and easily implemented, making them a popular solution. Plastic will not contain radiated EMI, but plastic enclosures that receive an internal conductive coating do well at reducing radiated EMI.

Devices put into a plastic enclosure should include the capability to easily add a conductive internal coating. In this manner, the low-cost option of a plastic enclosure can easily shift strategy to an FC if radiated EMI requires it.

An enclosure needs to be air-vented without using slots, which can allow EMI to leak. Hole patterns are commonly used and work well. Air filters using fine metal mesh screens are also available for air passage without EMI leakage.

Enclosure seams need to be carefully thought out for proper sealing. Electrically conductive gasket materials are available to reduce EMI leakage at the seams. EMI-absorbing materials are also available for lining the internal walls of an enclosure if necessary.

Connection ports going out of an FC can be problematic. Some connection plugs were designed with EMI leakage in mind, but most were not. When passing signals through the wall of an FC, you must keep the noise in while getting the signal out (Figure 6-40).

Different methods of passing signals involve different amounts of filtering:

- Case A shows a straight through connection, which can pass EMI on the wire to potential external antennas.
- Case B provides an LPF and is suitable for control signals that can tolerate some series resistance. Most digital control signals can utilize this with good results.
- Case C uses an FB instead, and is better suited to a current carrying connection.
- Case D uses optical isolation and is popular with high power switching power supply controllers.
- Cases E and F show filters that attenuate in both directions. Remember that noise can travel on connections in both directions.
- Case G uses feed-through capacitors, which exhibit low series inductance (ESL) so that the filter remains effective at very high (typically >1 GHz) frequencies.

A suggested strategy for most situations uses Cases E and F for PCB layout. A first PCB assembly with components can leave the capacitors off. Subsequent radiated EMI testing can determine whether they are needed.

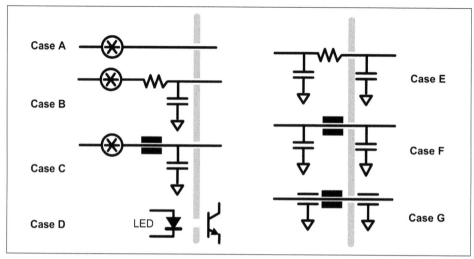

Figure 6-40. Connections into and out of an FC

Electrostatic Discharge Protection

Electrostatic discharge (ESD) is commonly experienced when walking across a carpet and touching something conductive, like a metal doorknob. Thunder and lightning is the most spectacular example of ESD.

On a smaller scale, all systems need to survive ESD events undamaged, and be functional afterward. ESD can destroy electrical circuits, and designers need to deal with the overvoltage/overcurrent incidents that ESD can cause to their designs.

A system can be designed to both survive and keep functioning properly through an ESD event. Medical systems are required to do exactly that. Understanding how to do this starts with understanding a model for an ESD event.

Figure 6-41 shows a simple model that approximates an ESD event quite well:

- Charge a 150 pF capacitor up to V_{ESD} (−8 KV to +8 KV is commonly used).
- Connect the charged capacitor to the victim device.
- A high-voltage transient occurs at the victim.

Typically, the voltage will transition to V_{ESD} in <2 ns and decay in <120 ns.

ESD events can be applied anywhere to the system exterior. Anything conductive on the exterior of the enclosure is especially vulnerable. The charge polarity can be positive or negative, so ESD protection needs to deal with both. Depending on the victim circuits, the sensitivity to a positive or negative strike can be very different.

The difference between the two most common models, the human body model (HBM) and the machine model (MM), is the series resistance. Human bodies (1500 Ω) don't conduct as well as metal machines (100 Ω).

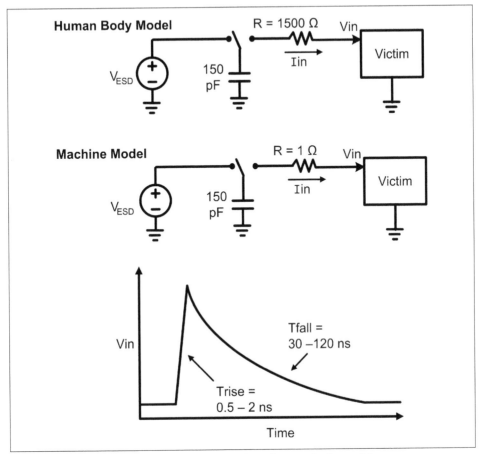

Figure 6-41. ESD as an electrical model machine/human body model

The best protection against ESD overvoltage damage is voltage limiting or clamping with a nonlinear circuit. Specialty transient voltage suppressor (TVS) diodes are commonly used. TVS devices have low impedance when forward-biased or in their avalanche breakdown region.

TVS diodes function as voltage limiters. The TVS is optimized for high-current transients over brief time periods. Two common variants are shown in Figure 6-42.

Unidirectional TVS devices perform similar to a high-current Zener diode. When forward-biased, the voltage and current are similar to a conventional silicon diode. When reverse-biased (V_{for}), the voltage will be limited to a specified clamping voltage

(V_{clamp}). In addition to a clamping voltage specification, a standoff voltage (V_{soff}) is also specified. The standoff voltage is the maximum voltage where the device does not turn on. As an example, a 3.3 V logic input could be protected using a unidirectional TVS with 3.6 V standoff voltage and 4.5 V clamping voltage.

Bidirectional TVS devices limit voltage equally in both directions. The standoff voltage (V_{soff}) is the same for both positive and negative inputs. Most ESD protection is done using unidirectional devices.

The standoff voltages are available in values optimized to protect common logic levels and power supplies. One size does not fit all, and the TVS should have the appropriate standoff and clamping voltages for the device being protected.

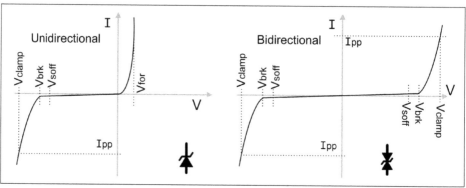

Figure 6-42. TVS device characteristics

Any port on a digital IC has a very limited functional voltage range. Outside the functional voltage range, permanent damage to the IC can occur. On-chip ESD is limited in capability. Digital devices with off-board connections need protection to avoid burning out.

Introducing a TVS quickly causes something else, namely a large current surge (Figure 6-43) through the TVS. TVS use prevents overvoltage damage. But the high-voltage transient is replaced with a multi-amp current spike that can cause other problems. Depending on the total impedance of the connection, a 4–16A surge current into the device is common. That huge transient current surge then becomes the newly created problem in ESD protection. Limiting the voltage was straightforward, but the resulting current can upset circuitry and grounds everywhere in the system.

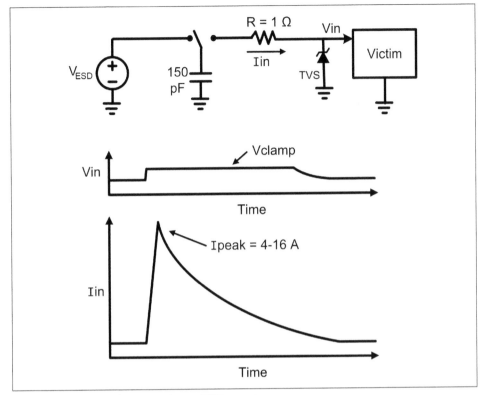

Figure 6-43. Injected current due to ESD event

Amps of current forced into the ground by the TVS cause inductive ringing of the system ground, as shown in Figure 6-44. Injecting a large surge current into the ground through the voltage limiter will cause the PCB ground to bounce due to connection inductance. A board with a low-impedance ground plane causes the entire ground plane to bounce together.

Power connections should have extensive high-frequency decoupling to ground, and consequently the power bounces with the ground. With P&G bouncing together, proper functionality on the PCB should continue through the ESD event.

Airplane turbulence and tightening your seat belt is a suitable analogy. Airplane or PCB, it will all stay functional if it all bounces together.

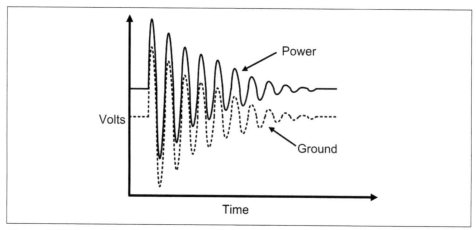

Figure 6-44. Ground and power bounce due to ESD

Establishing P&G integrity allows the PCB to maintain signal integrity across the board during an ESD event. However, external control lines are defined relative to an external ground and can get corrupted.

An ESD event in one location can cause an input elsewhere to be seen as faulty. The solutions for this are multiple:

- Inputs have to travel with the PCB ground, using an LPF to do so.
- Inputs can be nonresponsive to fast transients via software state sampling.
- Inputs use an error detection/correction protocol.

The elapsed time for the ESD event is under 150 ns for the initial current injection, and the ringing typically settles out in under 100 ms. Filtering, sampling, or error detection methods can readily deal with this.

Understanding what happens during an ESD event leads to the question of what needs to be protected from ESD, as shown in Figure 6-45. Some form of protective enclosure will be part of the device. Internally isolating the PCB is the first line of defense that needs to be carefully thought out. The device enclosure becomes the figurative fortress. With board-level ESD, building a fortress and establishing protected access points "across the moat" becomes the strategy. All items connected out beyond the "castle walls" need to be ESD protected.

When defining the enclosure, remember that ESD to any external part of the case can take multiple paths into the electronics. If there is a conductive path of any form to the PCB, it needs to be ESD protected as an external port, connected to ground, or removed from the conduction path.

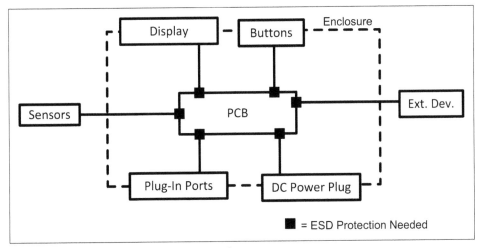

Figure 6-45. ESD strategy as a protective fortress

Protection circuits should be placed close to the entrance to the PCB and not downstream from the entrance point. ESD can be kilovolts and create possible arcing problems. This is best dealt with immediately where the connection enters the PCB. Many possible options for protection exist, and multiple options are examined in Figure 6-46.

Case A represents a basic TVS limiting circuit that provides overvoltage protection but creates surge currents into ground. For connections that can't tolerate any series impedance, this is better than nothing, but other problems may emerge. If possible, series impedance reduces the current injected into ground.

Case B represents a version that adds series impedance using an FB to the ESD signal path, and helps limit the surge current. The capacitor helps with high-frequency stability and the connection is still low impedance at DC. This is the preferred method for power connections. The FB also prevents the power wire from creating a radiating antenna.

Case C adds series resistance to the connection, which greatly reduces the current that gets pumped in the ground. For things like push buttons and low-frequency digital data, significant (1 KΩ to 10 KΩ) series resistance can usually be tolerated. High-speed data requires smaller resistance, but $R_{in} = 100\ \Omega$ on a fast data connection is often tolerable and helps reduce the ESD-created current.

Case D adds filtering capacitance to keep the input to the victim stable through the ground bounce of an ESD event. The RC time constant only allows slow (push-button type) responses, and can be made faster if the input needs more speed. This method works well to protect the large number of low-bandwidth inputs, which includes "set and forget" control lines, sensor inputs, and similar.

Although we are discussing input port protection here, output port protection is both similar and needed. TVS and additive resistance are appropriate. Voltage limiting helps prevent damage to semiconductors and other parts with voltage restrictions.

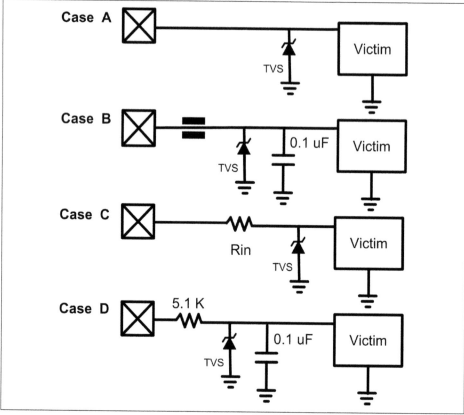

Figure 6-46. TVS strategy with supplemental R, L, C, and FB devices

For high-speed data, a properly designed communication protocol with error checking needs to be used. ESD would corrupt data, and bad data prompts a request to resend. Since an ESD event and its aftereffects last for less than 100 ms, the check, verify, and resend process deals with ESD invisibly. The end user is never aware that corrupt information was corrected. Some protocols don't check and correct data. If data goes off the PCB, it needs the capability to error-check and correct (FDI: Digital).

Summary and Conclusions

EMI has to be considered part of any electronic design. Noise can be both a regulatory and performance issue. Too much radiated noise can prevent device approval and commercial sale. Too much internal noise coupling can reduce performance and accuracy, or even render a device nonfunctional.

Many things that can be done to mitigate noise problems don't add significant cost to the product and should be included in any design. A simplified model for noise coupling sees three things: source, medium, and victim. This model helps explain how noise is conveyed between devices. Noise reduction techniques generally fit within these areas:

- Silence the source.
- Contain the source.
- Protect the victim.
- Deafen the victim.
- Separate the source and victim.

An initial design should approach noise reduction as an application of best practices. Recognize both noise sources and noise-sensitive victims. Then take actions to silence, separate, shield, or render them nonresponsive. Frequently, that is sufficient. After implementation, noise testing is best observed in the frequency domain where the frequency of noise sources provides well-defined clues to the circuit sources.

Transient switching is the predominant EMI source. Noise coupling paths include P&G connections, parasitic capacitance, inductive coupling, and radiated noise signals. Low-impedance grounding serves as a foundation for any EMI reduction strategy. But no ground is ideal, and small amounts of dynamic voltage variance in a ground are expected and need to be accounted for.

Open current loops should be avoided to minimize radiated EMI. Twisted wire pair interconnects, differential signal pairs, LVDS, and all PCB connections over a ground plane minimize current loops and their EMI.

Consider using slower clocks and slower transient edges for all digital signals and clocks. Frequently, devices can run more slowly, saving power, reducing EMI, and avoiding the need for impedance-matched transmission lines.

Any SMPS power regulator should include PCB locations for RC snubbers, ferrites to isolate the SMPS, and high-frequency filter capacitors. Heat sinks should be grounded out, and motors should receive RC snubbers with ferrites on their cables. Any DC reference voltage should have decoupling capacitors in close proximity to where that DC voltage is used. Creation of independent power sources for analog and digital devices is greatly preferred.

Designers should understand what parts are noise-sensitive victims and what parts are noise sources. Then they should implement a strategy that separates the victims from the sources. Liberal use of grounds between sources and victims aids in their separation.

Bandwidth limiting of amplifiers, comparators, and signal paths improves noise immunity. Lower-impedance connections and bias circuits are less sensitive to capacitive crosstalk.

Faraday cages can be used to reduce noise from an entire system, protect sensitive internal devices from noise, or contain noise emitted from a particular circuit. Frequently, multiple FCs are used within a single system to achieve needed noise abatement. In all cases, signals into and out of the FC need filters so that noise does not leak via electrical connections.

ESD protection will be required on all external connection paths out of the device enclosure. For some products, ESD protection will be defined and tested for as a regulatory requirement. For other devices, a lack of ESD protection becomes a liability as products fail in the field.

All ideas presented here on EMI and ESD are readily implemented into a design, many with minimal effort and expense. Schematic design, PCB layout, and enclosure implementation specifically for EMI and ESD issues should be reviewed.

Further Reading

- *Controlling Radiated Emissions by Design* by Michel Mardiguian, 1992, ISBN 0-442-00946-6, Chapman and Hall.
- *Noise Reduction Techniques in Electronic Systems* by Henry W. Ott, 1976, ISBN 0-471-65726-3, Wiley-Interscience.
- *Electromagnetic Compatibility Engineering* by Henry W. Ott, 2009, ISBN 978-0-470-18930-6, John Wiley & Sons.
- *Signal and Power Integrity – Simplified, 3rd Edition*, by Eric Bogatin, 2018, ISBN-13 978-0-13-451341-6, Pearson Education, Prentice Hall.
- *EMC for Product Designers, 2nd Edition*, by Tim Williams, 1996, ISBN 0-7506-2466-3, Butterworth-Heinemann.
- "Common Mode Filter Design Guide," Document 191-1, Coilcraft, revised November 8, 2007.
- Jerry Twomey, "Simple Grounding Rules Yield Huge Rewards," April 27, 2012, *Electronic Design Magazine, https://effectiveelectrons.com/articles-patents.*
- Jerry Twomey, "Quiet Down to Meet FCC Emission Standards," *Electronic Design Magazine*, July 10, 2013, *https://effectiveelectrons.com/articles-patents.*
- Jerry Twomey, "Protect Your Fortress from ESD," *Electronic Design Magazine*, August 9, 2012, *https://effectiveelectrons.com/articles-patents.*
- "EMC and System ESD Design Guidelines for Board Layout," AP24026 Infineon Corp., *https://www.infineon.com.*
- IEC 61000-4-2:2008 Testing and Measurement Techniques – Electrostatic Discharge Immunity Test, International Electrotechnical Commission, *https://www.iec.ch.*
- The ESD Association, *https://www.esda.org.*
- Lisa Linna and Christopher Semanson, "Exploring the Characteristics of Field Coupled Crosstalk," *Compliance Magazine*, March 31, 2016, *https://incompliancemag.com/article/exploring-the-characteristics-of-field-coupled-crosstalk.*
- James Kahl, "Fully-Differential Amplifiers," Texas Instruments Application Report SLOA054E, January 2002, revised September 2016, *https://www.ti.com/lit/sloa054.*

CHAPTER 7

Data Converters: ADCs and DACs

This chapter provides a brief introduction to data converters and then takes a look at some converter techniques widely used in embedded systems. "Further Reading" on page 262 includes resources for those who want to take a deeper dive into converter performance and internal circuit technology.

Within an electronic system, data converters create the needed bridges between digital control and nondigital peripheral devices. The conversion of analog information to digital data is fundamental to modern electronics (Figure 7-1). Modern embedded systems interface to peripheral devices using data converters. Digital signal processing and digital control systems require interfaces to the external world. Digital-to-analog converters (DACs) and analog-to-digital converters (ADCs) provide this interface and are also used as part of drive and sense circuits.

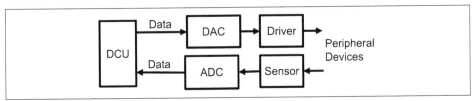

Figure 7-1. Data converters in a system

DAC Performance Basics

As a black box, DACs can be divided into two groups:

- Nyquist rate DACs update their output once per clock cycle.
- Pulse density DACs require multiple clock cycles to achieve a stable output based upon the averaging of repetitive digital patterns. Pulse width modulation (PWM) and delta-sigma ($\Delta\Sigma$) are the two methods used here.

Both Nyquist rate and pulse density DACs are commonly used in embedded systems. As of 2023, available DAC performance ranges up to 32 bits of resolution, and high-speed DACs that will settle out to their final value within 1–2 ns. High-performance devices can be power hungry and expensive. Many cost-effective offerings exist between 8 and 16 bits.

System-level designers will generally deal with these devices as a black box module, so the internal details are not covered here (FDI: Razavi, 1995). The one exception is the PWM DAC, which system designers often implement using a general-purpose input/output (GPIO) port on a microcontroller (MCU).

As shown in Figure 7-2, DACs can have one of three different types of output:

Voltage output, ground referenced
> The output signal is defined as the voltage between the output pin and ground. This is the simplest option and is widely used. However, it can suffer from noise due to both dynamic ground variance and electromagnetic interference (EMI) coupling to the output.

Voltage output, differential signals
> These devices have two voltage outputs, and the received signal is defined as the difference between them. The advantages of this are:

> *Larger amplitude signal*
>> The differential voltage is twice as large as the ground-referenced DAC. This provides 6 dB of signal-to-noise ratio (SNR) improvement.

> *Less sensitive to ground noise*
>> Ground noise is common to both outputs, and the receiver rejects common mode noise (20–30 dB common mode noise rejection is typical).

> *Less sensitive to EMI*
>> EMI is common to both outputs, and the receiver rejects common mode noise. Again, 20–30 dB common mode noise rejection is typical.

Current output, differential signals

These devices have two differential current outputs that are commonly terminated with external resistors, resulting in a differential voltage. By using resistor terminations on the printed circuit board (PCB), differential transmission lines can be easily configured. Often referred to as a *current-steering DAC*, these devices are capable of very fast conversion speeds.

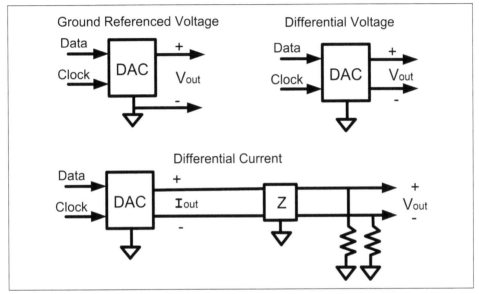

Figure 7-2. Nyquist rate DAC by output type

For an ideal DAC, loading the DAC sequentially with digital control codes should produce a staircase output, as illustrated in Figure 7-3. The resolution here is defined by the number of bits in the DAC design. Resolutions of 6 to 32 bits are commercially available. Typically, as the DAC resolution goes up, the conversion speed goes down.

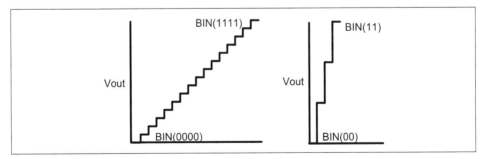

Figure 7-3. Nyquist rate DAC resolution

Nonideal characteristics can change DAC performance. Performance characteristics are divided into two groups: static behavior and dynamic behavior. Static behavior is based on device performance measured at fixed operating points for each digital input code, without transient switching effects.

Ideally the output staircase should follow a straight line. Nonideal internal DAC circuits can change that, as illustrated in Figure 7-4. The nonideal static characteristics of DACs include:

DC offset
> The origin of the staircase is shifted by DC offset; the origin of the staircase can be shifted up or down and the entire staircase shifts with it.

Gain
> Errors in gain illustrate themselves in the slope of the staircase, where high gain is illustrated by a steeper slope and low gain by a gentle slope.

Integral nonlinearity (INL)
> Instead of following a straight line, the staircase follows a curve.

Differential nonlinearity (DNL)
> Ideally, each step should be equal in amplitude. DNL errors are exhibited when the steps have an inconsistent size.

Monotonic versus nonmonotonic
> Poor DNL can lead to amplitude errors in steps, or even steps going opposite of the desired direction. This is known as a *nonmonotonic response*. Nonmonotonic behavior can be problematic when it is part of a feedback control system.

Problems with static DAC behavior can be traced to design limitations of internal DAC circuitry. Specifications can be reviewed to determine suitable devices, but devices that are commercially sold should have had these issues resolved. "Further Reading" on page 262 provides a deeper look at internal DAC characteristics (see Razavi, 1995).

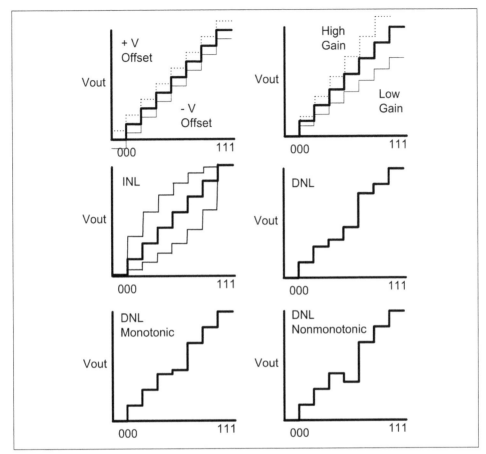

Figure 7-4. DAC: nonideal static behavior

Dynamic DAC behavior can show undesired transients as the output changes, as illustrated in Figure 7-5. As shown, an ideal staircase step can be replaced by ringing characteristics due to RLC components of the circuitry. Internal DAC circuits with poor switch timing can show undesired switching as part of the transition. These characteristics are transitory and are less problematic with a slower clock.

Conversion speed, aka clocking rate/frequency or conversion rate, denotes how quickly the DAC output can update to a suitable analog value. How quickly the DAC functions is dependent upon the internal circuitry design.

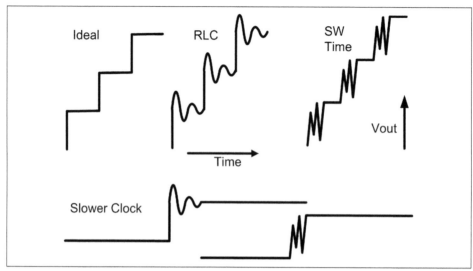

Figure 7-5. Nyquist rate DAC dynamic behavior

A common metric of DAC quality is spurious free dynamic range (SFDR), as illustrated in Figure 7-6. To measure SFDR, the direct digital synthesis (DDS) pattern for a sinusoid is repetitively clocked into the DAC and a spectrum analyzer measures the output. The desired output should be a single-frequency sinusoid, but nonideal behavior creates other spurious components readily seen in the frequency domain. The SFDR is measured as the amplitude difference between the sinusoid and the largest spurious signal component observed.

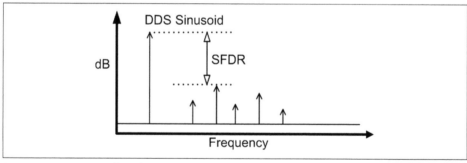

Figure 7-6. Spurious free dynamic range

Additional performance criteria also define DAC performance. This is a brief overview.

ADC Performance Basics

The basic performance criteria of offset errors, gain errors, INL, and DNL are also applicable to ADC devices. Instead of the output signal being shifted, the voltage where the ADC changes to a different code value is shifted. With an ADC, calibration can be done at min/max signal inputs to digitally compensate for offset and gain errors.

Most commercially available ADC devices use a differential input voltage. When dealing with ground-referenced input voltages one input gets attached to ground, near the voltage being measured. As of 2023, a broad ADC selection is available:

- Resolution from 4 to 32 bits
- Sampling rates up to 16 G samples/sec

Many top-speed devices are both expensive and power hungry. Typically these are designed for instrumentation, telecommunications, and other applications where cost is a low priority.

Most embedded designs use converters under 100 M samples/sec, and many devices are available that are both cost and power efficient. Designers should determine needed sampling rate, needed bits of resolution, and tolerable conversion latency for the ADC as a starting point for device selection.

Conversion latency is the time delay between input sampling and the associated digital data out. Depending upon the internal ADC structure, this can vary widely. Many ADC devices use multiple clock cycles with pipelining of both internal analog processing and digital data.

Quantization error, illustrated in Figure 7-7, exists in any ADC and is defined by the number of bits of resolution. As shown, the digital code out of the ADC (D-1, D, D+1) has a limited resolution, and the ADC output gives a numeric result where each binary value represents a range of input voltages.

For example:

> 10-bit ADC
> 1,024 possible outputs
> Max voltage range of 1 V
> LSB voltage step 1 V/1024 = ~1 mV
> Worst-case quantization error = ~0.5 mV

Quantization error is usually modeled as noise added to the ideal signal. As the number of bits goes up, the quantization error "noise" goes down. For an ideal system, the

SNR is a function of the bit count. Ideally, every 1-bit increase in resolution reduces quantization noise by 6 dB. The ideal equation is shown in Figure 7-7.

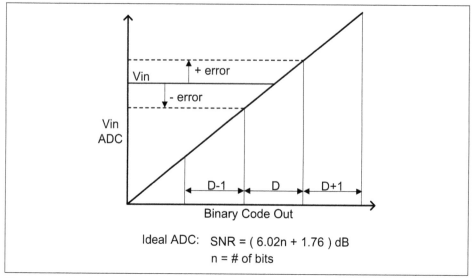

Figure 7-7. ADC quantization error

Aperture error deals with the timing uncertainty of the sampling point, as shown in Figure 7-8. Aperture error becomes an issue when the clock's rise/fall time becomes a significant part of the total clock period. The rise/fall duration of the clock edges can create a time uncertainty of when the signal is actually sampled.

Noise present on the clock signal can also cause aperture error. For signals that are simple and close to DC voltage measurements, aperture errors won't be a problem. Aperture error becomes an issue with high-speed signal capture. A wideband communication channel that uses an ADC as part of the signal processing can encounter this error source.

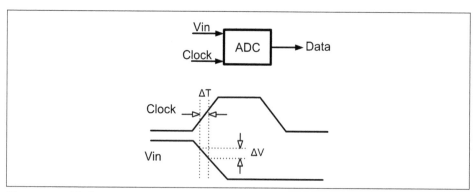

Figure 7-8. ADC aperture error

The needed ADC sampling rate is a function of the frequency content of the input signal. Remember, any periodic signal, no matter what the wave shape looks like, can be defined as a summation of sinusoids. That summation of sinusoids, also known as a Fourier series, defines the bandwidth and highest-frequency component of the signal. The highest-frequency (F_{sig}) component determines the needed sampling rate ($>2F_{sig}$). This is also known as the Nyquist sampling rate.

In Figure 7-9, the analog sinusoid (Input) is being sampled. The diagram shows sampling time points (ST3) of eight samples per one cycle of the sinusoid. The captured wave data (CD3) shows a staircase approximation of the original wave. With a suitable DAC and filter, this data could be used to re-create the original signal.

Reducing the sampling rate down to the Nyquist rate can produce problematic results. Sample times (ST1) are at the Nyquist rate, and the phase relationship between sample points and the original sinusoid shows samples at the sinusoid peaks. The captured wave data yields a square wave (CD1), which could be used to re-create the original sinusoid.

However, if the phase of the sample times is shifted (ST2), the amplitude of the captured wave will change. If sample points align with the zero amplitude points of the original waveform, the captured wave data (CD2) no longer represents the original signal.

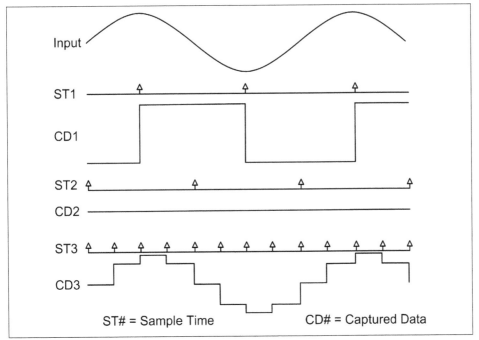

Figure 7-9. ADC Nyquist rate sampling

As the sampling rates go below the Nyquist rate, the captured wave data can falsely represent the original signal, commonly called aliasing, as illustrated in Figure 7-10. As shown, a sampling rate at 2F can have the same data points for sinusoids at 1F and 3F. If the original signal is 3F, then the data captured will appear as a 1F signal. To fix this, the sampling rate can be increased, or the bandwidth of the original signal can be reduced, removing higher-frequency input to the ADC. A common example of aliasing is seen in video capture of rapidly moving items, when the frame rate is too slow to properly capture motion. Wagon wheels rotating backward and helicopter blades appearing to rotate slowly are common examples of video image aliasing.

For slowly changing signals, aliasing issues are generally not a problem. For signals with high-frequency content, the relationship between the sampling rate and the signal's frequency content needs to be more carefully examined.

The signal BW, ADC sampling rate, and number of bits of resolution can be examined as part of a communication channel model if the need arises. For many common use applications, a simple RC LPF at the input of the ADC will suffice.

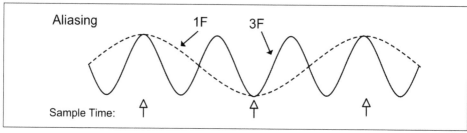

Figure 7-10. ADC aliasing

Depending upon the application, some sampling rates and resolutions are set by industry standards. The audio compact disc has a 44.1 KHz sampling rate and 16-bit resolution. This allows a sound frequency capture of >20 KHz and a signal to quantization noise of >96 dB.

Antialiasing Filters for ADC Inputs

Analog filters come in multiple forms, including:

- Passive RC
- Passive LC network
- Transconductor (gmC)

- Op-amp (gyrator)
- Switched capacitor

Filter structures that are readily implemented on a PCB include passive RC, passive LC network, and op-amp (gyrator) methods. Transconductor and switched capacitor devices are typically implemented internal to an IC and are not discussed here.

In addition to designing a filter, there are a number of system on a chip (SoC) integrated antialiasing filters designed to be inserted between the sensor and the ADC. For op-amp–based filters, the Baker (2015) paper on antialiasing filters is a suitable introduction; Baker also provides a useful discussion on how to define the filter parameters to fit the ADC and the input signals.

For many situations, a passive filter will suffice with either an RC LPF at the input or a higher-order LC network filter being suitable. The Coilcraft (2017) paper on reference designs includes values for both three-pole and seven-pole filters that can be readily implemented. In addition, a multitude of EDA tools for filter design are available. Williams and Taylor (2006) and Zverev (1967) provide a deeper dive into the topic (see "Further Reading" on page 262).

Pulse Width Modulation DACs

The PWM DAC is widely used in embedded systems. It costs little or nothing to implement, and most MCUs include internal digital PWM options. The PWM DAC is conceptually simple: a control pattern with a digitally defined duty cycle is fed into a low-pass filter (LPF) to create an analog voltage. These devices have limits that need to be understood to get the best performance. In addition, some useful and lesser-known methods to get optimal performance are described herein.

In Figure 7-11, a digital pattern is repetitively applied to an LPF. The output voltage (V_{out}) is the long-term average of the digital pattern. The PWM DAC requires many cycles of the control pattern to respond and stabilize.

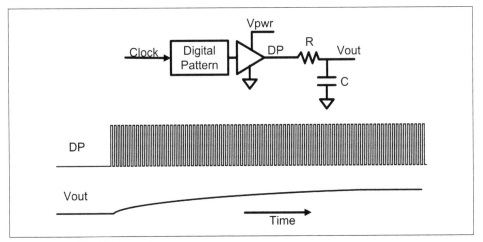

Figure 7-11. PWM DACs: basic concepts

A simple example of a 3-bit, eight-state PWM pattern is shown in Figure 7-12, with a state(000) of always off, a state(111) of always on, and the interim patterns shown. Creating a specific DC output voltage involves repetitive generation of the same pattern, and the RC LPF reduces the switching transients to an average DC voltage. For the 3-bit example, a pattern using eight clock cycles is needed. For a higher-resolution DAC, longer patterns are needed. A 10-bit device requires a pattern of 1,024 clock cycles. The necessary pattern length is 2^n clock cycles for a DAC with n bits of resolution.

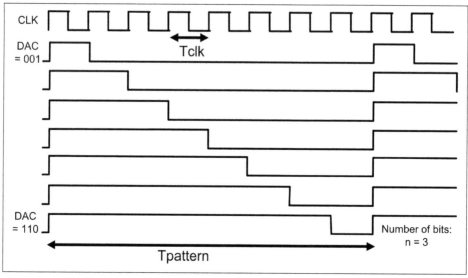

Figure 7-12. PWM DACs: PWM patterns, 3-bit example

The PWM DAC has important issues that limit the performance of the device. Ripple on the output signal is created by the non-DC content. Understanding the spectral content of the output (Figure 7-13) gives a better understanding of how to optimize the device.

As shown, the output has a very distinct spectral pattern. The average of the pattern is the DC value at f_0. The frequency of the first harmonic, f_1, is controlled by two things: the number of bits of resolution, n, and the frequency of the clock used. The pattern length $T_{pattern}$ is the inverse of f_1, and is controlled by the period of the clock, T_{clk}, and the number of bits, n, that the DAC is designed for.

As an example, n = 8 bits of resolution, with an f_{clk} = 10 MHz, creates f_1 at 39 KHz. The signal component at f_1 is the largest contributor to ripple on the output signal. Depending on the application, reducing the amplitude of this harmonic content may be necessary.

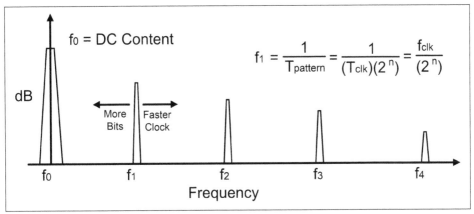

Figure 7-13. PWM DACs: spectral content due to clock and number of bits

Complete rejection of the harmonic components is an ideal goal. However, the RC filter (Figure 7-14) is not a perfect cutoff device. The RC filter provides 20 dB/decade of attenuation starting at f_{RC}. Consequently, attenuated versions of f_1 and higher harmonics will remain. If f_{RC} is lowered and/or f_1 is raised, more harmonic attenuation is created and ripple is reduced.

As illustrated in Figure 7-14, designers have some control over the placement of f_1 and f_{RC} to minimize in-band ripple:

Shorter patterns
> Pattern length is determined by how many bits of resolution the DAC needs. Shorter patterns will move f_1 up in frequency and achieve greater filter attenuation. If the design needs 4 bits (16 states) of DAC resolution, using a 12-bit (4,096 states) pattern needlessly moves f_1 closer to DC.

Higher clock frequency
> Both pattern length and clock periods determine f_1 location. A faster clock increases f_1, resulting in greater ripple attenuation.

Lowering f_{RC} increases harmonic attenuation and reduces ripple. However, DAC response time is also increased. For a lower f_{RC}, control pattern changes need a longer time period for output changes to stabilize at the new output voltage. By defining the needed bits of resolution, the available clock rates, the necessary response time, and the tolerable amounts of ripple, a suitable solution can generally be found.

Although the problem can be manually worked out, a simulator (SPICE), a math tool (MATLAB or similar), or even a spreadsheet with formulas can be used to work out the solution. The tool needs Fourier transform capability, but otherwise the math is straightforward.

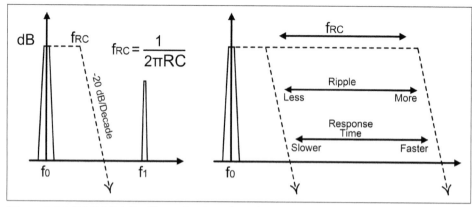

Figure 7-14. PWM DACs: filter rejection of ripple components

Other ripple reduction techniques are available. Using a logic inverter ideally creates an equal and opposite PWM signal. AC-coupling that signal creates an equal and opposite version of the ripple, as illustrated in Figure 7-15.

Subtracting the AC content is straightforward. In Figure 7-15, the two resistors are equal in value and the DC content passes through R_{DC}, but no DC current passes through C_{AC}. Summation of the two signals reduces the magnitude of ripple. Typical ripple reduction of 20 dB has been reported (see Noeman, 2020, in "Further Reading" on page 262).

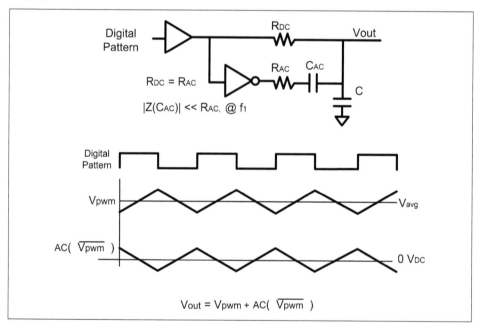

Figure 7-15. PWM DACs: ripple reduction by subtracting AC content

High-resolution (typically beyond 10 bits) PWM devices suffer from long patterns and either slow filter response or unacceptable ripple levels. Another approach includes the summation of two PWM signals of lower resolution, as illustrated in Figure 7-16.

In this example, two PWM patterns are used, one for the most significant bits (MSB) and the other for the least significant bits (LSB). The two PWM streams are summed through resistors that are not of equal value, but are a precise ratio of each other. The R_{MSB} has a smaller value and thus contributes greater I_{MSB} current than R_{LSB}, which contributes I_{LSB}.

As an example, a 6-bit PWM pattern is used for both digital patterns. This allows a PWM pattern that is only 64 clock cycles long. $R_{LSB} = 64\ R_{MSB}$ so that the full scale of the I_{LSB} is equal to a single bit of the MSB side.

This results in a DAC with 12 bits of resolution while using the shorter 6-bit PWM patterns (64 clock cycles) versus a 12-bit PWM pattern of 4,096 clock cycles. This provides a significant advantage (64X) in frequency location of the first harmonic, f_1, and response time. The system still needs to be examined for ripple magnitude to determine that the LSB performance is meaningful.

As part of a closed-loop control system, this technique is a low-cost and powerful method for achieving accurate feedback control. The LSB side can be set to run at half scale (50% duty cycle) while the MSB side provides coarse adjustments to a final operating point. When the MSB has achieved an approximate adjustment, the PWM pattern on the LSB side is then adjusted to the final stable value.

With this variant, the R_{LSB} is often made slightly smaller so that full variance of the LSB side covers two or three steps of the MSB side, thus assuring overlap even if the resistor accuracy is a bit off. The servomechanisms of millions of hard disk drives use this technique with excellent results.

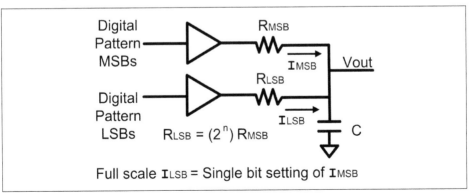

Figure 7-16. PWM DACs: high resolution by weighted current summation

The techniques of both weighted current summation and subtraction of AC content have also been combined to reduce ripple and get higher-resolution performance. Higher-order analog filters can be used to reduce ripple, but if the design needs op-amp filters or LC network filters, seeking out a better-quality integrated DAC is probably a lower-cost solution.

As a final consideration for achieving a quality PWM device, the power and ground of the digital driver gates should be kept low-noise. Any power supply noise on the driver gate, or difference in ground voltages between the driver gate and the filter capacitor, will manifest as noise in the V_{out} signal.

Higher-performance oversampling DACs exist in the form of delta-sigma devices. Some vendors are including delta-sigma devices as an internal feature in MCUs, or they can be implemented as an external device controlled through a serial port. PWM devices are given an in-depth examination here due to their widespread use and negligible cost of implementation.

Most oversampling DACs are limited to creating signal outputs under 100 KHz. PWM devices dominate the electromechanical control systems market, and delta-sigma devices are commonly used for creating audio signals. If a DAC with greater speed is needed, a Nyquist rate DAC probably needs to be added to the design.

Arbitrary Waveform Generation by Direct Digital Synthesis

DACs are frequently used in the creation of arbitrary waveforms. Arbitrary waveform generation (AWG) devices using direct digital synthesis (DDS) are useful design tools and are simple to implement. At the top of Figure 7-17, a clock is used to increment a counter. The counter output is used as an address to a digital look-up table that contains the digital amplitude data for a sampled waveform. The waveform data is resynchronized to the clock and used as a DAC input. The DAC output is the desired waveform. If needed, a smoothing filter can be added to remove the transition-switching staircase, resulting in an analog waveform.

Single-chip components are available with a complete DDS AWG system. Clock periods under 1 ns are readily available, and lower-performance, lower-cost variants can also be found.

Some fully integrated DDS devices include rapid frequency change capability (Figure 7-17, bottom). These devices do not change clock frequency to change output frequency. Instead, a delta phase (DP) number is loaded, where the DP value is repetitively summed in a counter. A larger DP number causes the address (ADR) to advance more quickly. The higher-order ADR bits (m) are used as the look-up table address, and the entire output of the summing function (n) is fed back to have the DP value repetitively added to it.

A small DP number results in a low-frequency signal at V_{out}. A larger DP number gives a higher output frequency. Changing the DP value allows rapid changes in output signal frequency. This technique is widely used in wireless systems to quickly change frequency.

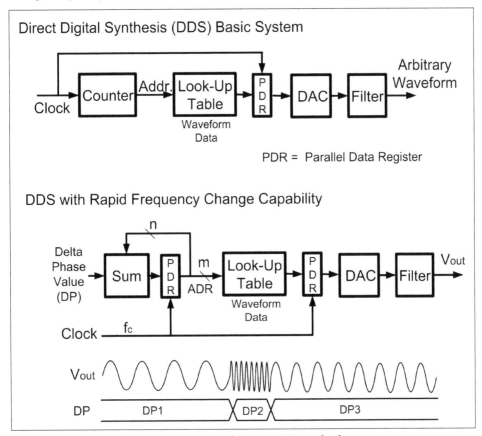

Figure 7-17. High-performance DACs and DDS/AWG methods

Summary and Conclusions

Data converters are an important part of any embedded system. Standalone converters are available that interface through a serial port, and many MCU vendors offer a variety of converters within the MCU chip.

The ADC sampling rate of a signal needs to be more than 2X faster than the highest frequency contained in the signal (aka Nyquist rate sampling). If the sampling rate of the signal is not faster than the Nyquist rate, the signal data will be corrupted by aliasing. A faster ADC can be considered, or reducing the signal bandwidth using an LPF is another option. Both are viable solutions.

A low-pass filter at the input of any ADC can be used to reduce the frequency content of the input signal and avoid aliasing. A simple RC LPF at the ADC inputs should be used as a minimum for all ADCs in the system. Signals requiring removal of more high-frequency content can be created with an LC network filter or an SoC integrated LPF.

PWM DACs are readily implemented with a GPIO port on the MCU and suitable code to generate the needed control patterns. Device implementation needs to be analyzed for in-band ripple to make sure the ripple is smaller than the LSB of the DAC.

Further Reading

- Bonnie Baker, "A Glossary of Analog to Digital Specifications and Performance Characteristics," Texas Instruments Application Report SBAA147B.

- "Understanding Data Converters," Texas Instruments Application Report SLAA013, 1995.

- *Data Conversion System Design* by Behzad Razavi, 1995, ISBN 0-7803-1093-4, IEEE Press.

- Ahmed Noeman, "Designing High Performance PWM DACs for Field Transmitters," *Analog Design Journal*, ADJ 3Q2020, Texas Instruments.

- Bhargavi Nisarge, "PWM DAC Using MSP430 High Resolution Timer," Texas Instruments Application Report, July 2011.

- Dennis Sequine, "Enhanced PWM Implementation Adds High Performance DAC to MCU," Cypress Semiconductor, September 2015, *https://www.elec tronicdesign.com/technologies/embedded/digital-ics/processors/microcontrollers/arti cle/21800934/enhanced-pwm-implementation-adds-highperformance-dac-to-mcu*.

- "Build Your LC Filter with Coilcraft Reference Designs," Coilcraft, 2017, *https://www.coilcraft.com/getmedia/25110d2a-de33-443e-a33a-fcd38fa36380/Doc124A_LC_Filter_appnote.pdf*.

- Bonnie Baker, "Designing an Anti Aliasing Filter for ADCs in the Frequency Domain," Texas Instruments, 2015, *https://www.ti.com/lit/pdf/slyt626*.

- *Handbook of Filter Synthesis* by Anatol Zverev, 1967, ISBN 0-471-98680-1, John Wiley and Sons.

- *Electronic Filter Design Handbook, 4th Edition*, by Arthur Williams and Fred Taylor, 2006, ISBN 978-0071471718, McGraw Hill.

CHAPTER 8

Driving Peripheral Devices

Drive and sense circuits connect the analog world of nondigital devices to digital control signals. *Driver circuit* is the generic name for circuits that provide suitable voltage and current to peripheral devices. Depending on the device, the needed currents and voltages are often beyond the capability of any digital control unit (DCU) output. As well, driving many devices requires signals that are of an analog nature, necessitating some form of digital-to-analog conversion (DAC) strategy. A *sense circuit* monitors real-world analog information and converts that information to digital data suitable for DCU processing. Chapter 9 is dedicated to exploring sensors and the circuit methods used to support them.

Figure 8-1 illustrates sense/drive concepts. A driver circuit provides high power to a motor and a sensor circuit determines the rotational speed of the motor. Devices can be operated without feedback sensing (aka *open loop*), or in conjunction with feedback within a *closed-loop* control system (FDI: Control).

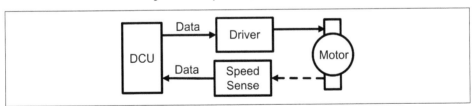

Figure 8-1. Drive and sense systems

This chapter focuses on driver circuitry. Other chapters focus on sense circuitry and closed-loop control methods.

WARNING: Safety First! As designers start controlling motors and other servomechanisms, safety must be a priority. Always consider what could possibly happen with an active system and where it can cause damage or harm.

Switched Driver Circuits

The simplest type of driver circuits are on and off devices that require current to go in one direction only through the device being driven. This requires the use of power transistors, suitable control of the power transistor, and techniques to avoid transistor destruction due to overvoltage or thermal overheating.

High- and Low-Side Switching

Typical digital devices can source or sink continuous currents less than 20 mA. Consequently, switching power on and off often requires higher currents than typical digital devices can provide. Figure 8-2 looks at some of the circuits used.

There are two ways to switch power on and off to high-current loads. Case A, commonly called *high-side switching*, places the switch between the power supply and load. Case B, known as *low-side switching*, places the switch between the load and ground. Both high- and low-side switching require an interface circuit and a power transistor. This text focuses primarily on negative/positive field-effect transistors (NFET/PFET) for power transistors due to their ease of use and diverse availability. (See "Power Transistor Selection" on page 267 for alternatives.)

Frequently, two different power supplies are needed. The LV_{pwr} in Figure 8-2 is associated with the digital power supply (1 V to 5 V typically), and the HV_{pwr} (5 V to >24 V) supplies power to the peripheral device, shown here as R_{load}.

Case C in Figure 8-2 illustrates high-side switching that commonly uses a PFET as the power transistor and an interface circuit to create an appropriate transistor control voltage. The interface circuit, known as a *high-side driver*, is available from multiple vendors.

Case D shows NFET low-side switching to provide a switched ground and a buffer driver to control the gate of the NFET. The NFET may be driven directly from the DCU, but a larger NFET may require a buffer circuit due to the capacitance associated with the NFET gate.

Case E illustrates the use of a high-side driver discussed in Case C, illustrating the single-chip solution to driving a PFET attached to a high-voltage power supply.

Case F integrates both the high-side driver and the PFET into a single chip, commonly known as a *load switch*, available from chip vendors with various current ratings.

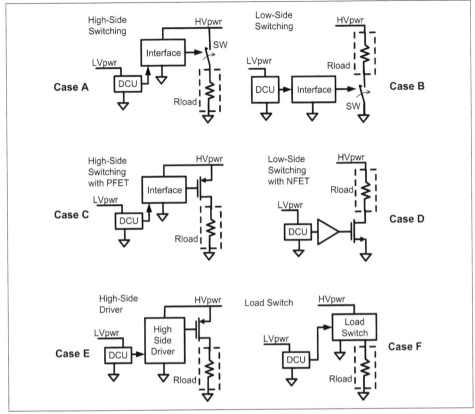

Figure 8-2. High- and low-side power drivers and load switches

Load switch devices are available for both high- and low-side switching and a variety of current ratings. Many load switch devices also include safety circuits to protect from excess current and temperature shutdown.

Either high- or low-side switching can be used in many cases. If the load must remain attached to chassis ground, high-side switching is needed. Low-side switching is simpler, often requiring a single transistor, and if the load can safely be connected to power and disconnected from ground, it is preferred for simplicity.

High-Power Load Isolation

Some high-power devices may need to be electrically isolated from the digital controller, as illustrated in Figure 8-3. High-current loads that create large dynamic current changes can cause the ground to bounce with the current surges. If the ground can be kept stable relative to the DCU side of the system, isolation isn't needed. If one stable ground is not viable, establishing two individual grounds may be needed. To do that, the DCU needs to communicate to the load without a common ground.

The opto-isolator (aka opto-coupler) is most commonly used to communicate without any electrical connection. Digital control passes between the two sides as light, using an LED and phototransistor. When using an opto-isolator, check the specification for:

- Isolation voltage limits
- Current leakage across the isolator
- Capacitance across the isolator
- Insulation resistance
- Switching speed capability

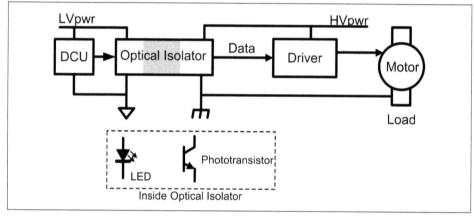

Figure 8-3. Isolation between controller and driver

Drive Signal Strategies

Device driving is commonly done via three different methods: two-state (on-off) switching, linear (aka proportional or analog) control, or pulse width modulation (PWM), all shown in Figure 8-4.

Two-state switching is straightforward. Any of the high-/low-driver methods outlined earlier are suitable. Some devices will have internal driver circuits allowing on/off control with a digital signal.

Linear control uses an analog control signal suitable to the device. Using a voltage output DAC to drive the peripheral is the prevalent method. The DAC may require a buffer circuit depending upon the needs of the peripheral.

PWM uses the duty cycle of on/off switching to achieve a suitable control signal. This method relies on the limited bandwidth of the device being driven to filter the PWM to a desired response. Two common examples of PWM drive control include:

- Motor drive, where the motor's rotational momentum provides a low-pass filter (LPF) for the PWM signal.
- LED display, where apparent brightness is due to the on/off duty cycle of the LED. The eye–brain interface interprets the duty cycle change as a variance in brightness.

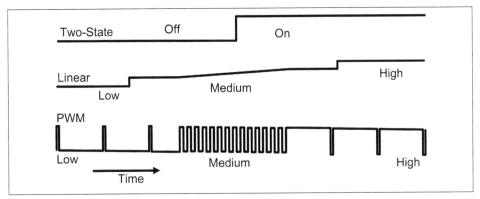

Figure 8-4. Proportional drive methods

A PWM drive should not be confused with a PWM DAC. A PWM DAC includes an LPF creating an analog voltage. A PWM drive requires LPF characteristics within the device being driven.

Power Transistor Selection

For high-current devices, an appropriate power transistor needs to be selected. There are many options (Table 8-1), but for most embedded systems, MOSFETs are easy to work with, cost effective, and widely available in different performance specifications.

Following is a brief look at the different types of power transistors:

Silicon carbide (SiC) transistor
   Designed for high current and voltage, with over 1 KV voltage capability and maximum currents up to hundreds of amps. SiC devices are a valuable part of the high-voltage and power conversion world, but they don't see much use in embedded system design.

Gallium nitride (GaN or GaNFET) transistor
   Well suited to power conversion and other high-current applications. GaN devices exhibit very low resistance, and in some situations they are useful for high-current design.

Insulated-gate bipolar transistor (IGBT)
   Used for power conversion applications.

Bipolar junction transistor (BJT)

Widely available since the late 1950s and still widely used in analog amplification, RF, and other applications. The BJT can be used as a switching device, but its most common modern application is for use with linear amplifiers of many different forms.

Junction field-effect transistor (JFET)

Requires virtually no input current and was popular in instrumentation and audio amplification applications that took advantage of the zero current feature. The JFET is still used in some commercial designs, such as operational amplifiers and audio microphones within an integrated circuit (IC). Its use as a single-transistor device has faded in favor of the MOSFET or a fully integrated device. The JFET has never been widely used as a power-switching device with most commercial devices under 50 mA.

Metal-oxide-semiconductor field-effect transistor (MOSFET or FET)

Available since the late 1960s with many variations in voltage and current capability. For most switching applications, the MOSFET is easily implemented at a low cost. Consequently, MOSFET use is the focus here.

Table 8-1. Transistors: Semiconductor type versus typical performance

Semiconductor type	Maximum Vds or V_{ce}	Maximum Ids or Ic[a]
Silicon carbide (SiC)[b]	600–3,300 V	3–600A
Gallium nitride field-effect transistor (GaNFET)[b]	15–650 V	0.5–90A
Insulated-gate bipolar junction transistor (IGBT)	1–4,500 V	1–550A
Bipolar junction transistor (BJT)	10–1,500 V	10 mA–100A
Metal-oxide-semiconductor field-effect transistor (MOSFET or FET)	10–4,700 V	5 mA–660A
Junction-gate field-effect transistor (JFET) non-SIC	15–50 V	1–50 mA
Values reflect typical range of devices based upon vendor survey		

[a] Continuous current

[b] N channel–type devices only

The electronics industry has not standardized field-effect transistor (FET) device symbols, as illustrated in Figure 8-5. Examples (1), (2), and (3) are negative metal-oxide-semiconductor (NMOS or NFET) devices, and three different symbols are used to represent the same transistor. Examples (5), (6), and (7) are positive metal-oxide-semiconductor (PMOS or PFET) devices and also represent the same device. Examples (4) and (8) show a flyback diode implemented with the power transistor, which is useful when switching current through an inductive load (more later).

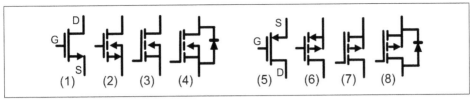

Figure 8-5. MOSFET symbols in common use

Understanding the basic performance criteria for power transistors is essential to selecting a suitable device (Figure 8-6).

Drain current, or *drain to source current* (I_d, I_{ds}), will have a maximum current. Design specifications will define instantaneous surge current capability, as well as continuous currents sustainable by the device. Both maximum current and maximum temperature criteria need to be met to avoid transistor damage.

Voltage drain to source (V_{ds}) is the maximum specified voltage that can be sustained without damage. In most circuits, maximum V_{ds} is seen when the transistor is turned off and the full power supply voltage is across the transistor.

Voltage gate to source (V_{gs}) serves as the control for turning the transistor on and off. For NMOS, when the voltage is raised above a device-specific *threshold voltage* (V_{th}), current will flow from drain to source. Typically, V_{gs} is set to twice V_{th} or more to be fully turned on. PMOS devices work in the opposite manner, with the voltage at the gate being lowered, relative to the source, to turn on. Each device has unique performance parameters, and specification sheets should be checked to determine suitable voltages.

Gate current (I_g) in steady state is typically zero. Inherent capacitance within the transistor (gate capacitance or C_{iss}) will cause I_g surges to charge and discharge the gate capacitance.

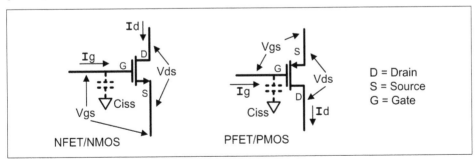

Figure 8-6. Design criteria of power transistors

Some simple examples in Figure 8-7 illustrate the switched characteristics of NFET and PFET devices.

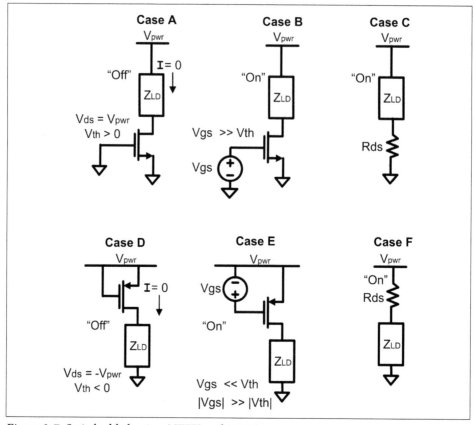

Figure 8-7. Switched behavior: NFET and PFET

Case A shows a low-side driver circuit, where the NFET creates the ground path for the load, Z_{LD}. With the gate voltage, V_{gs}, set to zero, no current flows in the load and the transistor is off. With no voltage drop across the load, V_{pwr} appears across the transistor. The maximum V_{ds} specification for the transistor needs to survive this.

Case B shows the same circuit with a V_{gs} that has been raised well above the threshold voltage. The NFET turns on and current flows through the load.

Case C further illustrates Case B by showing the equivalent resistance, R_{DS}, of the NFET while turned on. This resistance is a function of transistor size, and many different options are available. Knowing the impedance of the load, Z_{LD}, an estimate of the current and a determination of the transistor's maximum current can be made.

Case D shows a high-side driver circuit where the V_{gs} is set to zero and the PFET is off. No current flows in the load. Again, with no voltage drop across the load, V_{pwr} appears across the transistor. The maximum V_{ds} specification for the transistor needs to survive this.

Case E shows a situation where the transistor is on. The voltage at the gate has been lowered so that V_{gs} has exceeded the threshold voltage.

Case F illustrates the on resistance of the PFET and the equivalent resistance, R_{DS}, of the PFET.

These are simplified models, but they suffice for switching power transistors under 1 MHz. With this information, several criteria can be selected for the power transistor:

PFET or NFET
Whether the driver is high side or low side determines the choice of PFET or NFET. Although an NFET can be used as a high-side driver, it requires specialized circuits to create the gate control voltage. Common usage is PFET as a high-side driver and NFET as a low-side driver.

Maximum voltage drain to source (VDS-max)
A transistor specification with a larger $V_{DS\text{-}max}$ allows the device to survive higher voltages, but also leads to a larger R_{DS} for the same size transistor. Consequently, to get the same R_{DS} with a higher $V_{DS\text{-}max}$ the transistor needs to be larger and thus costs more. Therefore, to be cost-optimal, a "just enough" mentality is required when selecting a $V_{DS\text{-}max}$.

$V_{DS\text{-}max}$ for the transistor can be determined by the power supply (V_{pwr}) used. The transistor $V_{DS\text{-}max}$ should be greater than the power supply with room for error and transient behavior. As a conservative starting point for power transistors:

- $V_{pwr} = 1$ V to 3.3 V   Select $V_{DS\text{-}max} > 2 \times V_{pwr}$
- $V_{pwr} = 4$ V to 12 V   Select $V_{DS\text{-}max} > 1.5 \times V_{pwr}$
- $V_{pwr} > 12$ V   Select $V_{DS\text{-}max} > V_{pwr} + 6$ V

Maximum current ($I_{DS\text{-}max}$)
The maximum current for the transistor will be defined by the load. Generally, determining the peak load current and selecting a maximum continuous current twice that peak value is a conservative approach. Manufacturers of semiconductors also produce guide charts for "safe zones of operation" for devices, showing sustained current versus V_{DS} when turned on, and what the transistor is capable of.

Resistance drain to source (R_{DS})
Selecting R_{DS} requires determining the tolerable resistance in series with the load, as well as how much thermal heating occurs in the transistor. A larger R_{DS} causes

hotter transistors and reduces power supply voltage across the load. Also, lower power efficiency occurs due to energy lost in heating the power transistor. Hotter transistors may need a heat sink to keep the transistor within thermal limits. A smaller R_{DS} leads to higher device cost due to the need for a larger transistor.

Gate capacitance (C_{iss})

Large power transistors can have a sluggish response due to large amounts of internal capacitance of the gate connection. If rapid repetitive switching is used, the response time should be checked. When used as a switch, power transistors should be either on or off, with minimal transition time. If a large portion of time is spent in transition, the power transistor can overheat.

Power Transistor Thermal Performance

Understanding thermal heating of power transistors is necessary to determine tolerable values of R_{DS} and whether a heat sink is needed. Transistor heating is a function of the current, R_{DS} resistance, and duty cycle.

Figure 8-8 shows that using duty cycle (D_{cyc}) as part of thermal calculations requires distinct on/off switching of the transistor. When a transistor spends significant portions of time in transition, a simple duty cycle can't be assumed. Slow switching transitions result in periodic higher drain-to-source resistance and a hotter transistor.

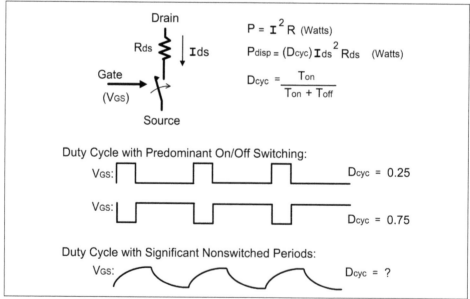

Figure 8-8. Thermal energy losses in switched power transistors

Figure 8-8 shows three cases: two cases show a well-defined duty cycle and the third case shows a problem where the gate driver circuit should probably be revised. Knowing load current, worst-case duty cycle, and R_{DS} allows an approximation of the power being dissipated, P_{disp}, within the transistor. Most power systems and motor devices use switching under 1 MHz, and this simplified model suffices for a first-order approximation.

How hot the transistor gets is also dependent upon the thermal paths available to remove heat from the transistor (Figure 8-9). Depending on the device, most power transistors are functional up to 150°C at the silicon. This is commonly referred to as the *junction temperature* or *die temperature*.

Thermal paths out of the packaged transistor are primarily through the mechanical attachment to the printed circuit board (PCB) or to a heat sink. However, air is a poor thermal conductor, and in some situations large amounts of thermal conduction area created by heat sinks are necessary to dissipate heat into the air. As an alternative, mounting the transistor to the PCB's copper ground plane can function as a heat sink. Depending on the amount of heat generated, both are useful methods. Also, make sure air ventilation is provided to remove heat from the enclosure.

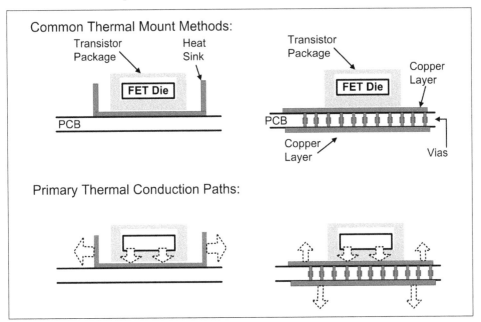

Figure 8-9. Thermal mounts, heat sinks, and conduction paths

Many low-current switches don't require heat sinks, but being able to determine when they are needed is useful (Figure 8-10). Determining junction temperature is a function of thermal elevation above the ambient environment temperature.

Temperature differences across the heat sink and across the transistor package are added to the ambient temperature to determine the temperature of the semiconductor and its transistor junction.

Temperature differences are determined by thermal resistance, often portrayed as a voltage drop across a resistance as an illustrative analogy. Thermal resistance is measured in degrees per watt, and good conductance of heat through a package or heat sink is illustrated by a small temperature difference per watt of power.

Junction Temperature Due to Power Dissipation and Thermal Resistance:

$$T_{jun} = T_{amb} + \Delta T_{hs} + \Delta T_{pk}$$

$$\Delta T_{hs} = P_{disp}\, R_{hs} \qquad \Delta T_{pk} = P_{disp}\, R_{pk}$$

R_{hs} = Thermal Resistance of Heat Sink ($^{\circ}$C / W)

R_{pk} = Thermal Resistance of Package ($^{\circ}$C / W)

P_{disp} = Power Dissipated in Thermal Loss of Transistor (W)

T_{amb} = Ambient Temperature

T_{jun} = Junction (Die) Temperature

ΔT_{hs} = Temperature Difference Across Heat Sink

ΔT_{pk} = Temperature Difference Across Package

Figure 8-10. Thermal calculations for power transistors

If thermal calculations show that the temperature is too hot, there are multiple methods for lowering the junction temperature:

- Select a lower R_{DS} transistor.
- Use a copper PCB layer as a heat sink with vias to multiple copper layers (FDI: PCB).
- Use a heat sink on the transistor.
- Use a lower duty cycle.
- Use two FETs in parallel, splitting the current between transistors. Identical MOSFET devices will split current evenly.

Once implemented, two lab checks should be conducted:

- Thermal imaging of the entire PCB to see if anything is running hot
- Inspecting gate control voltages to determine that on/off switching doesn't include significant soft transitions, as illustrated in Figure 8-8

Driving LEDs and Buzzers

Powering up LEDs and piezoelectric buzzers is commonly done with a low-side driver circuit. Piezoelectric buzzers are straightforward and come in many voltage ratings, and an NFET used as a low-side driver should complete the needed circuit.

Figure 8-11 shows that proper LED driver setup requires attention to appropriate bias currents versus brightness. The first priority is to understand LED bias curves and to get the proper operating point.

LEDs do not have the same bias curves as a silicon rectifier diode. As shown in Figure 8-11, the bias curves for different colored devices are also different. When selecting a device, the data sheet needs to be consulted for bias curves and brightness information.

In addition, achieving equal LED brightness for different colored devices can require different currents. Device specifications will provide brightness versus current information, and tricolor RGB devices often include data to get similar brightness from the three LEDs.

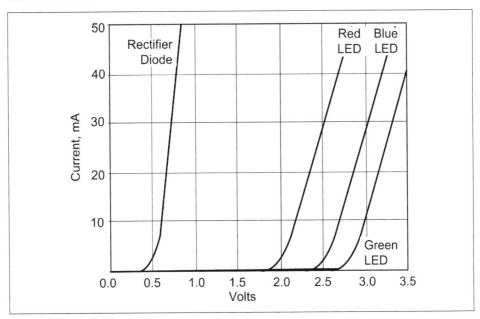

Figure 8-11. LED bias curves

Software adjustment of LED brightness using PWM switching is readily done. Rapid on/off LED switching is interpreted by the eye and brain as a variance in brightness, and not as a flickering/flashing device. Typically, a frequency over 2 KHz for the PWM pattern avoids any flicker effect.

Some common circuits used for LED power are shown in Figure 8-12.

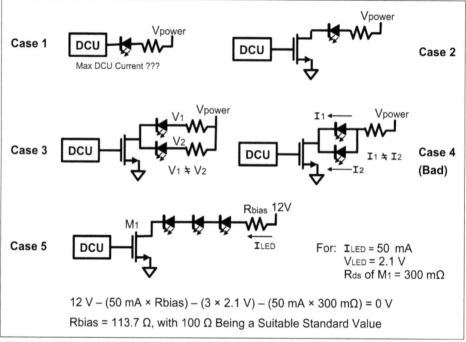

Figure 8-12. LED and multi-LED drive methods

Case 1 shows an LED fed off the power rail with a series current-limiting resistor. The current required for proper brightness and voltage is determined from the LED specification, and the resistor is sized to fit the current and voltage requirement. The possible circuit problem is that the DCU is being used as the grounding switch. LED current can range from 5 mA to 50 mA, and DCU GPIO ports often have maximum current limits under 50 mA. Check the max current capability of the GPIO versus the needed current of the LED.

Case 2 uses an NFET as the grounding switch. This approach doesn't burden the DCU with current loading and allows a higher LED current if needed. This would be the preferred method of driving a piezoelectric buzzer as well.

Case 3 expands on the NFET as a ground switch for multiple LEDs. With separate resistors for each LED, the currents can be selected for different LEDs. This may be useful for running red and green LEDs, set at different currents, to balance brightness.

Case 4 saves a resistor, but due to slight differences in the LEDs, it may result in different currents in the parallel devices and different brightness for the two devices. Device-to-device mismatch can be a problem here. This method is not suggested.

Case 5 explores multiple LEDs in series and puts some specific numbers on a typical circuit. If multiple LEDs need to be powered, stacking them in series minimizes both components and current consumption.

Many different LED driver ICs are also on the market, and some examples are shown in Figure 8-13. Cases 1 and 2 illustrate two of the constant-current drive ICs available. These devices generally have a digital on/off control and a connection for an external bias resistor to define the LED current. Some are high-side drive, some are low-side drive, and many multi-LED variations are also available.

Case 3 illustrates another common option in LED driver ICs where a data pattern is clocked into a shift register that controls multiple LED current sources. The pattern for the LEDs is shifted into a daisy-chained set of chips for large LED arrays, and when latched, the new LED pattern controls the current sources.

Various other chips exist to control arrays and strings of LEDs. Some have intelligent interfaces and serial ports, some include power regulators, and some have programmable PWM brightness capability.

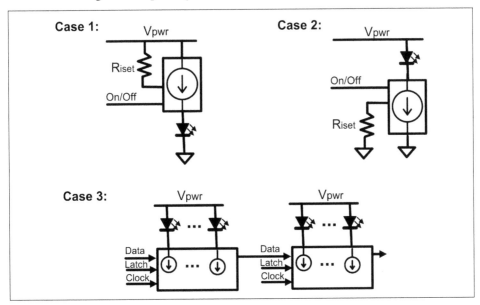

Figure 8-13. Examples of LED driver ICs

Frequently, LEDs and driver chips are integrated into display modules.

Selection of Static Displays

Static displays come in many forms and sizes (Figure 8-14). Color, monochrome, lit, unlit, low-power liquid crystal display (LCD), vivid color OLED, and RGB LED are among the many options.

A typical interface is a serial port and power connections. Displays range in size from freeway billboards to simple segmented number displays. Since displays are static or they update slowly, loading data through a serial port is within the capability of most microcontroller (MCU) and microprocessor (MPU) devices.

Some displays are configured for text or numbers only. The seven-segment display, used for numbers, is still widely used for meters and other simple applications. Dot matrix characters and 16-segment displays are more suited to alphanumeric characters.

Many displays now opt for a full dot matrix of the entire display, allowing either simple monochrome images or alphanumerics. A color display needs data loading for red-green-blue overlays, and higher pixel count devices can support high-quality graphics.

Things to look for when picking a display include:

Viewing angle
> Some displays have a limited viewing angle range. This can be a feature for security systems such as bank ATMs, but it can be a detriment in general-use devices.

Brightness
> Many devices used outdoors don't have displays that are viewable in bright sunlight. In other cases, a display's brightness can be blinding. Consider all expected application environments as part of the selection, and consider whether adjustable brightness should be used in conjunction with ambient light detection.

Versatility
> Pixel array and dot matrix displays are more versatile than alphanumeric character displays. Since devices evolve and change over a long period of use, opting for versatility could avoid hardware revisions over the product's life.

Power consumption
> Displays with bright lights, LEDs, and other features that consume power won't be a good option in battery-powered devices. LCDs use minimal power and are commonly selected when using battery power.

Interface and memory

Different displays can use different serial port protocols to load data. Determine whether the MCU and display have compatible interfaces that are easy to implement, and sufficient memory to support the needed display images.

Update rate

How quickly can the serial port update the screen contents? If the display is used in an interactive manner, screen update time needs to be considered.

Integration with touch screen capability

Touch screens (FDI: Sense) come integrated with displays, so if they are required for the application, a touch screen methodology needs to be considered in conjunction with the display methodology.

Custom versus off-the-shelf (OTS) display

Using an OTS display will get a product to market more quickly and at a lower development cost. Custom designs will be the path of choice when large product volumes can support the up-front costs of a custom design or when a suitable standard product can't be found.

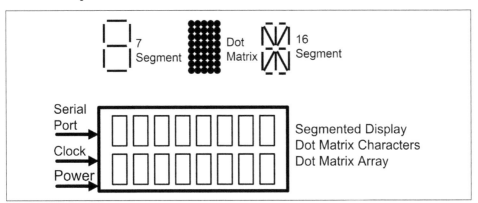

Figure 8-14. Typical LCD/LED/OLED static display

Most controllers readily provide the typical data rates needed to support static images. If the design needs to support streaming video, data processing becomes much more demanding.

Streaming Video Output

Data rates needed to support streaming video are considerably higher than those for static images. For example, the industry-standard digital video interface, HDMI 2.1, streams data at over 40 Gbit/s. Systems supporting both streaming video and MCU task management can be designed, but a simpler approach is available.

A single-board computer (SBC) with operating system software that supports streaming video is easier to implement and lower cost. An MCU to support the embedded system control is linked through a serial bus (CAN bus, RS-485, or other). This approach (see Figure 8-15) uses OTS items for everything except the MCU and peripherals. The effort to develop video software (SW) is sizable, and using an OS makes more sense. A video SW driver is part of any general-use OS, and there's no need to reinvent the wheel.

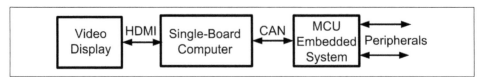

Figure 8-15. Common option to support HR video

Driving displays, lights, and piezoelectric buzzers is straightforward, and digital data control or simple switches get it done. Many other loads have important issues associated with their transient behavior that need to be better understood.

Driving Inductive Loads

Many switched loads have inductance as part of the load. This includes motors, solenoids, relays, speakers, and others.

Transient Current in a Switched Inductor

Switching current through an inductor (Figure 8-16) can create destructive transient voltages. To illustrate this, (A) in Figure 8-16 shows a low-side driver with an inductor as the load, and (B) shows a simple model for the circuit. Resistance sources are R_{DS} (transistor turned on) and R_L (resistance of the inductor).

Using the simplified model of (B), the circuit behavior can be readily examined. Examining the behavior at inductor (D), when the switch is closed (t_1) the current builds in the inductor (t_2) until it maximizes (t_3) due to series resistance. When the switch is opened (t_4) the negative di/dt on the inductor creates a large transient voltage across the inductor.

At transistor (E), the transient voltage created by the inductor results in a large positive voltage spike, V_{DS}, across the transistor. Typically, the transistor is destroyed due to excess voltage.

The (C) illustration introduces a diode across the inductor. The diode is off during normal operation of the circuit and does not conduct current. When the transistor turns off (E, t_4), the resulting voltage spike turns the diode on, thus restricting the voltage created by the inductor.

The V_{DSD} (with diode) becomes limited to (V_{pwr} + V_{diode}), thus protecting the transistor from overvoltage damage. This diode is referred to as a *flyback diode* (aka snubber diode, freewheeling diode, commutation diode, or clamp diode). For a single inductor with current flowing in one direction, a single diode suffices. Bi-directional current in an inductor may require diode protection for both positive and negative spikes. Any inductive load requires flyback diode protection for the associated driver transistors.

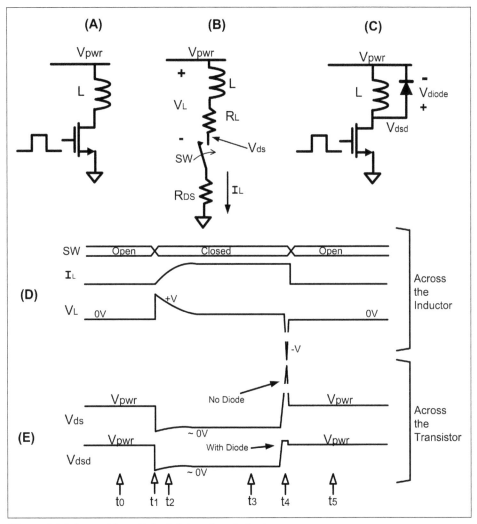

Figure 8-16. Low-side switched drive of an inductor

Driving Solenoids and Relays

Relays and solenoids utilize similar driver circuits. Both are electromagnets that create mechanical motion. Solenoids move a mechanical actuator, and relays open/close a set of electrical contacts.

Recognize the need for a flyback diode, and the controlling relays and solenoids are straightforward, as shown in Figure 8-17. When selecting the components:

- Determine the static current of the inductor when turned on.
- Select a power transistor with suitable current and open circuit voltage, V_{DS}.
- Check the R_{DS} of the power transistor and determine if heating is an issue.
- Check that the maximum current of the flyback diode can handle the static current that was in the inductor.
- Include some design margin in the selections.

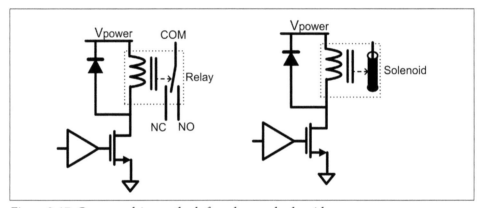

Figure 8-17. Common drive methods for relays and solenoids

Relays and solenoids require constant current to maintain an active state. There are strategies to apply full current during the transit time (pull-in current) and then reduce the current (hold current) while maintaining the active state. Relay-solenoid driver chips exist with fully integrated capability for pull-in current and controlled holding of PWM current, completed with the necessary power transistor.

PWM techniques can be used to reduce the average current, but there are often other options that can be used. For example, consider removing the relay from the design. Relays can suffer from mechanical failure, often due to burnt electrical contacts. Also, relays require constant current to keep them activated. With any mechanical device, always determine if there is an electronic method instead. Frequently, an electronic method is both less expensive and more reliable.

To switch DC power, a transistor switch will usually get the job done better than a relay. Any switching DC power supply can be configured to use a low-current on/off control line. If AC power needs to be switched, a Triac circuit can be utilized (see ON Semiconductor, 2006, in "Further Reading" on page 297). Solid state relays are another alternative. However, these devices tend to be more expensive than the power transistors and isolator circuits inside them.

The constant current of a powered relay/solenoid is not attractive for low-power systems. If they can't be designed out, consider using a bi-stable (aka latching) device (Figure 8-18). Bi-stable devices don't use continuous current to maintain state. Rather, a brief (typically <100 ms) pulse of current changes the state. With zero current needed to hold state, these devices consume less power. However, there is no "default state" when the control electronics shut down. That needs to be considered as part of the safety review of the device in the system.

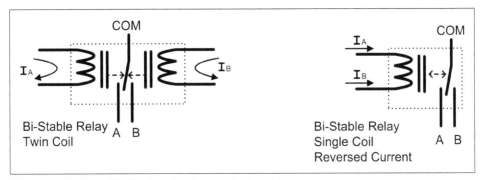

Figure 8-18. Bi-stable (latching) relays and solenoids

To function, bi-stable devices require either two coils, or the capability to reverse current in one coil. The capability to reverse current requires using an H-bridge drive circuit.

H-Bridge Drive Circuits

All prior circuits are capable of creating current in the load in only one direction. Many devices require the capability to reverse the current direction through the load. This requires an *H-bridge drive circuit*.

H-bridge drivers are widely used in the control of brushless motors, reversing the rotation of DC motors, stepper motors, servomechanisms, latching relays/solenoids, speaker drivers, and others. It is a very useful circuit in motion control, and a closer look is valuable.

As shown in Figure 8-19, the capability to reverse current in a load hinges on the sequencing of four switches:

- When A and D are closed and B and C are open, positive current flows through Z_{LD}.

- When B and C are closed and A and D are open, negative current flows through Z_{LD}.

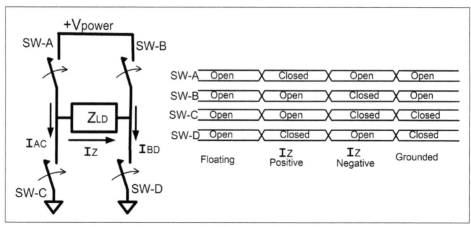

Figure 8-19. Bi-directional current in load using H-bridge driver

Problems can occur near the time points when the switches transition. In Figure 8-20, the phase relationship of the switches is skewed slightly:

- At t_0, switches are lined up ideally.

- Between t_0 and t_1, current flows properly through the load.

- At t_1, both A and C are on together and a current spike flows through the I_{AC} path.

- At t_2, B and D are closed at the same time, causing current through the I_{BD} path.

This phase relationship is known as *make-before-break switching*, and the undesired current spikes are referred to as *shoot-through*. At t_3, the switching phase relationship closes switches slightly later than opening them, thus eliminating the undesired current spike. This *break-before-make switching* is essential.

The phase relationship between the top and bottom switches needs to be carefully controlled to avoid shoot-through. Considering that the power supply is being short-circuited to ground through two power transistors, the outcome can be destructive.

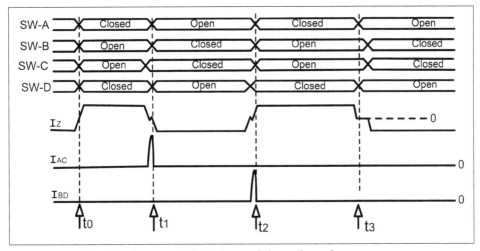

Figure 8-20. H-bridge driver: switch timing and shoot-through

Using FET switches to implement the H-bridge is shown in Figure 8-21. Implementing the driver circuit with NFETs on the bottom and PFETs on the top is straightforward. Some supplemental items are shown here to complete the circuit.

Typically, a pair of top and bottom transistors (M_A and M_C or M_B and M_D) are controlled together to maintain a proper break-before-make phase relationship. A control chip, known as a *half-bridge driver*, is used. Common features in bridge driver chips include:

- Sufficient drive strength to quickly charge/discharge the gate capacitance of large power transistors
- Control signal level shifting up to the positive power rail, suitable for PFET gate control input
- Adjustable phase delays to define break-before-make timing

Two half-bridge driver circuits are often bundled into a single IC, and are then known as a *full-bridge driver circuit*. Also, complete H-bridge devices are now offered with bridge drivers and power transistors on a single chip.

Following are other important circuit considerations:

- An inductive load (e.g., a motor) is shown at (3), with an RC snubber (4) to reduce EMI caused by switching transients (FDI: EMI & ESD).
- Flyback diodes are placed across all four transistors. This limits the voltage at both ends of the inductor (3) to protect the transistors from overvoltage spikes. Without the diodes, both positive and negative voltage spikes can occur due to current switching in both directions.

- Depending on the application, a current sense resistor (5) may be used to support the feedback control circuit of the device. The typical value for current sensing is under 1 Ω, which allows creating a current sense voltage—usually under 100 mV, seen at (6)—while not interfering with operation.

If current sensing isn't needed, the source pins of the two NFETs go to ground.

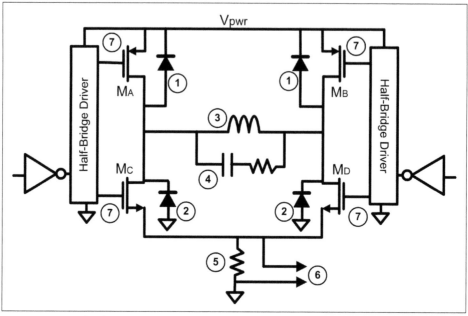

Figure 8-21. H-bridge circuit

Driving DC Motors

DC motors are available in multiple formats: motors with commutation brushes, motors that require external electronics to commutate the motor, stepper motors, and voice coil motors are the most common. Depending on motor type and application, different drive circuits are suitable.

Motor Selection

Brushed DC motors (BDCMs) have been in common use for many years, and are inexpensive and easy to control. BDCMs are available with supply voltages from 1.5 V to 32 V, with power ratings from 0.1 W to 35 W commonly available.

BDCMs can be electrically noisy due to arcing brushes, and brushes can wear out with long-term use. For consumer products where low cost is the priority, brushed motors will frequently suffice. Industrial and commercial applications where reliability is a top priority should probably review long-term wear issues when considering a BDCM.

Brushless DC motors (BLDCs) use electronic switching for motor commutation instead of brushes. This improves motor reliability. Many BLDC motors use a three-phase structure and have low torque variation, which can be useful in many applications.

AC motors are designed to run off AC mains power. Both single- and three-phase motors are widely manufactured. Mains voltage from 48 to 575 VAC is available. Motors up to 500 kW power are produced in volume. Unless using a specialized power inverter to change the AC frequency, AC motors are designed for fixed rotational speeds, dependent upon the AC mains frequency.

Some motors will be power-rated in watts; others use horsepower as a rating factor. Typically, a one-horsepower motor equates to 746 watts. Adding to the confusion, an actual horse can generate up to 15 horsepower for brief periods. Torque specifications are a better indicator of suitability than either watts or equestrian-related measurements.

Items to consider when selecting a motor include:

- Type—AC single phase, AC three phase, BDCM, or BLDC
- Rotational speed (RPM)—fixed, variable, or min/max
- Rotational direction or bi-directional capability
- Continuous torque
- Starting torque
- Torque variation as a function of rotation angle
- Supply voltage
- Rotational encoder for angle and speed
- Integrated or independent control electronics
- Braking system
- Mechanical issues—size, mounting, shaft type, and/or reduction gearing
- Environment issues—sensitivity to corrosion, moisture, and/or temperature
- Internal bearings—ball bearing, brass sleeve
- Expected life

Brushed DC Motor Driver Circuit

Driving a brushed DC motor is straightforward, as shown in Figure 8-22. A BDCM requires an NFET as a low-side power switch. Frequently, a rotational speed sensor is added, allowing speed feedback to the controller. The NFET serves as a low-side driver, and speed is controlled via PWM control of the NFET. The mass momentum of the motor acts as an LPF to the PWM switching, and speed becomes proportional to the PWM duty cycle (FDI: Control).

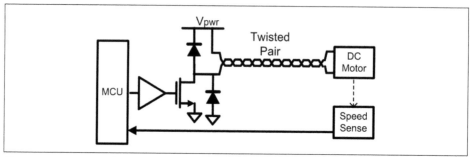

Figure 8-22. Typical brushed DC motor driver

Low-cost motor circuits commonly allow putting energy into the motor and rely on frictional losses and shaft load to slow the device. Consequently, there is no need to reverse motor current. The wires to the motor are configured as a twisted pair to minimize EMI (FDI: EMI & ESD), and flyback diodes protect the NFET from going too far beyond the power/ground voltages.

Brushless DC Motors: Single and Three Phase

BLDC motors come in single-phase and three-phase variants, shown in Figure 8-23. Both require the capability to reverse current in the windings. A single-phase motor uses an H-bridge circuit, as described before. Three-phase motors use a variant on the H-bridge, using three sets of power transistors instead of two.

In addition to power transistors driving the motor windings, information is needed to determine where the motor is rotationally. This phase information comes from an encoder of some form, and is commonly acquired with a Hall sensor (FDI: Sense) built into the motor. This is used to ensure that the proper motor winding is powered during specific phases of the motor rotation.

Three-phase motors can be implemented in either a delta or a Y (aka Wye) configuration. Smaller motors are usually implemented with a Y configuration, and the delta configuration is more commonly used with large industrial motors powered directly from AC power mains.

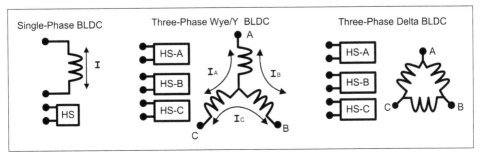

Figure 8-23. Typical single- and three-phase brushless DC motors

The motor is generally controlled and driven with a *motor driver controller*; a dedicated IC designed to interface with the motor. A typical motor with support electronics is shown in Figure 8-24. Three sets of power transistors provide the power switching. A power transistor gate drive is provided by three half-bridge drivers built into the IC. The motor controller chip monitors the rotational phase through three Hall sensors installed in the motor. Current sensing is through a grounded set of sense resistors.

Motor control by the host digital controller comes through a digital serial port and other digital control connections. Most motor control ICs use this architecture or a similar variant (FDI: Hughes & Drury, 2019).

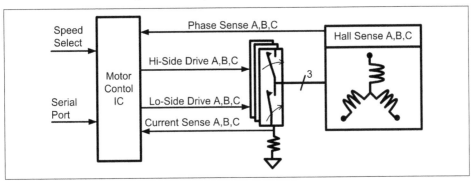

Figure 8-24. Brushless three-phase DC motor with support electronics

Motors with Integrated Control Electronics

Figure 8-25 shows another option: using a motor with fully integrated electronics. In this example, a motor system includes a PCB bolted onto the motor. The motor controller IC and power transistor circuitry has been included in the motor enclosure, and the interface is digitally controlled from an MCU.

Although a fully integrated motor simplifies the design effort, it comes with two drawbacks. First, a high cost for full integration is typical. These devices cost well beyond the added expense of the electronics. A lower-cost solution puts motor driver circuitry on the MCU PCB. Second, motor selection without integrated electronics is extensive and diverse. There are considerably fewer options with fully integrated electronics.

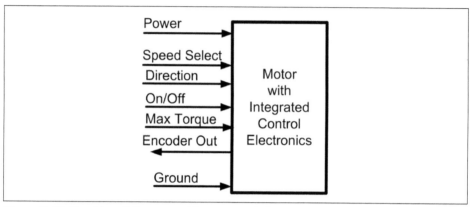

Figure 8-25. Brushless DC motor with integrated electronics

 For a high-volume product, a fully integrated motor is not cost effective. As a general rule, always ask motor vendors for their documents on suggested drive and control electronics. Typically, vendors have electronic design solutions ready to go for their motor.

Stepper Motors

Stepper motors (aka steppers) make discrete individual steps of motion. One example is a rotary stepper motor designed with 200 steps per rotation. This motor will move 1.8 degrees per step and is very consistent and repeatable. This well-defined motion is popular for use in robotics, industrial automation, and any application requiring motion steps that are easy to control.

Stepper motors also come in linear versions to provide straight line motion, but rotary stepper motors predominate. Linear motion from a rotary stepper is often accomplished with mechanical methods, such as a lead screw, rack and pinion, or other technique.

Figure 8-26 shows a simple illustration of how steppers operate internally. The rotating body (rotor) in the center of the stepper motor can be either a permanent magnet, or a ferrous element that is attracted by a magnetic field. Selective powering of the stator coils creates the magnetic fields needed to move the rotor.

The example in the figure shows a four-state sequence and how the rotor moves depending on which stator coils are powered. As shown in the example, the stator coils require the capability to reverse the magnetic field. To do so, the current needs to be reversed in the coil (unipolar), or a second coil (bipolar) on the stator needs to be used.

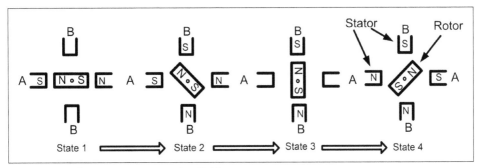

Figure 8-26. Stepper motor rotating field

The differences between unipolar and bipolar winding of a stepper are illustrated in Figure 8-27. Unipolar devices typically connect the center tap (CT) to the positive supply voltage while selectively grounding either the A1 or A2 connection to activate the stator. Bipolar devices require the use of an H-bridge driver circuit to reverse the current and thus reverse the magnetic field.

Bipolar devices will generally offer more torque than unipolar devices, since the entire coil is used when activated. Unipolar devices only use half of the coil winding at a time. Bipolar stepper motors occupy about two-thirds of the market as of 2023.

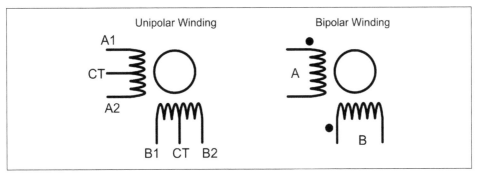

Figure 8-27. Unipolar versus bipolar winding

Stepper motors are very repeatable and accurate in their position, but the transitions can exhibit mechanical resonances and ringing effects, as shown in Figure 8-28. When taking a step, the motor can overshoot and a time period is needed to settle to the final position. For many applications, this is not a problem. Typically, the magnitude of the ringing is much smaller than the step size. Reduction of stepper ringing is commonly done via two methods. When viable, waiting for the device to settle before taking further action is one method. Also, inertial damper devices are available to bolt onto steppers and soften the step response.

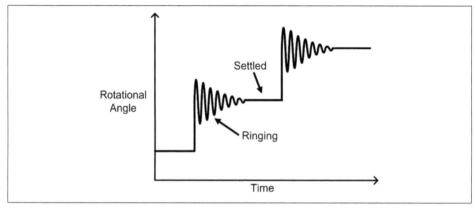

Figure 8-28. Typical stepper motor motion

Figure 8-29 depicts the process of driving a bipolar stepper motor. H-bridge driver circuits are used and the rest of the control system is digital. Although stepper motors can be run without feedback control, an encoder (FDI: Sense) is often added to confirm motion, determine a home position, or determine whether enough time has passed that the motor has settled into the new destination location.

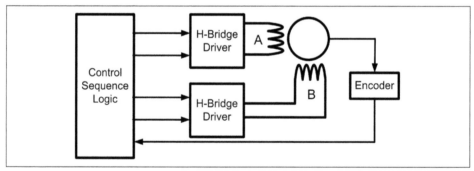

Figure 8-29. Typical stepper motor system

If high-speed motion with no stepping is needed, a voice coil motor may be necessary.

Voice Coil Motors

The linear voice coil motor is very similar to the coil magnet setup used within an audio speaker, as shown in Figure 8-30. Voice coil motors come in both linear and rotary variants. These devices can be very fast and precise, but they require a feedback and control system that monitors location, velocity, and acceleration to give optimal performance (FDI: Control).

The rotary voice coil motor has been used as the head positioning servomechanism within hard disk drives since the 1980s. The linear voice coil motor has been used in fast-focusing optics, and any application where small size and fast adjustments are needed.

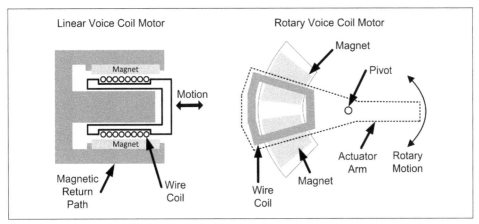

Figure 8-30. Linear and rotary voice coil motors

H-bridge driver circuits are typically used to drive a voice coil motor.

Stall Currents and Protecting from Self-Destruction

Before leaving the topic of motors and other electromechanical actuators, a brief review of safety issues is appropriate. Some devices are low power and safety is not a concern. Others can be high power and destructive if not properly controlled.

In addition, mechanical stalling out or jamming situations can lead to power overloads, smoky electronics, and burned-out motors. Consequently, some attention has to be given to safety controls. Conceptually, Figure 8-31 shows the controls to consider including.

Forcing a safety shutdown before damage occurs can be accomplished in multiple ways. A power cutoff is a first possibility. Power cutoffs (fuses, circuit breakers, or PTC devices) respond to sustained overcurrent situations and respond with an open circuit. However, this method is too slow in some situations.

Safety sensors on the motor and power circuitry can include:

Limit switches for servo actuators
    Going beyond the allowed range of motion triggers a shutdown.

Fast-response current sensors
High current often indicates a stalled motor.

Temperature sensors
High temperatures often indicate problematic bearings or other mechanical loading that is not part of normal functionality.

Rotational speed detection
An encoder to monitor motor speed is a simple tool for determining proper functionality.

All safety sensors can send shutdown/safety requests to the DCU, or directly shut down the power drive circuits. Many motor controller ICs include safety shutdown methods inherent to the chip design.

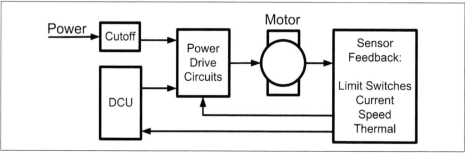

Figure 8-31. Protecting motors from self-destruction

If the mechanics of the system don't work properly, the outcome needs to be analyzed for possible safety problems. Methods to avoid mechanical self-destruction or damaging the power system need to be part of the design. In addition, if software control is lost, direct shutdown methods that are not DCU dependent can be invaluable.

Figuratively or literally, throw a wrench in the gears and make sure the system can safely shut down.

Audio Outputs

Audio outputs are frequently a system requirement. Modern devices are commonly configured with a "digital to the speaker" strategy. This entails sending digital data to a chip designed to drive a speaker or headphones. Analog signals only occur within a single chip and the speaker.

If the application is low power, such as headsets and small speakers, the architecture shown in Figure 8-32 is common. There are variants on this, but a digital data stream, processed through a DAC, filter, and amplifier, is commonly integrated onto a single chip. Many of these use a digital audio data stream format known as the I²S standard (Inter-IC Sound standard).

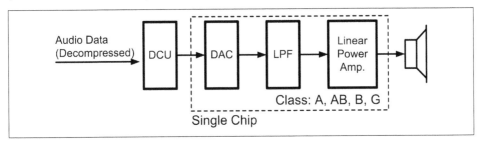

Figure 8-32. Typical data to audio, classic linear

Audio amplifiers can consume large amounts of power, and various amplifier types are used to improve efficiency. Low-power amplifiers use linear methods (most commonly Classes AB and G) while higher-power devices resort to a switching amplifier (Class D) approach, as illustrated in Figure 8-32.

Conceptually, the top diagram in Figure 8-33 shows the internal functions of a digital audio power amplifier. The approach used involves a data stream for the audio signal, which is converted to a PWM format. The PWM signal controls a switching amplifier, commonly implemented with an H-bridge driver circuit. The switching amplifier output drives an LPF connected to the speaker. A single-chip implementation is shown in the bottom part of the figure, with an I²S data stream input and the use of an LC LPF driving the speaker.

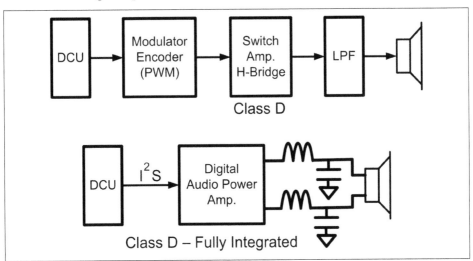

Figure 8-33. Typical data to audio, digital/switched methods

Summary and Conclusions

This chapter is a survey of some of the many devices an embedded system can be called on to drive and control peripheral devices. For any of these devices, consulting with the device vendor can often yield documentation on suitable driver circuits. As well, dedicated chips for various driver tasks are often supported with application notes on how to set up and configure these devices for proper functionality.

Remember that:

- High-side driver and bridge driver circuits are available as an SoC, with no need to design from scratch.

- Many transistor options are available, but MOSFET devices can be utilized for most embedded system applications.

- If an R_{DS} transistor creates a hot transistor problem, seek a lower R_{DS} device. If a suitable MOSFET doesn't exist, a GaNFET may have the desired low R_{DS}.

- Examine the gate control signals of all power transistors to determine that duty cycle switching is primarily on/off with minimal time spent in the linear regions.

- Check power transistors for acceptable temperature elevation, both during design (calculations) and after implementation (thermal imaging).

- Diode bias curves are not the same among LED devices, and a rectifier diode has different voltages than an LED.

- Inductive loads of solenoids, relays, and motors all require flyback diodes.

- H-bridge circuits require careful timing coordination to create break-before-make control and avoid high-current shoot-through.

- Stepper motors may require settling time or mechanical damping to mitigate mechanical ringing.

- Safety shutdown methods should be considered for all motor systems to avoid self-destructive behavior when mechanically stalled or jammed.

Further Reading

- "Power MOSFET Selecting MOSFFETs [*sic*] and Consideration for Circuit Design," Toshiba Electronic Devices & Storage Corporation Application Note, 2018, *https://toshiba.semicon-storage.com/info/docget.jsp?did=13416*.

- H. Lin, C. Chung, and J. Chen, "Understanding Power MOSFET Parameters," AN-1001, Taiwan Semiconductor, *https://taiwansemi.com*.

- James Bryant, "Choosing Discrete Transistors," Analog Devices Wiki, May 19, 2014, *https://wiki.analog.com/university/courses/electronics/text/choosing-transistors*.

- Jerad Lewis, "Common inter-IC digital interfaces for audio data transfer," EDN, July 24, 2012, *https://www.edn.com/common-inter-ic-digital-interfaces-for-audio-data-transfer*.

- *Electric Motors and Drives: Fundamentals, Types and Applications, 5th Edition*, by Austin Hughes and Bill Drury, ISBN-13 978-0081026151, 2019, Newnes-Elsevier.

- *Switching Power Supply Design, 3rd Edition*, by A. Pressman, K Billings, and T. Morey, 2009, ISBN 978-0-07-148272-1, McGraw Hill (see Chapter 9: "MOSFET and IGBT Power Transistors").

- "Choosing between Brush and Brushless DC Motors," Allied Motion, *https://www.alliedmotion.com*.

- "Thyristor Theory and Design Considerations Handbook," HBD855/D Rev. 1, ON Semiconductor, 2006, *https://www.onsemi.com*.

CHAPTER 9

Sensing Peripheral Devices

Sense circuits convert real-world analog information and events to digital data suitable for digital control unit (DCU) processing. Figure 9-1 shows an example of using a motor under digital control. This figure illustrates the sensor possibilities for just a rotating motor. Depending upon application and system needs, a motor can have multiple sensors to determine the state of the motor, including:

- Power supply voltage
- Rotational speed
- Output torque

- Current
- Rotational phase
- Temperature

Sensors are used to both control the system and monitor its health.

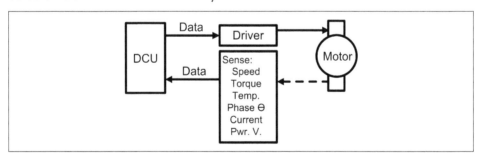

Figure 9-1. Sensor systems on a motor

Suitable sensors are available for all of these motor performance monitors.

Sensors for Everything

Sensors are the path through which real-world physical phenomena are represented by equivalent electrical signals. For embedded systems, sensor information will be converted to digital data.

Many older electronic systems would directly use an analog sensor output as part of a control system. Modern systems primarily digitize the sensor output for use by a digital control protocol.

Sensors have been developed for many different things, including:

- Fluid levels in storage tanks
- Flow rates in pipes
- Physical force/strain/torque
- Specific gas detection
- Environment humidity
- Magnetic field
- Digital compass orientation
- Acceleration/velocity/motion
- Avionics: pitch/yaw/roll
- Tilt
- Vibration
- Image sensors: still image/video
- Distance ranging
- Optical: light, infrared, specific frequency/color
- Particle-dust counters
- Smoke detection
- Gas pressure
- Angular orientation
- Linear position
- Proximity
- Shock force
- Temperature
- Radiation
- Acid-base pH scale
- Sound

Some sensors include internally implemented signal processing and quantization to create a digital data output. Other sensors provide an analog output, requiring support electronics to produce data useful to a digital system.

Sensor Output Types

Although sensors exist that respond to a diverse multitude of stimuli, most sensors respond with a limited number of output types. Depending on the output type, different sensor devices can use similar signal processing electronics. Figure 9-2 summarizes the different output types.

Two-state sensors are simple in concept. If some real-world stimulus exceeds a sensor-defined boundary, then the output changes state. Examples here include motion beyond a limit, exceeding a flow rate maximum in a pipeline, air pressure going below a defined minimum, a tilt sensor that indicates a machine has tipped over, and others. With two-state sensors, depending on the nature of the event, the DCU may want to observe the output transition for a limited time period to determine that the output is stable and not transitory. This weeds out glitches, noise, and electromechanical switch contact bounce.

Pulse counting of events is another common sensor output. Some type of event is sensed as singular events, and the useful information is embedded in how many events occur over time. The events may or may not be evenly spaced. The Geiger counter used to detect radiation is an obvious example where counting discrete ionization events yields radiation information. Another example is a sensor used to control loading devices into a container, such as counting a set number of boxes to put onto a pallet in a factory automation scenario.

Frequency sensors output a stream of pulses or a continuous frequency that can be correlated back to the stimulus. This requires both an event counter in the DCU and a reference clock to create a defined time period. With both, accurate frequency can be determined.

Sensors producing a frequency output have better noise immunity than devices that produce an analog voltage. Some older-technology sensors used a voltage-to-frequency converter within the sensor itself for better noise immunity. Other sensors may create a change in inductance or capacitance, which can be inserted into an oscillator circuit. The output of the oscillator frequency then represents the stimulus change.

The output of *proportional analog-linear* sensors is proportional to the stimulus. A linear response of stimulus to output is fairly easy to process. The output can be a voltage, current, resistance, inductance, or capacitance (VIRLC). These devices need signal processing to create useful data for a DCU. Variable L or C sensors are often converted to a frequency using resonant oscillator methods.

Devices that output a current are often converted to a voltage using an accurate reference resistor. Variable resistance sensors can be converted to a voltage through use of a reference current, or can be connected into a voltage divider with a fixed-value reference resistor.

Voltage output devices can be directly connected to an analog-to-digital converter (ADC), or may require amplification for sensors with low-amplitude outputs before connecting to an ADC.

Proportional analog nonlinear at extremes sensors are proportional to the stimulus but only over a limited range. These devices are dealt with in a similar manner to the linear devices while recognizing the limited range of the sensor. Data is no longer accurate (or destructive to the sensor) beyond certain limits.

Proportional analog nonlinear devices respond on some form of a curve, so a straight line approximation will not yield useful information. Consequently, using these devices requires curve-fitting to multiple data points or some form of look-up table for converting analog data to physical data.

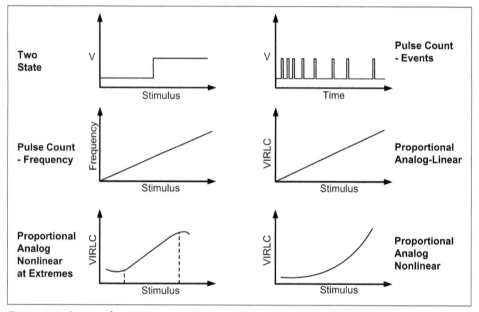

Figure 9-2. Sensors by output type

Sensor Data Capture and Calibration

The goal is to get accurate and reliable digital information from the sensor back to the DCU. Typically the sensor is connected to an ADC, and the following sensor discussions illustrate many of the useful techniques available to do this.

Data Capture Method

Sensor data capture is dealt with in several different ways:

- Interrupt-driven data capture is configured so that the sensor generates a request to be read by the DCU. This strategy is used when the stimulus to the sensor changes infrequently. An example would be some type of fault sensor that rarely asserts in normal device operation.

- Periodic polling of the sensor by the DCU can be used to create a data stream when the exact timing is not critical. An example would be a temperature sensor where data is collected every five minutes. Since exact timing is not critical, the DCU can request the data as a low-priority task.

- Streaming continuous data is used for faster sampling, or when sample point timing needs consistent accuracy. A communication channel, audio input, or any situation where the input is constantly changing requires continuous sampling and ADC conversion controlled by a reference clock.

How time-critical the data is or how frequently the data changes will determine the strategy here.

Sensor Calibration

Some sensors will need no calibration, but others require some way to determine one or more data points. Many sensors vary with use, age, and temperature. Depending on the needed accuracy, variance can be calibrated out, or it can be tolerated within an error budget. Understanding sensor variance and the tolerable error will define whether calibration is needed.

Sensors with a linear response often calibrate using two data points at min/max extremes of operation and then interpolate the data points in between. As an example, flatbed optical scanners calibrate black and white image extremes as part of their startup strategy. Many other devices also have calibration built into their systems.

When implementing any sensor in a system, the need for calibration and the method to be used must be examined as part of the design.

Sensor Response Time

Depending on what's being measured, physical phenomena have different response times. Any sensor system should be designed to respond in a manner similar to the physical phenomena. If the stimulus changes in a matter of minutes, a sensor system response of nanoseconds is not needed, and often not desirable.

Overly fast devices can respond inaccurately to out-of-band noise. In many cases, data output can be averaged over multiple digital samples to mathematically reduce sensor bandwidth and improve noise immunity where needed. Conversely, if the stimulus includes rapid transients, the sensor needs to be fast enough to track the stimulus, or stimulus information will be lost.

Two-State Devices: Switches, Optical Interrupters, and Hall Sensors

The mechanical switch as a sensor (Figure 9-3, Case A) can be problematic, especially with high-speed use or large amounts of repetitive use. The problem with switch contact bounce was covered earlier (FDI: Essentials), and switches have a limited life span due to wear. Issues of corrosion and burnt contacts are also problems. For a sensor, there are often more reliable options.

The mechanical switch can often be replaced by an optical interrupter. As shown in Figure 9-3, Case B, an LED is used to illuminate a phototransistor and turn it on. A physical light interrupter can be moved between the LED and phototransistor to turn the device off. The interrupter is attached to the mechanics of the system, but otherwise there are no mechanical pieces to fail. These devices are widely used, are very reliable, and can function at higher rates of operation than mechanical switches. Optical interrupters can have problems in dirty environments when particulate matter blocks the light path. Also, the optical interrupter requires the use of power while being operated, so it may be a consideration in ultra-low power consumption.

The reed switch (Figure 9-3, Case C) is a variation on the mechanical switch, where the electrical contacts are encased in a sealed glass tube. The connection is controlled by externally applying a magnetic field that moves the contacts. Devices are available with both normally open or normally closed contacts without the magnetic field.

Reed switches were used in old analog telephone networks to route phone calls. More recently, the use of reed switches has included laptop computers and flip phones to allow on/off power control with no static power consumption. Other uses of magnetic reed switches include alarm security systems for doors and windows. The reed switch is still useful as a low-cost device where zero power consumption is important.

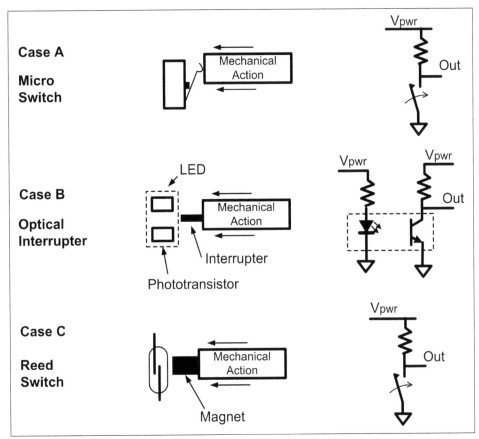

Figure 9-3. Switches, optical interrupters, and reed switches

As shown in Figure 9-4, the Hall effect sensor can also be used as a two-state position sensor, but it is completely electronic, with no mechanical parts to break. A Hall sensor works by using electromagnetic methods. When a current (I_{Bias}) flows through a conductor placed in a magnetic field, the path of the current is bent due to the Lorentz force. The bent current creates a voltage (V_{Hall}) across the conductor that is proportional to the strength of the magnetic field. Hall sensors can be used to quantify a magnetic field (denoted as Analog Output in the figure) or amplified and turned into a digital control signal (denoted as Digital Output in the figure).

A fully integrated Hall sensor appears very simple from the outside: power, ground, and output are the only needed pins. The Hall sensor can be used to detect the presence of a moving magnet, or detect the presence of a piece of ferrous metal such as a steel gear tooth.

Hall sensors are simple, reliable, inexpensive, and widely used. Most electric motors that need to provide rotational position information use Hall sensors. Because Hall sensors work reliably in dirty environments, such as automotive and other machine systems, they enjoy widespread use where optical sensors can't function.

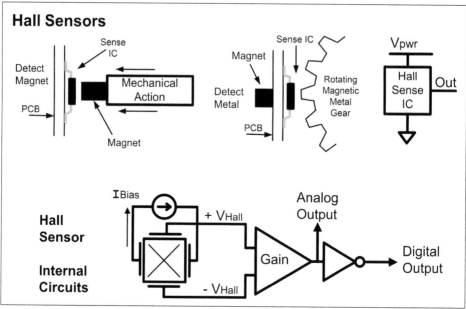

Figure 9-4. Hall effect sensors

Optical sensors and Hall sensors are suitable for most switch-type sensing situations while providing a reliable and low-cost solution. Depending upon the application, any mechanical contact switch should be evaluated for repetitive-use failure. Frequently, mechanical switches are more expensive and less reliable than electronic sensors. Typically, all electronic sensors respond more quickly than mechanical options, but any sensor needs to be checked for response time when used in a high-speed on/off scenario.

A two-state detector is often employed in an encoder that yields position or rotation information.

Position and Rotation Encoders

An *encoder* is a device that yields digital data showing its position: either along a linear axis or the rotational position (Figure 9-5). That real-time position information can be used with a time reference to also determine speed and acceleration.

The encoders shown in Figure 9-5 use a single sensor, which is the most common implementation of an encoder. However, the single sensor is limited to relative position only. To determine absolute position, a method to figure out the start location (aka home position) can be used while counting sensor pulses to see how far the device has moved from the home position. Many electromechanical systems will drive an actuator against an end stop as part of a start location calibration process.

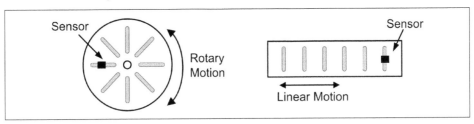

Figure 9-5. Encoders for linear and rotational data

When absolute position needs to come from the encoder, other options are available. Figure 9-6 shows three options. Case A (Encoder – Relative Position) uses a single sensor (aka incremental encoder) that indicates relative motion. When used with some method that determines home location, it can also find the absolute position by counting transition pulses as the device moves away from the home location.

Case B (Encoder – Absolute Position, Binary Code) uses four sensors to create a binary code showing the absolute position. A binary output like this can be problematic, because multiple sensor outputs can change at the same location. This can result in some unpredictable outputs, especially near a transition point. Consider the state change from 0111 to 1000 where all four outputs are changing at the same location. If data is sampled at the transition point, the location information can be in error.

Case C (Encoder – Absolute Position, Gray Code) circumvents the binary code problem by using a Gray code sequence instead of a binary sequence. A Gray code sequence is constructed such that any transition involves only a single bit change. By doing this, location errors are limited to a pair of adjacent locations. Encoders capable of absolute position are commonly done with Gray code methods.

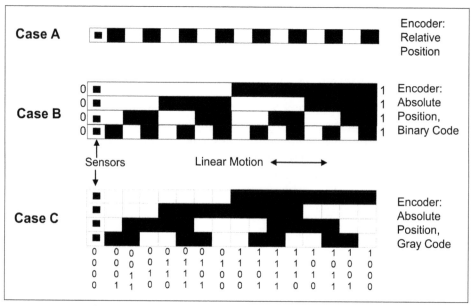

Figure 9-6. Encoder position: Gray code versus binary

Encoder technology is available in numerous devices from many vendors. Rotary encoders are available with resolutions beyond 10,000 steps per revolution. Encoders for linear motion often use mechanical conversion to rotary motion followed by a rotary encoder.

Analog-Linear Sensors: A Closer Look

Analog sensors do not always offer voltage output. Although voltage output is the most common, some devices (Figure 9-7) output a current, and some can be a variable resistance, capacitance, or inductance device.

Typically, analog sensor outputs are converted to a voltage suitable for sampling by an ADC. Current outputs are put through an accurate resistor to yield a proportional voltage. A variable output resistance can be driven with a precise current and a suitable voltage is created. Variable L and C devices are often put into an oscillator structure and the resulting frequency of oscillation is used to derive the sensor output value.

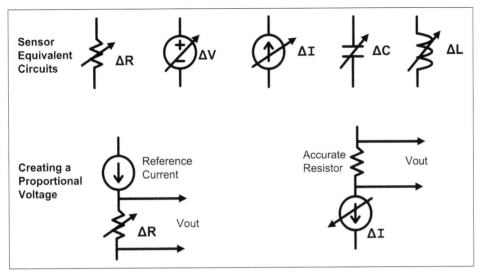

Figure 9-7. Sensor characteristics, equivalent circuits, and proportional voltage

Characteristics of Analog Sensors

Certain characteristics should be considered (Figure 9-8) when selecting a sensor:

Usable range

Sensors are designed to function over a defined stimulus range. As a general rule, fitting a sensor's usable range to the expected stimulus provides more reliable results. Using a temperature sensor as an example, a sensor output will have a limited min/max voltage, and getting a sensor that fits the needed input stimulus range allows more accurate data. Therefore, for a 1 V min/max output, a temperature sensor with a 500°C range will give 2 mV/°C, and a sensor with a 100°C range will give 10 mV/°C. Typically, the larger voltage variance will be easier to accurately monitor.

Ideally, the sensor's usable range should be centered on the stimulus range and have enough usable range to encompass all working scenarios. Many sensors exhibit nonlinear characteristics beyond their usable range, and some can be damaged when operated beyond design limits.

Linearity

Ideally, a straight-line response between stimulus and output is desired. How much deviation is tolerable depends upon the accuracy needed. Either curve fitting or a look-up table to translate a nonlinear response can be used to improve accuracy.

Repeatability

As the same stimulus is applied multiple times, it should always produce the same response. Variation in the response determines the limits of repeatability and thus the usable resolution of the sensor.

Sensitivity

For a given change in stimulus, how much the sensor output changes is a measure of sensitivity. Sensitivity is not a "more is better" scenario, but rather a matter of fitting the range of the stimulus, to the usable sensor range and its output, to the voltage range of the ADC it attaches to. Depending upon the sensor, some devices may require amplification to make full use of an ADC's input range.

Stability and durability

Some sensors can change their output response with the number of on/off stimulus cycles, the temperature of the environment, or other aging factors. Some devices vary thermally, and a temperature reference or other method to compensate for the environment temperature is needed. Some sensors require periodic replacement to remain accurate. Ideally, any sensor that ages without variance, has no temperature variance, and can function through a high stimulus cycle count is easier to implement and maintain.

Hysteresis and backlash

Hysteresis is most commonly associated with magnetic devices or electronic circuits that use feedback. It is exhibited when a rising stimulus gives a different output than when the stimulus is falling. Backlash (aka play or slop) is created by the mechanics of a sensor, where different results are seen due to looseness in the mechanics. This is most often experienced in mechanical motion when the stimulus reverses direction.

Response time and bandwidth

Response time (aka latency, group delay, or phase delay) is the time delay between the application of a stimulus and the output showing a response. A sensor's response time can affect the characteristics of a control system that depends on the sensor for timely information on control status.

Finally, sensor bandwidth is important when tracking a quickly changing stimulus. Errors in measurement will occur as the stimulus frequency approaches the bandwidth of the sensor. In most cases, a sensor can be thought of as a low-pass filter (LPF), and errors in measurement will occur as the stimulus frequency approaches the bandwidth limitations of the sensor. High-frequency stimulus events can be missed because the sensor does not have the bandwidth to respond.

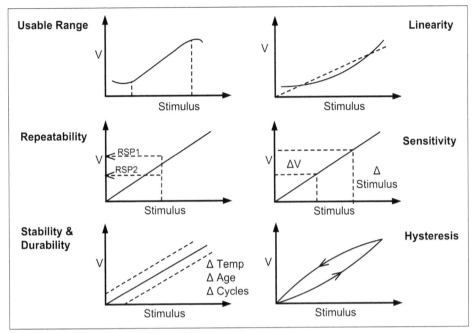

Figure 9-8. Sensor characteristics: key parameters

Signal Processing for Analog Sensors

Frequently, a sensor is placed out on a set of wires that connect back to the main printed circuit board (PCB). Figure 9-9 shows some additional parts to supplement the sensor connection.

A typical sensor connection includes a twisted/shielded pair of wires to minimize noise coupling. The shield ground is connected only at the receiver end (FDI: EMI & ESD). Sensors that use ground as a connection often suffer from noise problems due to variation in the ground voltage between the sensor and the ADC.

The RC components at the ADC create an LPF to also minimize undesired noise into the ADC. Transient voltage suppressor (TVS) diodes are included to protect the ADC inputs from electrostatic discharge (ESD) events that may occur at the sensor or along the cable (FDI: EMI & ESD).

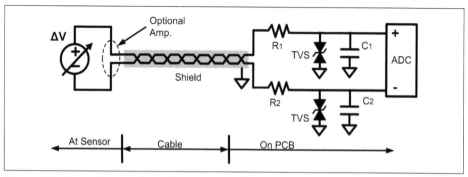

Figure 9-9. Typical sensor installation with cabling, ESD, and bandwidth limiting

Placing an amplifier near the sensor may be necessary to improve noise performance. This may be needed where the sensor signal is small (mV), the cable is long (beyond 10 cm), or the environment has a lot of electromagnetic interference (EMI). Transporting analog signals over distance is often problematic, and there is a better method, as illustrated in Figure 9-10.

Introducing a programmable gain amplifier (PGA), an ADC, and a serial port all placed at the sensor eliminates the need for noise-sensitive analog signals to be transported over distance. Signals on the cable are composed of low-voltage differential signaling (LVDS) digital data, which has robust noise immunity. Multiple vendors produce single-chip devices to provide these combined functions.

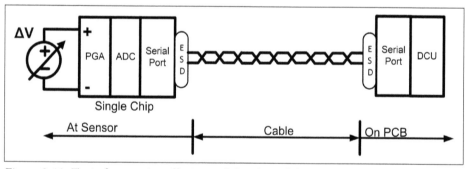

Figure 9-10. Typical sensor installation with local amplification and ADC

Sensor Calibration

Depending on required accuracy, some sensors may need to be calibrated. If the sensor is linear over the range, a single data point in the middle of the range may suffice, but calibration at extremes of the functional range is often implemented, as illustrated in Figure 9-11.

Whether calibration is needed requires a quantitative analysis. Items to consider include:

- Needed data accuracy
- Device-to-device variance of multiple sensors
- Variance due to stability and durability issues
- Offset and gain variance of circuits between the sensor and ADC
- ADC accuracy

For a linear response sensor, two calibration points at min/max operating points often suffice. By using two points, both offset and gain errors are calibrated out.

Calibrating a sensor with a nonlinear response is more complicated. Calibration of multiple points along the response curve is common. Interpolation between calibration points is then used. In some cases, a detailed look-up table can be created to cover the sensor response range. This is a common strategy used for nonlinear thermocouples.

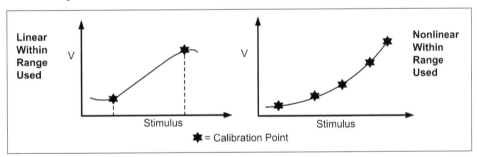

Figure 9-11. Sensor calibration: two-point linear and curve fitting

Frequently, analog sensor outputs can be converted to a two-state device. Many systems don't need quantitative data, and a trip point will suffice. Some examples include:

- A temperature sensor that asserts above 80°C
- An air pressure sensor that asserts over 120 PSI
- A tank fill sensor that asserts when the fill level goes under 10%

An ADC can be used with a sensor, followed by a software-defined trip number to assert at. Alternatively, a comparator that asserts at a specific voltage level can be used. A set trip voltage without calibration (Figure 9-12) needs to be carefully analyzed for errors and accuracy. As shown conceptually on the left side of the figure, comparing a sensor voltage against a trip voltage is straightforward. On the right side of the figure are the items that need to be considered to do this with success.

One of these is the trip point accuracy that the system requires. For example, a temperature trip at 30°C is ideal, but a temperature trip of 29.5°C (min) and 30.5°C (max) would be acceptable. In a similar manner, a pressure sensor that trips at 120 PSI would be ideal, but 115 PSI (min) and 124 PSI (max) would give acceptable behavior. The physical limits defined here will then be transformed into defined voltage limits associated with the sensor.

If trying to use a sensor without calibration, the sensor-to-sensor variance (sensor output variance) needs to be understood. This is especially important in a high-volume product with many sensors being deployed. Useful sensor data includes the nominal voltage at which the system should trip, how much the nominal trip point voltage changes between devices, and the change in voltage as a function of the change in stimulus.

For the preceding temperature example, the nominal V_{sens} = 1.5 V at 30°C, and the sensor changes 5 mV per 1°C of temperature change. For this example, 5 mV of voltage error creates 1°C of temperature error. Error sources include the sensor output variance from nominal, accuracy of V_{ref}, voltage divider (R_1, R_2) accuracy, and input offset (V_{off}) of a nonideal comparator.

Using a high-accuracy voltage reference and accurate resistors, and selecting a comparator with low input offset voltage specifications, will minimize the resulting temperature error.

 Sensor variance among multiple devices is a red flag for needing calibration, or the designer needs to select a sensor with a more consistent response.

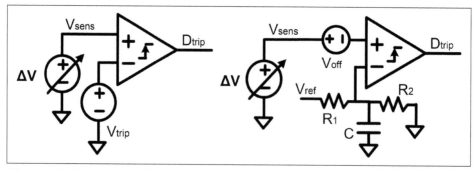

Figure 9-12. Comparators for a single decision/trip point

If better accuracy is needed, use an ADC instead of a comparator to create an adjustable system. Include a calibration method and digital adjustment of the trip point as defined by a programmable ADC output value. Avoid using analog adjustment potentiometers or changing resistor values. Such methods are not suitable to high-volume production.

Current Sensing Methods

Having the capability to determine circuit current is useful in many applications. For circuits like the H-bridge driver or the low-side driver circuit (Figure 9-13), this is often done with a resistive shunt. The voltage across R_{shunt} is proportional to the current in the circuits above it. The value of R_{shunt} is small, typically under 1 Ω, and V_{shunt} is usually under 100 mV.

Current sensing using shunt resistance is suitable for most small-scale systems. Since R_{shunt} is a small-value resistance, V_{shunt} should be measured directly across R_{shunt} and should not include PCB connection resistance.

A Kelvin connection (FDI: PCB) is needed so that the voltage across the shunt resistor is what the ADC measures, and resistive PCB connections are not part of the measurement. In addition, when selecting R_{shunt}, a high-accuracy resistor with a suitable power rating and minimal thermal variance should be used. A larger value of R_{shunt} wastes energy in heat lost, and reduces the power supply voltage available to the load.

Connecting a differential input ADC directly across the shunt resistor is readily done. If needed, a zero current measurement can be implemented to eliminate ADC offset errors. For noisy switching devices, an RC filter at the ADC input is suggested.

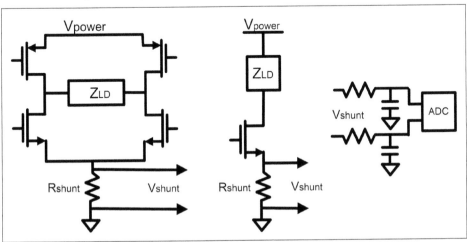

Figure 9-13. Current sensing with shunt resistor

Use of a shunt resistor to measure current fulfills the need in most situations. Where that is not viable, sensors that use electromagnetic methods—specifically, a current transformer, flux gate, and the Hall effect—are available, as shown in Figure 9-14.

Current sensing using a current transformer puts the unknown current through a magnetic core, and a proportional current is created in a secondary winding that is processed and quantified.

In a flux gate, the unknown current ($I_{unknown}$) creates magnetic flux in the core. A measurement circuit controls a known reference current (I_{ref}) that creates an opposing magnetic field in the core. The reference current is adjusted until the magnetic fields null (cancel) each other out. Knowing I_{ref} and the number of turns of wire that I_{ref} goes through allows determination of $I_{unknown}$. Null sensing is often done using a Hall sensor to determine where the fields cancel out.

The *Hall effect* current sensor uses the unknown current to create flux in a magnetic core. The flux passes through a Hall sensor placed in a gap of the magnetic core. This sensor uses the proportional analog response of the Hall sensor in a calibrated manner to define the current.

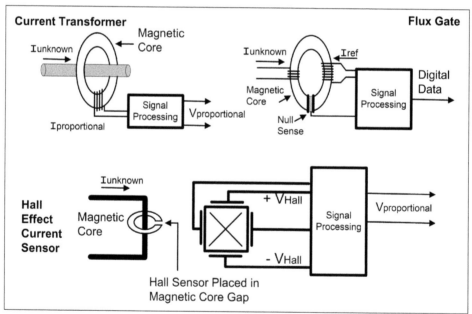

Figure 9-14. Electromagnetic current sense methods

These specialized current sensors are available as integrated modules. For most design scenarios, measuring voltage across a shunt resistor should suffice.

Voltage Sensing

Scaling signal amplitude up to better fit an ADC was covered earlier. Scaling signal amplitude down by voltage division should be straightforward (Figure 9-15). High-voltage connections should have a couple of circuit additions for reliability and safety.

Voltage division using R_{top} and R_{bot} reduces the voltage to a suitable amplitude for ADC input. Typically, the input voltage to the ADC should be selected so that it is at half scale when at its normal operating point. This allows the ADC to convert both over/undervoltage signals.

Following are additional things to consider:

- Both R_{top} and R_{bot} need to be accurate if a meaningful ADC value is needed. Check the tolerances of the resistors.

- R_{top} can be a large value. Resistors above 100K can have apparent value changes due to the environment with both dust and humidity being a factor. Keeping an accurate resistor value may require controlling the environment of the resistor. A clear, insulating, and waterproof overcoat (aka conformal coating) of electronic components is often used to do this.

- R_{top} can have a lot of voltage across it. Physically small surface mount components often have maximum voltage limitations. High-voltage (HV) arcing across a device can occur when violating voltage limits. R_{top} may need to be divided into a string of resistors to spread the voltage drops among multiple devices.

- HV power supplies can be electrically noisy or poorly filtered. The capacitor (C_f), in conjunction with R_{top} and R_{bot} creates an LPF into the ADC. If the HV supply is noisy, this should help reduce noise variance in the ADC output.

- Frequently, HV power can have switching transients well beyond the nominal voltage. This is common when the power supply is used for any switched inductive load. Use of a TVS diode with a V_{max} limit value selected similar to the power supply of the ADC limits (typically 1.8 V to 5 V) will protect the ADC input from overvoltage damage (FDI: EMC-ESD).

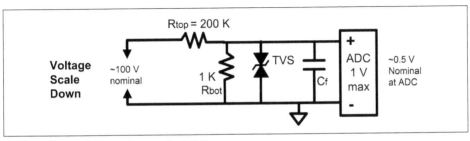

Figure 9-15. Voltage scaling to fit ADC range

Specific Sensor Applications

So far in this chapter, the discussion has been about generic sensors, where the stimulus can be many different things. This section examines different common sensor uses and some of the items to consider when implementing them. Specifically, methods for measuring and sensing of pressure, temperature, strain, sound, imaging, and touch panels are discussed.

Pressure Sensors

As a black box, electronic sensor pressure gauges come in two variants:

- Two-state pressure sensors with a conditional trip point. An example is a pressure sensor that asserts logic high when the input pressure exceeds 35 PSI.

- Quantitative output sensors, typically linear-analog output, which is quantized with an ADC.

Internally, most pressure sensors use strain gauge technology. Some sensors will output an analog signal, while others will include internal signal processing and provide a digital data output.

Pressure is measured against a reference:

- *Absolute pressure* references the pressure to a vacuum.

- *Gauge pressure* references the pressure to the surrounding environment pressure.

- *Differential pressure* measures the difference in pressure between two input sources.

- *Compound pressure* measures positive or negative pressure (aka suction).

Units of pressure can be inconsistent, depending largely on the application's environment:

- Metric pressure is measured in Pascals (Pa), where $1 \text{ Pa} = 1 \text{ Newton/meter}^2$.

- Pressure in imperial units is measured in pounds per square inch (PSI), and the units PSIA (pounds per square inch absolute) and PSIG (pounds per square inch, gauge) are often used.

- Other pressure units are still used within specialty industries. Most of these are based on older liquid column measurement methods, namely inches of water (in H_2O), millimeters of mercury (mmHg), and others.

Operating pressure range will generally be specified over a min–max range, and this is the useful linear range of the pressure sensor. Typically, a sensor that works over a narrower range will give more accurate data than a wide-range device. Since pressure sensors use thermally sensitive strain gauge techniques, the specifications should be examined for accuracy versus temperature changes. Many pressure sensors include internal thermal compensation methods to improve accuracy.

In terms of physical considerations, tubing or other plumbing type interconnects will be part of the mechanical definition. Depending on the medium being measured (air pressure, water pressure, hydraulic fluid pressure, caustic liquids, etc.), a mechanical gauge isolator between the medium and the sensor may be needed to protect the sensor.

Temperature Sensors

Sensing temperature (Figure 9-16) is commonly done using four different methods:

Thermistor

A thermistor is a resistor that varies with temperature. Depending on the materials used, the resistance can go up or down with temperature. Positive temperature coefficient (PTC) devices increase resistance with rising temperature and negative temperature coefficient (NTC) devices decrease it. Most thermistors have a nonlinear resistance response to temperature.

Resistance temperature detector (RTD)

The RTD is also a thermally varying resistor. The RTD is fabricated as a long metal conductor wrapped around an insulator. The conductor is commonly made of copper, nickel, or platinum. Many older RTD devices were fabricated as a coil of wire, and newer manufacturing methods use thin film lithography on an insulating substrate. Modern devices can be very small and are available in standard surface mount packages, similar to an SMT resistor.

Thermocouple

Thermocouple devices monitor the voltage created in the contact of two different metals. This "hot junction" voltage varies as a function of the temperature. Thermocouples are suitable to both high temperatures and applications where a fast response is needed.

Silicon diode

The voltage across the PN junction in silicon reduces 2 mV for every 1°C increase in temperature. A hotter junction is a smaller voltage. PN diode junction temperature devices take advantage of the thermal characteristics of the silicon diode junction.

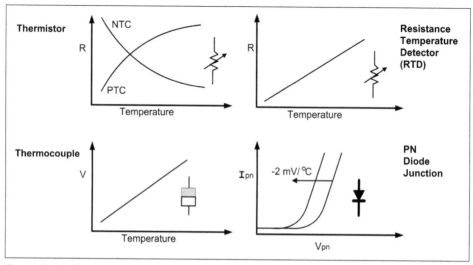

Figure 9-16. Common temperature sensors

Picking a temperature sensor to use starts with looking at the temperature range it covers (Table 9-1). Thermistors and RTDs cover a similar temperature range (−50°C to 300°C). PN diode methods are fully integrated onto an IC and are limited by the temperature range of the semiconductor (−55°C to 150°C). Thermocouples offer the broadest temperature range (−270°C to 1,100°C) but offer small voltage changes that require specialized signal processing and calibration to yield accurate results.

Table 9-1. Typical thermal range for temperature sensors

Sensor type	Min (°C)	Max (°C)
Thermistor	−50	300
RTD	−55	300
Thermocouple	−270	1,100
PN diode	−55	150

Figure 9-17 depicts a typical thermocouple. As shown, the hot junction is used to measure temperature. However, the device actually has three junctions of dissimilar metals and all three will create some form of thermally changing voltage.

The two cold junctions (CJ) are held at the same temperature, and the temperature of the CJ is independently measured. Knowing the CJ temperature allows the CJ voltages to be determined from a data table. With that, and the voltage out of the thermocouple structure, the CJ voltages can be subtracted out to find the hot junction voltage, which is then converted to temperature.

As an example, a standard K type thermocouple creates a voltage of about 40 uV per degree C. With such small voltage variance, measurements can be difficult to keep exact, and accuracy within 3°C is typical.

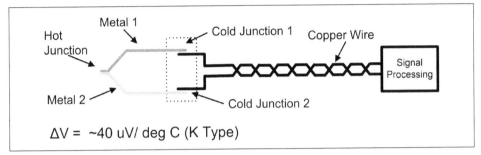

Figure 9-17. Thermocouple structure

Thermocouples are useful for broad temperature range and fast response, but accuracy limits and sensitive signal issues can be a problem. For a narrower temperature range, a silicon diode temperature sensor is more easily implemented (Figure 9-18). Silicon diode devices are provided as a fully integrated IC with the temperature sensing diode junction in the IC. A serial port provides digital temperature data to the host.

Some devices use an external PN junction to allow remote temperature sensing. Modern high-power MPUs (Intel Pentium and newer) have a thermal diode internal to the MPU that can be attached to for monitoring the MPU die temperature. Typically, this is done using a positive-negative-positive (PNP) bipolar transistor, which is available on most CMOS semiconductor processes.

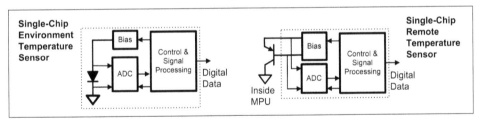

Figure 9-18. Fully integrated PN junction temperature sensors

Fully integrated temperature chips (Figure 9-19) exist to support thermistor and RTD devices, as shown in Figure 9-18. Chips designed for resistive temperature sensing usually include certain features:

- Dedicated bias system with calibrated currents
- Differential input ADC with Kelvin connections to the thermal resistance
- Digital signal processing (DSP) with memory and programmable look-up tables

- Serial port access for temperature readback
- Serial port access to load configuration data from the host

For most scenarios, a silicon diode sensor or an RTD will provide reliable temperature data. Thermistors are nonlinear and consequently require more work to configure and compensate than RTD sensors. Thermocouples are necessary when wide temperature ranges or high temperatures are involved, but they are limited in accuracy and can be fussy to get properly configured and cold junction–calibrated.

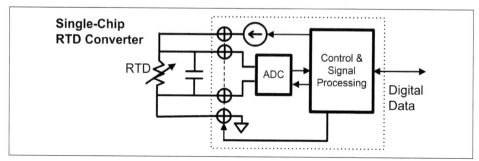

Figure 9-19. Single-chip processor for RTD and thermistor temperature sensors

Strain Gauges

Strain gauges (SGs) are used to measure deformation (strain) of the material that the gauge is mounted on (Figure 9-20). Common examples include:

- Strain of a structural support beam (buildings and bridges)
- Support cable tension
- Shifts or deformity in concrete structures
- Weight measurement apparatus
- Pressure gauge devices
- Sensing rotational torque and shearing stress
- Strain measurement of structures and control surfaces for airplanes and rockets

Strain gauges can be set up to measure both tension and compression. As illustrated in Figure 9-20, when the metal conductor of a strain gauge is compressed, the resistance reduces. When the strain gauge is stretched, the resistance increases. To increase the magnitude of this resistance change, multiple connections are put in series, and all are strained along the same axis.

To properly measure a strain gauge, an accurate bias current is applied and the resulting voltage is measured. This Kelvin connection has zero current in the sense connections and thus eliminates connection resistance from the measurement.

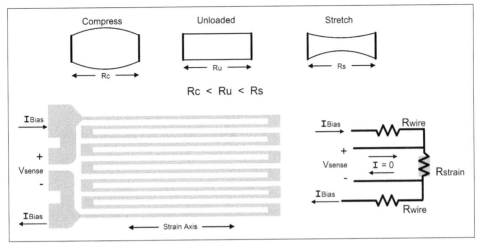

Figure 9-20. Strain gauges: structure and measurement

Many older texts advocate using a Wheatstone bridge method when measuring an SG. With modern electronics, that should not be needed. Modern ADC converters are available with 32 bits of resolution, many low-cost 24-bit devices are widely available, and sufficient accuracy can be obtained through direct measurement, select calibration, and careful implementation of support circuitry.

Temperature variance of an SG can be significant and needs to be calibrated out or numerically removed. On the left in Figure 9-21, an accurate reference voltage, V_{ref}, drives a voltage divider. An accurate and thermally stable reference resistor, R_{ref}, completes the voltage divider with the SG resistance, R_{sg}. With the capability to multiplex the ADC input, three voltages can be measured: V_{ref}, V_{div}, and V_{gnd}. The R_{sg} can be accurately derived.

This setup lacks a direct method to account for temperature variance. In a mechanical system where the strain can be turned on/off (mechanical engagement of a motor or other device), the system can be measured with and without strain applied, and thus the temperature effects can be accounted for. An alternative approach is to place a second SG in the same thermal environment, without the physical strain, and measure it to determine a zero-strain, at-temperature baseline. With these methods, the compensation is done in the DSP math.

Yet another method is shown on the right of Figure 9-21. Two strain gauges, in the same thermal environment, are part of the voltage divider: one is under strain and the other is not. Voltage variance at V_x is a function of the lower device strain, because thermal resistance variance in both devices should be equal for matched devices.

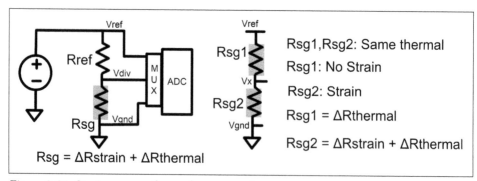

Figure 9-21. Strain gauges: thermal compensation

Can a Wheatstone bridge method be useful in applications? Perhaps. As shown in Figure 9-22, Case 1 puts an SG into a bridge circuit for measurement. Knowing the three fixed resistors (R_1, R_2, R_3) and the ADC voltage, the value of SG_{R4} can be derived. There is no compensation for thermal variance, and finding the value of SG_{R4} requires measuring R_1, R_2, R_3 to get useful data.

Measuring the SG directly without the bridge circuit is simpler. Case 2 inserts an SG in the R_3 position, and SG_{R3} does not have strain applied. This approach allows a hardware method of cancelling out the thermal variance of the SG. The V_1 node changes voltage with strain, but (ideal case) not with temperature. Measuring V_1 to V_2 does not have an advantage over measuring V_1 to ground in most situations.

Case 3 uses four SGs in all positions. In situations where both compressive and tensile strain is present (i.e., opposite side of a bending beam), this may be useful.

Ideally, all four SGs thermally track each other, cancelling out thermal variance. By combining SG_{R1} (in compression) and SG_{R2} (in tension), the strain-induced change of V_2 is doubled. As well, SG_{R3} and SG_{R4} are combined with the compression/tension roles reversed, thus doubling the change of V_1. Taking the differential voltage between V_1 and V_2 has increased the signal output by 4X.

To summarize, Case 1 yields no useful improvements over direct measurement. Case 2 offers a hardware method to compensate for thermal variance, but direct measurement without the extra two resistors should suffice. Case 3 introduces a large amount of complexity to increase the amplitude of the measured signal.

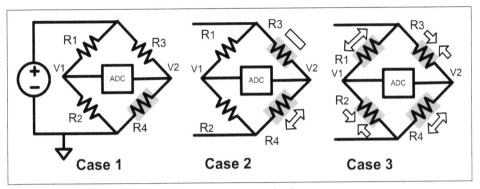

Figure 9-22. Strain gauges: Wheatstone bridge methods

Using a Wheatstone bridge to do DC measurements is a legacy practice. Instead, using an accurate and stable bias source, Kelvin-structured sense/drive paths, and an accurate ADC should suffice without needing extra components. The capability to multiplex the ADC allows measurement of sensor voltage, supply voltage, voltage across reference resistor, and offset voltage. The resulting data can be processed digitally to get the desired strain information.

Best practices when working with SGs include:

- Balanced interconnects, shielded cables, and twisted pairs to reject noise
- Kelvin connections that avoid bias currents in the ADC input connection
- Placing the ADC near sensors to minimize cable noise pickup
- Low-pass filtering at the ADC input for noise immunity
- Multisample averaging of the ADC output to digitally reduce noise
- Low bias current magnitudes to avoid self-heating issues
- Temperature compensation using a reference SG or temperature monitor and digital adjustment

Sound and Microphones

Microphones (Figure 9-23) are necessary to turn audio into a data stream. Multiple methods have been used and several techniques prevail in modern use.

Carbon microphones utilize crushed carbon granules under a sound-sensitive diaphragm. Sound on the diaphragm creates varying pressure on the carbon, which results in a sound-dependent variable resistance. The carbon microphone was widely used in landlines and is still deployed in these legacy telephone systems.

With piezoelectric microphones (aka crystal microphones), an acoustic diaphragm puts force on a crystal, which generates a voltage. The crystal microphone has been in use since the 1930s. Modern versions, using microelectromechanical systems (MEMS) technology, enjoy low cost, small size, and rugged resistance to mechanical shock. Their response is not ideally linear, and consequently they don't see much use in high-quality recording.

Dynamic microphones (aka moving coil microphones) use a sound-sensitive diaphragm to move a wire coil through a magnetic field, creating an output voltage. Although less sensitive than a condenser microphone, the dynamic microphone is often used for live performances, recording percussion instruments, and general use.

Condenser microphones (aka capacitors, electro-condensers, electret-condensers) use a sound-sensitive diaphragm with a conductive surface. The diaphragm is placed beside a solid conductive plate so that the two conductive surfaces form a capacitor. Sound hitting the diaphragm causes motion, and the distance between plates changes with the sound. The distance variation causes the capacitance to change, which modulates the voltage across the capacitor. Since the only physical moving part is the low-mass diaphragm, these microphones can be both very sensitive and very responsive to high-frequency sound. Modern condenser microphones use MEMS technology to create extremely small versions, and the condenser microphone is widely used in cell phones, music recording, and elsewhere.

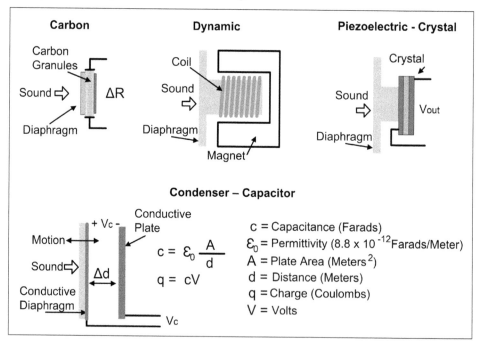

Figure 9-23. Internal electronics of common microphones

Microphones for use with computers, small-studio podcasting, and similar applications have largely adopted a digital output strategy via USB interfacing. Many options are available off the shelf. Cabling of a USB microphone is limited to 5 meters, so wiring up something beyond a few closely placed microphones won't be possible over USB.

Older analog microphones for computer use depended on a converter within the computer to digitize the sound. These are inexpensive and still widely available, but are not encouraged for new designs.

Professional audio has changed slowly. As of 2023, cabling of standalone microphones still largely consists of differential analog signals (aka balanced audio) based upon the XLR3 connector (aka XLR, Cannon connector), which has been in widespread use since the 1950s. The connector uses three pins: audio+, audio−, and ground/shield. The mechanics of the connectors are designed such that the ground/shield connection makes contact first upon mating and separates last when disconnecting.

Frequently, microphone power is provided through these wires. *Phantom power* (Figure 9-24) involves attaching a positive DC voltage (typically 12 V or 48 V) through resistors to both audio+ and audio−.

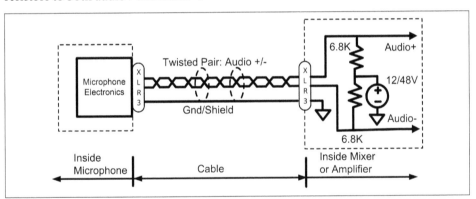

Figure 9-24. Phantom power for microphones

Microphones that use digital methods to communicate are slowly gaining market share in professional audio. Large amounts of legacy analog audio equipment remain in use, slowing down widespread adoption. Two different methods are competing for the digital output microphone market.

AES-42 is the Audio Engineering Society standard that has been adopted for digital interface microphones. The AES-42 audio data stream is also known as AES-3, IEC 60958, or S/PDIF. AES-42 is designed as a point-to-point protocol, such as a microphone that connects to a sound mixer board, where the sound information is passed as digital data, not analog signals.

AES-67 and Ravenna can be described as an "audio over network" methodology. For AES-67, a microphone is configured to communicate using Ethernet methods. Ravenna is an open source standards group that has adopted AES-67 and expanded upon it to include all media sources. The Ravenna standard, widely supported by industry vendors, is growing as the predominant method for digital-based media communication.

Professional audio will get to a mostly digital approach, but the methods used, and their final adoption, are still in flux as of 2023.

Different sampling rates and ADC bit resolution structures are in use:

- High-quality audio uses a resolution of 16 to 24 bits and a sampling rate of 44 KHz or higher.

- Recording studios (circa 2023) use 24 bits and a sampling rate of 44.1 KHz or higher.

- Uncompressed audio CDs (*.wav* format) use 16-bit resolution at 44.1 KHz.

- Lower-quality audio with less resolution and slower sampling is commonly used for nonstandardized audio applications when quality isn't a priority.

A typical set of internal electronics for analog and digital output microphones is shown in Figure 9-25. For the analog microphone, a variable capacitor forms the microphone sensor. The bias circuit creates and keeps a fixed charge across the capacitor. Sound variation causes diaphragm motion, thus creating a dynamic change in capacitance, resulting in voltage proportional to sound into the amplifier. The amplifier inputs have negligible input current so as not to interfere with or load down the sensor capacitor. A differential filter is often included as part of the amplifier.

For an analog-only microphone, the amplifier outputs are AC coupled and exit the microphone as audio+, audio−. A phantom power system takes the DC voltage (12–48 V) present on the two audio connections and creates an internal power supply for the microphone circuitry.

For a digital output microphone, further signal processing is needed. An LPF eliminates aliasing and an ADC quantizes the signal. This is followed by a digital controller and a serial port system to arbitrate the incoming control data and the outgoing audio data stream. This is a generic device for illustration, but many variations on this exist.

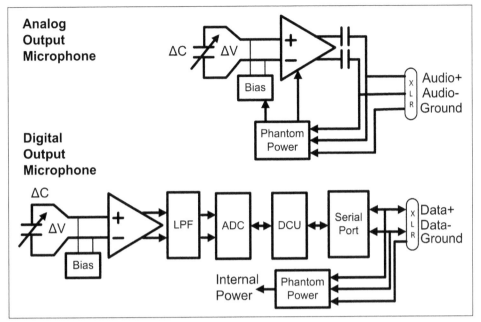

Figure 9-25. Modern condenser microphones

Since human hearing has such wide dynamic range, noise is of prime concern. Minimizing the use of analog signals will minimize noise corruption.

Image Sensors and Video Cameras

In addition to video cameras for image streaming, many specialized image sensors exist. A few examples include sensors for high-speed motion capture, thermal-infrared imaging for temperature profiles, low-light (night vision) devices, and X-ray sensing.

The image sensor within a camera is the heart of the system. Modern image sensors are fully integrated onto a single chip, with a streaming digital output to convey the image data. Figure 9-26 shows a typical color camera setup. At the input is a camera lens, followed by an RGB color filter (aka Bayer filter) and the image sensor IC. The image sensor contains a pixel array and support electronics.

The electronics include an analog front end (AFE) that features row/column multiplexing, and a buffer-amplifier to drive the ADC. DSP follows for corrections to color, linearity, and brightness. The image data is formatted into row and column data that exits via a serial port.

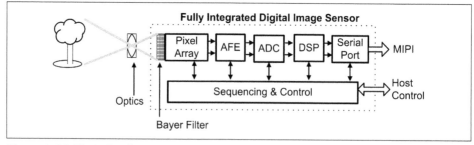

Figure 9-26. Typical video camera system

Multiple video interface protocols are used, including:

- MIPI-CSI (Mobile Industry Processor Interface – Camera Serial Interface)
- DVI (Digital Visual Interface)
- SDI (Serial Digital Interface)
- HDMI (High-Definition Multimedia Interface)
- DisplayPort
- HDBaseT – Ethernet
- USB
- And others

Some camera interfaces are designed to be cable friendly (USB, HDMI, Ethernet, DisplayPort), while others are designed to be an interface to another IC (MIPI-CSI). Most fully integrated cameras use cable-friendly interfaces. Older camera systems use analog signal outputs, based upon BNC connectors or RCA phono plugs. To be useful in an embedded system, some form of converter interface is needed. Analog output cameras are discouraged for new designs.

Many cameras are not designed for conventional video but are optimized for industrial inspection and automation systems. Purpose-specific cameras are readily available, with many variants. These cameras often use USB and Ethernet interfaces and are the solution of choice for most situations. Custom cameras can be developed, but OTS designs are the quickest path to a product.

Following are things to consider when selecting a camera system:

Monochrome versus color

For many industrial imaging and automation applications, color imaging isn't needed. Simpler electronics and better resolution can result using a monochrome system. One application where color cameras are often preferred is automated systems that handle and inspect food, where coloration is often part of the inspection criteria.

Frame rate

High frame rate cameras are useful for motion capture, high-speed inspection, automation handling, and similar applications. Specialty cameras are readily available with frame rates into the millions of frames per second. Ultra-fast cameras come at a price premium and require high-speed data capture. Consequently, the necessary frame rate should be defined for optimal camera selection.

Light sensitivity (aka ISO sensitivity)

This is an indicator of an image sensor's capability to detect light and define an image. High sensitivity provides the capability to function properly in low-light and dark environments.

Pixel resolution

Early digital cameras were low-resolution, low-pixel-count devices. Consequently, an early selling point of digital cameras was the pixel count. As of 2023, camera sensors capable of sensing over 50 million pixels are commonplace. Sufficient pixel resolution is rarely an issue with modern cameras.

Frame aspect ratio

The frame aspect ratio is the horizontal to vertical ratio of the sensor image. Classic television systems (NTSC/PAL) use a 4:3 aspect ratio. Modern HDTV video uses a 16:9 aspect ratio. The 3:2 aspect ratio is also in common use and has an origin based in 35 mm film photography that used a 36 mm × 24 mm image size.

Many industrial and medical specialty cameras diverge from these ratio standards, depending on application need. Medical endoscopes are often designed with circular image fields or a 1:1 (square) image to optimize use of the image sensors within the miniature cameras used.

Frequently, a camera sensor is designed for a particular aspect ratio, and applications needing a different ratio adapt the image by stripping away pixels or by stretching the image to fit the desired frame shape.

Lens selection

Frequently, digital cameras will allow selection of an application-specific lens to be mounted to the camera. For this, suitable optical considerations need to be made. Following is a quick overview:

Field of view (aka angular field of view)

For a defined working distance from the camera, the field of view is the size of the image area covered. The angular field of view removes the working distance from the definition and defines this as an angle of view. A wide-angle lens (i.e., 160 degrees) has a larger field of view than a telephoto lens with a narrow field of view (i.e., 20 degrees).

Depth of field
>This is the min-to-max distance from the camera where the image remains in focus.

Lens aperture (aka F-stop, lens speed)
>This is how much light the lens allows in.

These considerations are similar to those for conventional cameras and are lightly touched on here. For those seeking more information, these topics are well covered in many sources focused on the technology of photography and video imaging.

Streaming high frame rate video is data intensive and can load down a computer system. Consequently, estimates or experiments need to be conducted to determine suitable processor capability to support the camera system. High performance video processing may require a dedicated MPU or specialty DSP ASIC, to provide high speed image processing. Use of a single-board computer system with an operating system and an embedded controller peripheral is often the simplest implementation.

Touch Panels

A touch panel typically comes integrated with a display panel. To make touch and display work together, the display needs to be fed image data and touch sensing needs to be processed. Several different methods are used for touch screen sensing, with resistive touch, capacitive touch, and projected capacitive touch being the most common.

Resistive touch screens are widely used for simple applications. The most common is the four-wire touch screen version shown in Figure 9-27. As shown, when a user touches the screen in response to display information, the system needs to determine the touch point (X, Y) location. The cross-sectional view of the touch screen shows two conductive layers, both transparent, placed over the image display. Both conductive layers are resistive in nature, and when they are pressed together, a four-resistor circuit is created. To complete the circuit, the four edges of the display contain electrical connections to the conductive layers.

If the four resistance values can be determined, the (X, Y) locations can be determined. Selective switching of a voltage source and ADC to the resistor network can determine R_3, R_4, and then can change connections and determine R_1, R_2. The problem is simplified because the voltage at the ADC, divided by the input voltage, directly yields (X, Y) values.

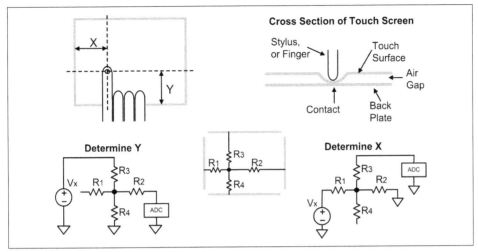

Figure 9-27. Resistive touch screens, four wire

Designers can configure a resistive touch screen controller out of discrete parts, but many dedicated chips have been designed to do the job, as shown in Figure 9-28.

A resistive touch screen controller IC will include a multiplexer to make the needed connections, a voltage source, ADC, and the needed sequencing and control logic. These devices determine (X, Y) position and provide data back to a host through a serial port.

Resistive touch control is simple, low cost, and widely utilized. The limitations of resistive touch are:

- Limited product life due to flexing the touch screen surface
- Ability to sense only one contact at a time

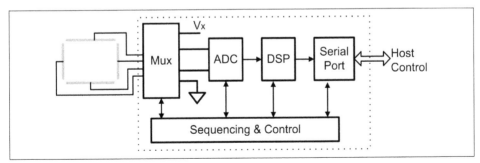

Figure 9-28. Resistive touch screen controller

Capacitive touch screen methods solve the durability problem by using a nonflexing touch screen. As shown in Figure 9-29, capacitive touch uses a nonflexing clear front panel, with a conductive coating underneath. When a finger contacts the screen, a capacitive connection between the finger and the conductive coating is created. The finger is at a different potential than the conductive coating, and current is injected through the capacitor into the conductive coating. The injected current can be positive or negative, depending on the potential difference.

The four corners of the conductive coating are metered for current sensing. Since the four resistances to the corners define a unique location, the inverse of the currents to the corners reflects those distance vectors. The distance vectors to the corners of the display can be determined from the currents. The distances (A, B, C, D) between the four corners and the contact point can be converted to an equivalent (X, Y) location. Other approaches to this measurement method exist, and all share the common method of impedance to corners and use various methods to measure it.

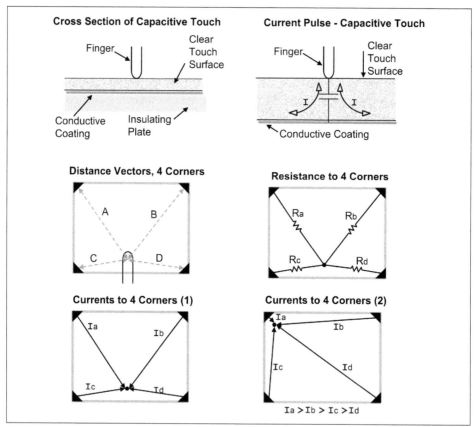

Figure 9-29. Surface capacitive touch screens

Surface capacitive methods suffice for single-touch systems. To have a touch panel that can process multiple simultaneous touch events, a projected capacitance touch screen is needed (Figure 9-30). Projected capacitance touch screens use two conductive layers, one with a row pattern and the other with a column pattern. Touching the front panel injects current by capacitive coupling into the closest row and column. Row and column data correspond to the (X, Y) location. Multiple touches at the same time can be processed. These touch screens are rugged and very reliable.

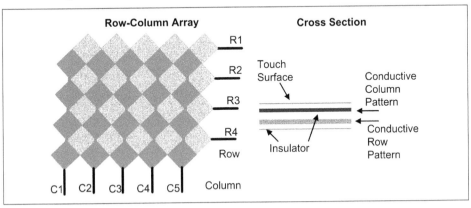

Figure 9-30. Projected capacitive touch screens

To summarize, following is a comparison of the touch screen methods covered here (other methods exist, but these are the three most popular):

Resistive touch
>   Low cost, thin physical implementation, simple perimeter connections, low power, long-term durability may be a problem, not multitouch

Surface capacitive
>   No moving parts, no flexing, high durability, fingers only (no gloved use), not multitouch

Projected capacitive
>   No moving parts, no flexing, high durability, supports multitouch, sensitivity can be adjusted to be responsive while wearing gloves

Most touch screen systems are available as a fully integrated module with an interface for the display and the touch screen, as shown in Figure 9-31. Typically, these modules come with a serial port interface for the display and a separate serial port for the touch screen. Vendors are constantly improving and adding new display and touch screen products, so researching current product offerings is suggested.

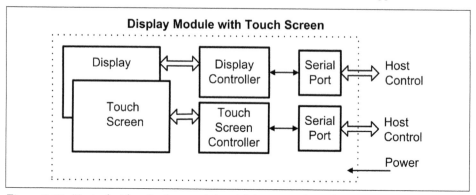

Figure 9-31. Typical video display module with integrated touch screen

Summary and Conclusions

Sensors are an important part of most embedded systems. Providing the capability to monitor real-world phenomena is a keystone in creating systems that interact with the application environment.

Following are some key ideas:

- There's a sensor for pretty much anything.
- Sensor outputs come in a limited number of formats: digital data, two state, event counting (events and frequency), frequency, and proportional to stimulus (linear and nonlinear).
- All sensor outputs can be converted to digital data usable by the host.
- Data capture is usually by interrupt request, periodic polling, or streaming.
- Sensor calibration may be needed depending on accuracy needs and sensor repeatability, stability, and durability.
- Avoid mechanical switches where possible. Optical interrupters and Hall sensors are often a good substitute.
- Sensor specifications should be examined for usable range, linearity, repeatability, suitable sensitivity, stability, durability, and response time.
- Signal processing for analog output sensors should be filtered and ESD-protected coming into the ADC. Cables should be kept as short, twisted pair differential wires with a shield connected at the ADC end.
- If long cables are needed from a sensor, consider putting an ADC at the sensor, and putting data on the cable, not analog signals.
- Temperature is easily monitored with an RTD sensor or a PN diode junction sensor. Thermocouples should be avoided unless monitoring of wide temperature ranges is needed.

The following are necessary when implementing a sensor design:

- Early in the design, determine whether calibration is needed. Adding calibration support circuitry late in the design process can be costly.
- Check component accuracy versus needed data accuracy.
- Always ESD-protect any cable port onto a PCB.
- Always include RC filters on an ADC input to limit EMI sensitivity.

Further Reading

- Chris Wolf, "Powering Microphones," MicPedia, *https://micpedia.com/powering-microphones*.

- David Potter, "Measuring Temperature with Thermocouples – A Tutorial," National Instruments Application Note 043, 1996, *https://users.wpi.edu/~sulli van/ME3901/Laboratories/03-Temperature_Labs/Temperature_an043.pdf*.

- "Engineer's Guide to Accurate Sensor Measurements," National Instruments, 2016, *https://download.ni.com/evaluation/daq/25188_Sensor_WhitePaper_IA.pdf*.

- Jerad Lewis, "Common inter-IC digital interfaces for audio data transfer," *EDN Magazine*, July 24, 2012, *https://www.edn.com/common-inter-ic-digital-interfaces-for-audio-data-transfer*.

- "Image Sensor Terminology," ON Semiconductor Technical Note TN6116/D, July 2014, *https://www.onsemi.com/pub/collateral/tnd6116-d.pdf*.

- Jack Ganssie, "A Designers Guide to Encoders," DigiKey, April 19, 2012, *https://www.digikey.com/en/articles/a-designers-guide-to-encoders*.

- *Handbook of Transducers* by Harry Norton, 1989, ISBN 0-13-382599-X, Prentice Hall.

- Joseph Wu, "A Basic Guide to RTD Measurements," Texas Instruments Application Report SBAA275, 2018, *https://www.ti.com/lit/pdf/sbaa275*.

- Bonnie Baker, "Temperature Sensing Technologies, AN679," Microchip Technology Inc., 1998, *https://www.microchip.com/en-us/application-notes/an679*.

- Bonnie Baker, "Precision Temperature Sensing with RTD Circuits, AN687," Microchip Technology Inc., 2008, *https://www.microchip.com/en-us/application-notes/an687*.

CHAPTER 10

Digital Feedback Control

Digital feedback control is used in a plethora of devices. The device under control (DUC) can be a motor, temperature, servo actuator, system pressure, flow rate, and many others. Many control textbooks will use the term *process* or *plant* instead of DUC due to the historical use of control systems by industrial processes and manufacturing plants.

Before diving into this chapter, readers should be familiar with Chapters 7, 8, and 9, as that material will be utilized here.

Control system textbooks usually take an approach that is heavily mathematical with well-defined system models. Many have found problems with this approach in application:

> ...the mathematics of control involves such a bewildering assortment of exponential and trigonometric functions that the average engineer cannot afford the time necessary to plow through them to a solution of his current problem. (Ziegler & Nichols, 1942)

A minimal math approach is desirable for expediency and ease of design. Also, a "mostly digital" control system is highly preferred over analog methods due to cost, consistency of performance, and ease of adjustment. However, a DUC has analog behavior that's being controlled. Consequently, a control system design needs to deal with those analog behavior issues. Because of that, the approach used here can best be described as digital emulation of an analog system.

With this "analog functionality in a digital box" approach, some select topics from control theory are presented so that readers have the foundation they need to design digitally based controllers that are stable and work properly.

Industry implementations often avoid the rigorous math approach. Instead, proportional integral derivative (PID) controllers are often configured using empirically derived tuning rules. With over 80% of the market, PID methods dominate industry control applications and provide satisfactory performance in most cases.

Trajectory control methods forgo a mathematical approach for a cause-and-effect adjustment protocol. Various forms of trajectory control are widely used where high performance is a necessity. Both PID and trajectory control will be examined here.

In addition, a section is included on sequence control and the common issues encountered there. Most embedded control devices use a mix of sequence control and feedback control, depending on the devices in the system.

Overview of Sequence and Feedback Control

Sequence control is part of many embedded control systems. Figure 10-1 illustrates the concept. A simple sequence controller turns things on and off in a defined pattern on a fixed time schedule or in response to yes/no sensor prompts. One example is a dishwasher or clothes washer that turns valves, motors, and pumps on and off in the pattern defined by the washing process.

Older methods used switches, relays, and electromechanical sequencing devices. Modern methods use digital electronics, but electromechanical sequence controllers are still used in some simple, low-cost, and slow devices.

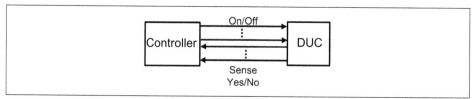

Figure 10-1. Basic sequence control

A feedback control system adds some complexity. An example of controlling the speed of a DC motor is shown in Figure 10-2. All three cases shown in the figure represent the same motor speed control system. Different texts will use varying levels of detail when representing a system. In all cases, a representation of the output is brought back to be compared to the set point reference in what is called a *feedback loop* or *control loop*.

Case A in Figure 10-2 shows some details of the system elements. The set point represents the desired operating point of the motor, which is within the DUC. A controller device provides suitable signals to a driver circuit (FDI: Drive), and the motor shaft output is connected to a rotational speed sensor (FDI: Sense) providing "recent speed" information to be compared to the set point reference.

Case B simplifies the illustration by putting the driver circuits within the DUC and removing details associated with the sense device. The difference between set point and speed information from the sense block creates the error signal, which the controller processes to create a corrective action for the DUC.

Case C minimizes the system representation down to just the DUC and control devices, with the difference circuit implicit within the control block.

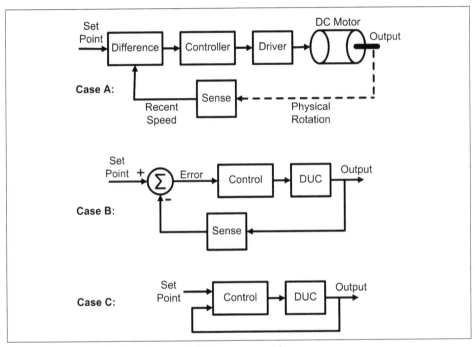

Figure 10-2. Basic feedback control, DC motor speed

At this level of detail, no information is shown on how the blocks are implemented. If electronic, these blocks can be analog or digital. Control systems predate electronics, and early-era devices were implemented with mechanical, pneumatic, or hydraulic methods. Some of these older techniques are still used in specialty scenarios, but most modern control systems are electronic.

Digital Versus Analog Circuit Methods

Electronic control can be analog or digital, as shown in Figure 10-3. Typically, five math functions are used in a controller structure: gain/multiply, sum/addition, difference/subtraction, derivative, and integral. These can be implemented using analog methods, typically using op-amps, or using digital methods with either logic gates or code executed by a microcontroller (MCU).

Analog control methods are still used within many integrated circuits for bias control, nonswitching voltage regulators, phase locked loops, alignment circuits, automatic gain control, offset control, and others. Embedded systems have largely migrated to digital control methods.

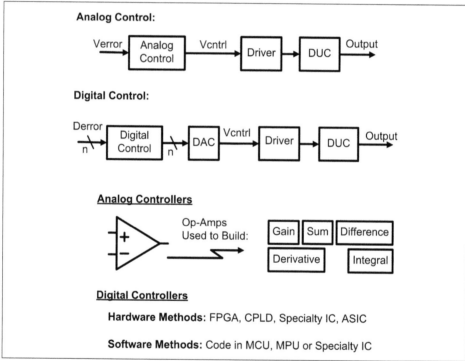

Figure 10-3. Examples of analog versus digital control circuits

Implementation methods differ greatly, but analog and digital feedback control can perform the same control loop functions. Figure 10-4 depicts the difference element and the creation of the error information. When done with analog methods, the set point is most often defined as a voltage and the sense output is rendered as a voltage. Finding the difference between the voltages and creating a proportional voltage error signal is commonly done using op-amps.

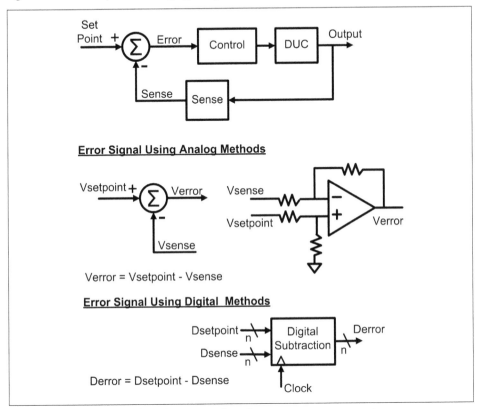

Figure 10-4. Examples of analog versus digital sense, error components

As another example, Figure 10-5 shows how a sense circuit for angular position can be implemented using digital or analog methods. For angle detection, a rotary actuator is connected to a sensor. The analog sensor is a resistive potentiometer, and the output is a voltage proportional to the shaft's rotary angle. The digital sensor is an optical encoder where a digital number indicates the shaft's angle (FDI: Sense).

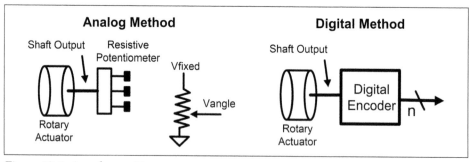

Figure 10-5. Angular position sense

For most parts of a control system, there's a digital way to do things. As will be shown in "Transition to Digital Control" on page 372, the controller structure has two digital approaches: either digital circuits or a digital code implementation. A circuits approach is used when the update rate of the system is very rapid or requires a tightly controlled time/phase relationship. This is rarely necessary with modern devices, and the bulk of digital controllers are implemented as code installed on a microcontroller.

Preliminary Definitions and Concepts

As a start, some commonly used ideas in control systems need to be defined. In Figure 10-6, closed loop control senses the state of the system output and responds to it. Open loop control has no sensors and is simpler to implement. Open loop control is typically nonvarying sequence control. In some situations, open loop control is viable where accuracy is not a necessity. The example in Figure 10-6 shows an open loop clothes dryer executed with a fixed period of drying time, and a closed loop version that monitors the humidity.

The closed loop dryer terminates when the humidity drops below a defined level. The open loop version stops at a predetermined time. Typically, the open loop dryer uses a time period that is longer than necessary, thus assuring dry clothing. This method uses more energy than necessary. The design trade-off here is more cost and complexity for a closed loop dryer, or an open loop dryer that wastes energy.

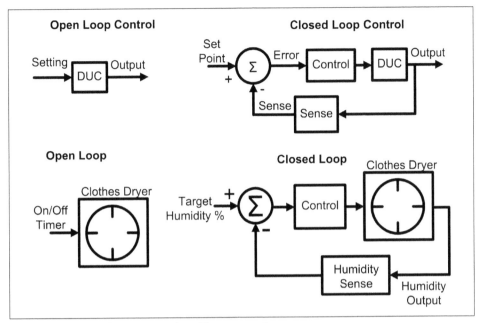

Figure 10-6. Open loop versus closed loop control

Transfer Functions, Block Diagrams, and Basic Feedback

Control systems, both digital and analog, are defined using transfer functions as part of system block diagrams. A brief overview of properties is illustrated in Figure 10-7.

A *transfer function* is a mathematical definition of the input to output relationship. The inputs and outputs can be in many forms. Binary numbers, voltage, and current are all commonly used, but things like speed, temperature, pressure, and other physical phenomena are also possible. As an example, a transfer function could also be defined for a pair of gears, where the input and output are the rotational speed defined by the transfer function of the gear ratio.

For control systems, transfer functions deal in values that represent system behavior in some way. Analog systems represent the behavior as a proportional voltage (or current), while digital systems use binary numbers.

Values can be manipulated and transformed by the transfer function through operations of summation, differencing, multiplication, cascaded operations, and other mathematical transformations.

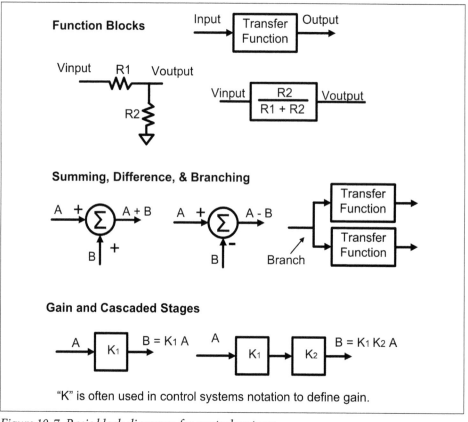

Figure 10-7. Basic block diagrams for control systems

Transfer functions are assumed to be ideal. Namely no loading or interaction between stages is assumed to be present, and all transfer functions are assumed to be linear without restrictions on operating range. In application, this may not be valid.

A functional definition of a feedback system can be mathematically represented as a simpler equivalent system, shown in Figure 10-8. This idea was first put forth in 1934 by Harold Black at Bell Labs, and it became the cornerstone of electronic feedback systems. For large values of G (forward gain), the transfer function mathematically approaches the inverse of the feedback path, namely 1/H.

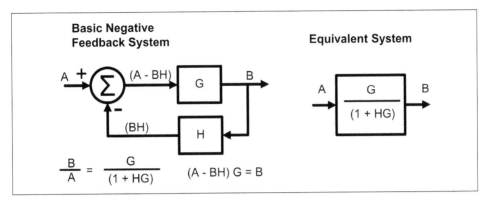

Figure 10-8. Basic feedback systems

Transient Response Terminology

Characterizing a DUC as it responds to a step in the set point input is a common test for a control system. Figure 10-9 shows the transient response of the system. Following is an explanation of the terms shown in the figure:

Lag time (t_{lag})
   The elapsed time between the transition of the step input and the point in time where the output starts to respond.

Rise time (t_{rise})
   Typically measured between 10% and 90% of the response step. This time period is a useful indicator of how quickly the system can respond.

Overshoot magnitude (M_{os})
   Indicates how far the output exceeds the final settled value.

Settling time (t_s)
   Defines the total time needed for the output to stabilize after the set point changes. This final value is typically specified to be within +/−1% of the final value.

Peak time (t_p)
   The elapsed time between the input step and the amplitude peak of the output.

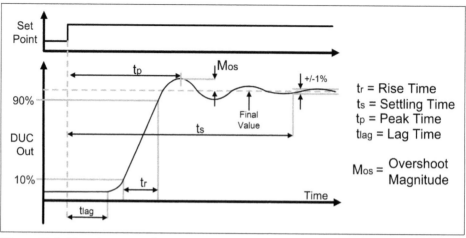

Figure 10-9. Transient control response

Transient behavior is used to characterize a control system, including the DUC, by injecting a step function into the set point. Characterization of just the DUC can also be done by stepping the DUC input and observing the output. Both techniques are used in the creation and analysis of control systems.

Figure 10-10 illustrates that a control system should minimize the difference between the set point and the output. This is called *minimization of the static error*. Here, the output (via the sensor) is compared to the set point. After the system has stabilized, a static error between the two may exist. This static error is typically due to a lack of DC gain and will be examined as part of PID control.

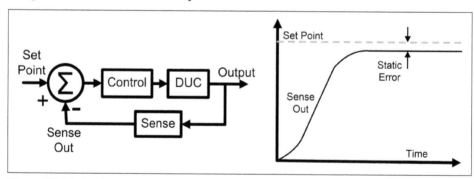

Figure 10-10. Static error

DUC Performance Selection

Before developing a control system, check the DUC for application suitability. The DUC should perform with sufficient margin beyond normal operating requirements.

Figure 10-11 shows that saturation occurs when a device input goes beyond its linear range of operation. Two examples are shown. The motor example illustrates the control voltage response while under load. Motor #1 operates primarily near the maximum input voltage and has little margin for changing conditions. Motor #1 is going into saturation and can't provide the needed torque/power for the application. Motor #2 has more torque and the midrange operating point is better suited to the application.

The thermal chamber example uses a heater with energy losses to the environment. The temperature is held constant by adding heat energy into the system. If the duty cycle of the controller is always on, the maximum heater power output is insufficient to compensate for the thermal losses. If the heater is mostly off, either the thermal losses are very small or the heater element may be too powerful for the system, causing a rapid temperature rise in the chamber. Sizing the heater so that the duty cycle is near 50% allows the controller to perform temperature control over a varying range of energy losses.

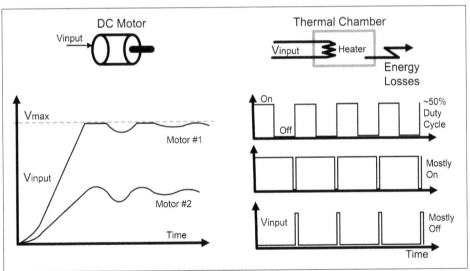

Figure 10-11. DUC selection to avoid saturation

These examples illustrate the need to properly "size" the DUC so that the controller can manage the system's operating point and not be hindered by limitations of the DUC.

Further understanding the DUC, the presence of an integrating function within the DUC itself needs to be determined. This is shown in Figure 10-12.

Introducing a step input to the DUC can cause a change in operating point, which indicates no integrator within the DUC. Alternatively, a positive step into the DUC can cause the device to increase without limit, indicating the presence of an integrator within the DUC.

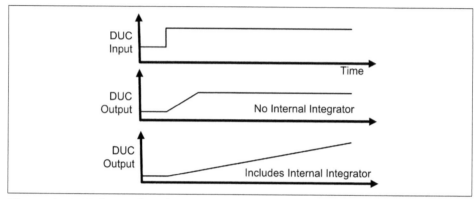

Figure 10-12. Understanding the DUC: internal integrator

The DUC can also exhibit low-pass filter (LPF) characteristics, as shown in Figure 10-13. In this example, the DUC input receives a series of pulses. The input duty cycle goes from ~10% to ~90% as shown. The DUC output does not respond to each individual pulse; rather, it slowly changes due to inherent LPF characteristics.

This built-in LPF characteristic is common to many electromechanical devices and can be useful in designing the control and driver circuitry. For such devices, a switched driver is used instead of a linear control voltage, and control settings are provided by the duty cycle. This is a common approach for many motors and other electromechanical actuators (FDI: Drive).

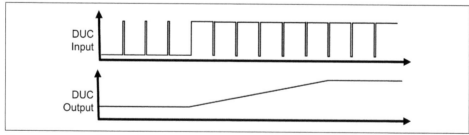

Figure 10-13. Understanding the DUC: inherent LPF

The DUC can exhibit other characteristics that define how the control system needs to be configured. Some examples of this are shown in Figure 10-14.

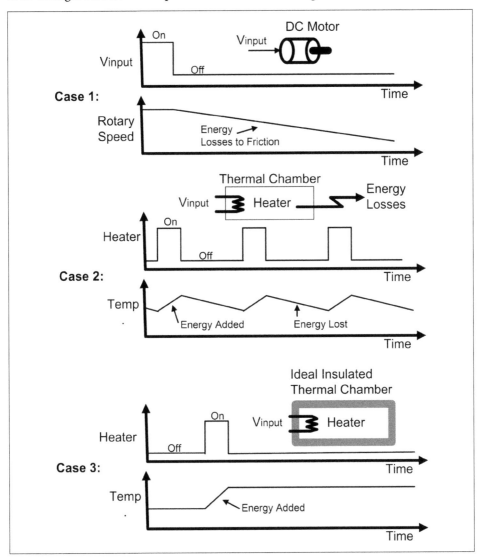

Figure 10-14. Understanding the DUC: lossy versus nonlossy

Case 1, the DC motor, exhibits energy loss due to friction and loading. If the energy loss is sufficient, it can be configured so that the circuitry only puts energy into the system. If rapid speed reduction is required, some form of active braking to take energy out may be required as part of the system.

Case 2, the thermal chamber, exhibits energy losses to the environment. If the set point temperature can be either above or below the environment temperature, or rapid temperature changes are needed, both heating and cooling control may be needed.

Both Cases 1 and 2 require passive losses to be inherent to the DUC. If the DUC does not exhibit loss, the system won't perform as needed. Case 3 illustrates this idea. The thermal chamber has been improved with high-quality insulation and thermal losses are negligible. Such a system requires both heating and cooling sources to function properly.

The important realization is that some systems require both "up and down" methods to function properly. As well, some systems depend upon the external environment for proper functionality.

Another example is a phase locked loop (PLL), shown in Figure 10-15. The DUC here is the voltage-controlled oscillator (VCO). For the PLL, there are no "losses" to the environment. Both pump up and pump down are needed.

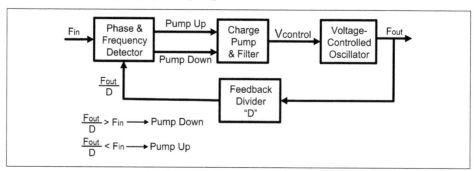

Figure 10-15. PLL with charge pumping

Whenever an up/down active control is implemented, the behavior at the crossover point needs to be carefully considered. Consider the ideal thermal chamber system shown in Figure 10-16.

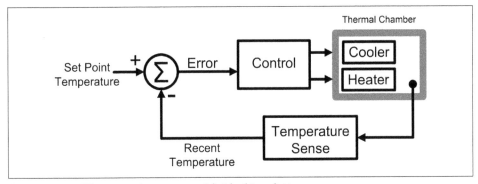

Figure 10-16. Heater-cooler system with ideal insulation

The crossover point is where the sense data is equal to the set point data. For this thermal management system, the sense temperature is equal to the set point temperature. For this system, crossover point behavior may have problems, as illustrated in Figure 10-17.

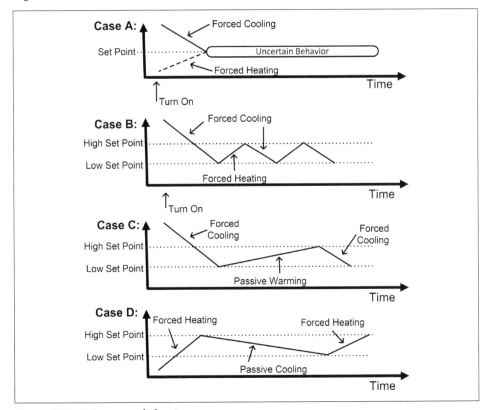

Figure 10-17. Crossover behavior

In Case A, upon starting up, the chamber could be either above or below the set point. The controller moves the output toward the set point, but the action at the set point is not well defined. How the system behaves at the crossover needs to be consistent and predictable, and that's not shown here.

Case B introduces a second set point into the system allowing the DUC to cycle between forced heating and forced cooling. However, if the system is designed to maintain a single temperature, this would be inefficient.

Case C and Case D show a commonly used method where the set point used changes depending on whether the system is forcing the temperature up or down. The system then enters a passive state of heating/cooling depending upon the temperature of the external environment.

Temperature control is used as an example here because it is easily understood. Many control systems exhibit problems at the crossover. Consequently, the important takeaway is that crossover behavior needs to be carefully reviewed and tested for potential issues.

Sequence Control

The two basic variants of sequence control are blind and verified sequencing, illustrated in Figure 10-18. *Blind sequencing* turns the DUC on and off with no sensors involved. To use this successfully, DUC functionality must be reliable and repeatable. Also, blind sequencing is suitable only when a possible malfunction can't cause damage or harm. *Verified sequencing* uses sensors to provide DUC performance data that confirms proper response as the event sequence is performed.

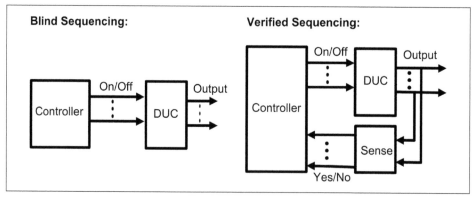

Figure 10-18. Basic sequence control systems: blind and verified

There are some potential issues with blind sequencing. Two examples are shown in Figure 10-19. First, accurately filling a tank for a fixed time period will be dependent on the fluid flow rate. The fill level will be variable and may not meet the required accuracy. Second, running a DC motor by turning the power on and off will result in a motor speed that varies with mechanical loading, variation of power supply, and bearing wear.

If accuracy is unimportant, blind sequencing might meet system needs. This approach is mostly used where accuracy isn't a concern and low implementation cost is a priority.

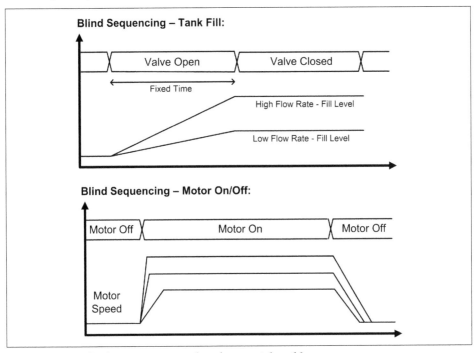

Figure 10-19. Blind sequence control and potential problems

Sequence control can be clock based or event based, as shown in Figure 10-20. A traffic light system is an example of sequence control. The top illustration in Figure 10-20 uses clock-based blind sequencing, where the traffic lights repetitively cycle yellow-red-green ad infinitum.

The bottom example uses event-based sequencing, by adding sensors into the roadway that detect vehicle presence. A vehicle arrives at the light and triggers sensor A, which initiates a light sequence allowing the vehicle to proceed.

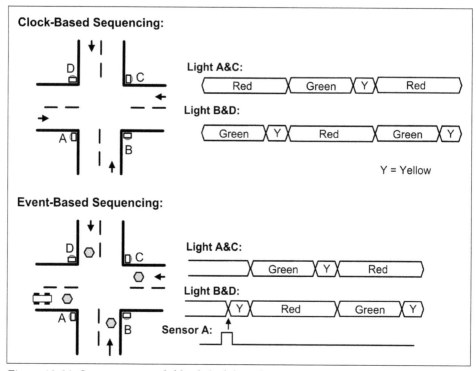

Figure 10-20. Sequence control: blind clock based versus event based

Verified sequence control is widely used. An industrial control example is shown in Figure 10-21. In this example, the sequence controller is used in manufacturing where two liquids are measured out, mixed together, and drained. Valves (V#) are operated as defined by the sequence controller, but control execution is conditional upon level sensor (LS#) status. Verification of fluid levels provides well-defined fluid quantities and safer system operation.

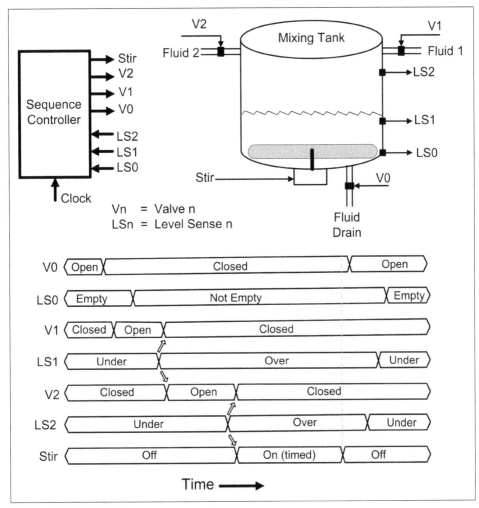

Figure 10-21. Verified sequence control

Digital controllers are generally faster than the electromechanical devices they control. Frequently, systems require time separation between events, allowing the system to complete the prior task. Consequently, definable wait times are an essential feature. Two examples are shown in Figure 10-22.

The "wait before measure" case turns on a power supply and then allows time for the voltage output to stabilize, before performing a test measurement. The "wait before drilling" action is common in automated machining systems. A stepper motor exhibits some oscillatory motion when going to a new location. The wait time allows a consistent and repeatable drill pattern.

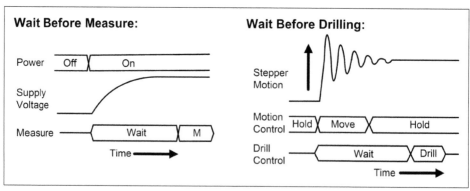

Figure 10-22. Sequence control: wait time in electromechanical systems

The code used to create a wait time should be linked to the cycle counting of a real-time clock (RTC). Wait routines not using an RTC reference are vulnerable to changes in hardware performance or may create problems if the system is changed and suddenly time delays are changed.

In addition to wait times, the phase relationship of switched devices should be checked. As illustrated in Figure 10-23, trying to create simultaneous events can sometimes generate problems. The time relationship between devices A and B in Figure 10-23 is ambiguous when trying to make them occur at the same time. Depending on the function of the devices, more predictable behavior is generally realized when defining a "make before break" or "break before make" phase relationship. Which to use depends on the devices involved. The motor-brake illustration releases the stopping brake on a motor before spinning the motor. Explicitly releasing the brake before rotating the motor avoids conflict.

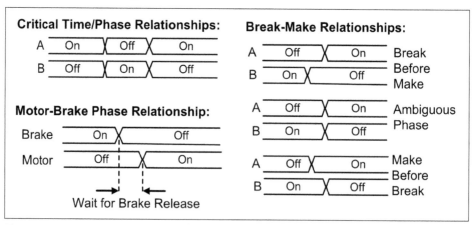

Figure 10-23. Sequence timing–critical time/phase relationships

Designing a sequence control system is usually straightforward. Using suitable state verification sensors, including intentional time delays that allow devices to stabilize, and creating well-defined event phasing helps create predictable and repeatable behavior.

Select Topics in Analog Control Systems

Invariably, when trying to vary the output of a DUC the device exhibits some form of analog behavior. Due to this, a control system design needs to address the analog characteristics of the DUC. Consequently, control system design is inherently analog in nature. However, a digital controller that mimics that analog behavior can be readily created. As a start, a better understanding of where analog systems are still used and what the problems are with that is useful insight.

Fully analog control systems are still used in some scenarios. However, the bulk of these applications tend to be implemented within an integrated circuit (IC). Figure 10-24 depicts a linear voltage regulator using fully analog methods. As a single IC, it's a minimal circuitry approach, with millions of them in existence. If the design requires no adjustments and can be put onto a single chip, it's a valid approach for an IC designer.

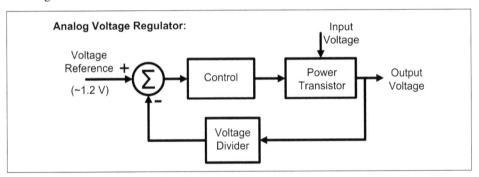

Figure 10-24. Modern uses of analog control systems

Potential problems with analog control systems can be better understood by looking at the analog implementation of a PID controller, shown in Figure 10-25. The functional drawing (top) is a generic PID controller. An error signal is created by subtracting the sensed output from the set point. This error difference is applied to three signal paths: proportional, integral, and derivative.

The proportional path has a gain multiplier (Kp) and no other signal processing. The integral of the error signal and the gain multiplication (Ki) occur in the second path. The third path takes the derivative of the error signal and multiplies it by a third gain coefficient (Kd).

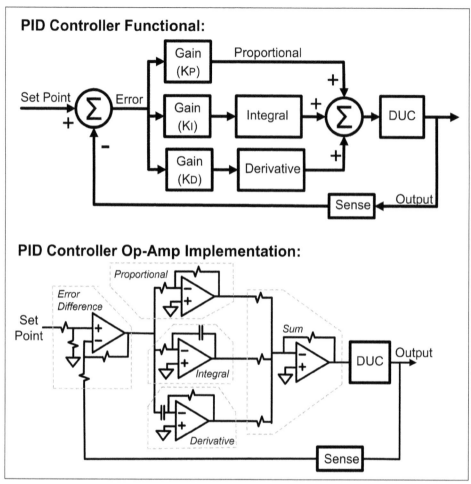

Figure 10-25. Analog control system weaknesses

The three signals are summed and used to control the DUC. Regardless of whether the PID is an analog circuit or is a sampled and quantized data stream using DSP, the mathematics of the control system remain the same. (Further details will be provided shortly.)

The analog PID controller can be implemented using op-amps to create the required math functions. The op-amp version is shown in Figure 10-25. These devices have limitations when the components are not ideal. For instance, op-amps have additive noise and offset issues, capacitors can leak current such that integration is not ideal, and resistors can be inaccurate, creating gain errors. In addition, the multitude of components adds to the implementation cost.

Most important, adjusting parameters requires component changes or resistive trim pot adjustments. None of that is friendly to high-volume production. In modern design, if adjustments are required, the capability to automatically do so under software control is essential.

A digital PID implementation can often be accomplished entirely within a single-chip MCU. For this, the needed data converters are internal to the IC and the rest of it is just code. Also, a digital PID is readily adjusted through parameter changes within the code. Between consistent performance, ease of adjustment, and minimal cost, this is the preferred method where possible.

Digital emulation of an analog system is the approach to be outlined for PID control. A minimal math approach uses the gain and phase response of the system to easily determine stability and make adjustments. Consequently, understanding certain aspects of linear control systems is useful to reach this goal.

Linear Systems and Approximations

Ideally, a linear device responds with an output that is proportional to the input. Plotting the input-to-output relationship should follow a straight line, as shown in Figure 10-26.

The slope of the line defines the gain of the device. The gain transfer function can be a dimensionless scalar, using the same units for in/out, or it can have units associated with the transfer function. For the voltage-controlled oscillator example in Figure 10-26, the units would be frequency/voltage (Hz/V).

This relationship is for a steady state input. A rapidly changing input introduces other considerations of frequency response that will be explored shortly.

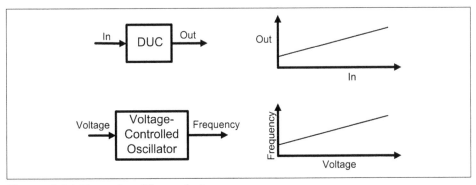

Figure 10-26. Examples of linear devices

Some devices exhibit a nonlinear transfer function (Figure 10-27), which can make them more difficult to implement in a control system. In this figure, the voltage in/out of the diode and resistor circuit has a distinct change in the in/out relationship as the diode turns on. A voltage comparator exhibits a distinct on/off characteristic. Neither device is linear over the full input range.

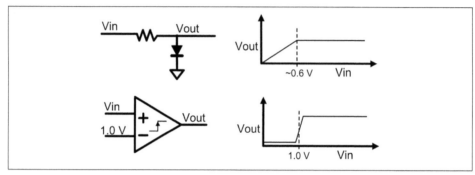

Figure 10-27. Examples of nonlinear devices

Many devices are not perfectly linear at extremes of operation, but they can be approximated as a linear transfer function over a limited range of use. This is known as a *linearization approximation* or a *small signal approximation* and is shown in Figure 10-28. Cases A and B show where there is a linear region of the transfer function. If the device is operated in this region, it can be approximated as linear. Cases C and D lack a linear region, but an approximation can be used, as shown.

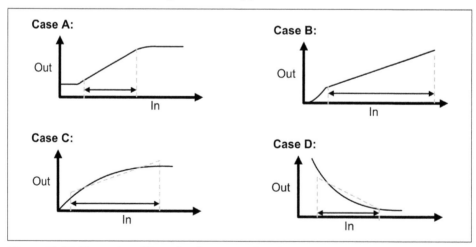

Figure 10-28. Examples of linearization approximations

For most well-behaved devices, a linear approximation can be used to analyze the system.

Bode Plots for Stable Control Loops

Since a DUC is invariably an analog device, control system analysis is often accomplished using differential equations. The math to do so can be cumbersome if the system has any complexity. However, by converting the time domain response (differential equations) to a frequency domain response (Laplace transforms), the math is simplified greatly.

The goal here is to eliminate most of the math analysis by transferring to the frequency domain and then using a graphical gain and phase analysis via Bode plots to create a stable control feedback system.

Consider the Laplace transform in Figure 10-29. The Laplace transform converts from the time domain to the complex frequency domain, allowing for evaluation of simpler algebraic expressions. Integration transforms to division by S; derivation transforms to multiplication by S.

By doing analysis in the frequency domain, a gain and phase response can be determined for any transfer function in the form of graphical Bode plots. The Bode plot can be used in most cases to determine the stability and adjust the control system. Complex math methods are thus circumvented.

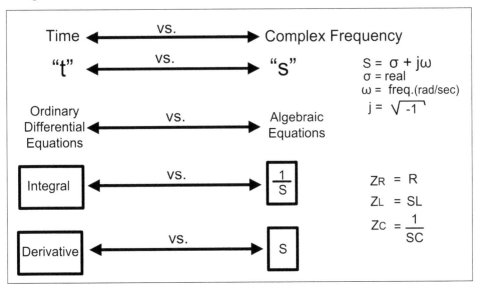

Figure 10-29. Overview of Laplace transforms

This is not a tutorial on Laplace transforms. Many textbooks exist for those interested in further reading. Understanding the gain and phase response of a transfer function, and the Bode plots thereof (Figure 10-30), is what a designer needs to work with control systems.

In Figure 10-30, a resistor and capacitor are configured as an LPF in a simple example of a transfer function with both gain and phase components. The actual gain and phase responses are a little more rounded in the bends, but the straight line approximations are easy to create and serve as a good estimate.

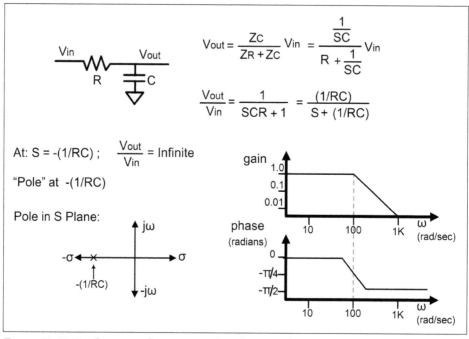

Figure 10-30. Laplace transforms: a simple pole example

The Laplace transform often uses radians/sec for frequency, radians for phase, and a logarithmic scale for gain. Most industry documentation will convert this to Hertz for frequency, degrees for phase, and decibels for gain.

Bode Plots for Gain and Phase Response

Figure 10-31 shows that using dB for the gain scale allows easy manipulation of linear scale graphs on the vertical axis. Most industry documentation will use dB for gain, degrees for phase, and Hertz for frequency. The capability to cascade multiple stages by adding gain profiles and summing up phase delays allows building an overall gain-phase response of a multistage system.

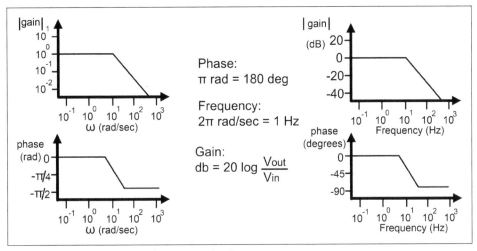

Figure 10-31. Bode plot, industry-standard units

As shown in Figure 10-32, a graphical addition of plots for gain and phase is easily done. In a similar manner, being able to easily "add up" the cumulative responses around the loop gives the gain and phase response for an entire control loop.

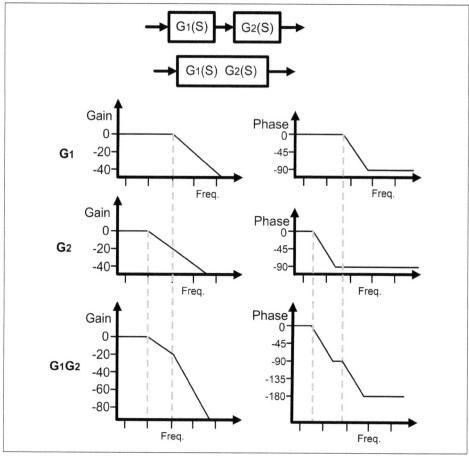

Figure 10-32. Bode plot: cascade of gain and phase responses

Bode Plots for Gain and Phase of a Control Loop

Figure 10-33 illustrates that the Bode plot of each part of the device can be determined via lab tests or device specifications. The total response of the forward path, G(S), includes the DUC, driver circuits, and controller electronics. The feedback path, H(S), includes the sensor circuitry. In the figure, the difference amplifier is assumed to have zero phase and unity gain.

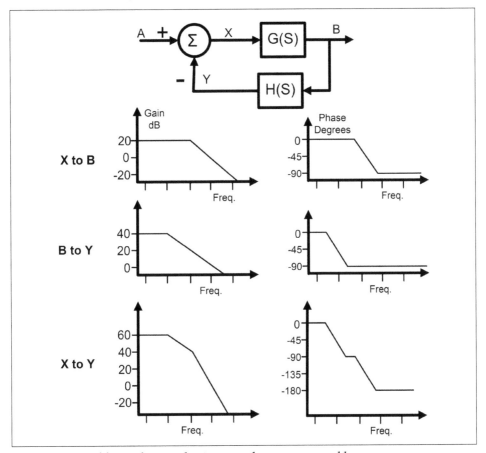

Figure 10-33. Additive phase and gain around an open control loop

With a gain-phase plot set for the entire loop, adjustments can be made to keep the control loop stable, as illustrated in Figure 10-34. A gain and phase response of each part can be determined in lab testing or found from specifications. Those are then combined for the total loop response. As shown in Figure 10-34, the additive phase is 135 degrees at 0 dB of gain and −10 dB of gain at 180 degrees of additive phase. With 10 dB of gain margin and 45 degrees of phase margin this control loop should be stable.

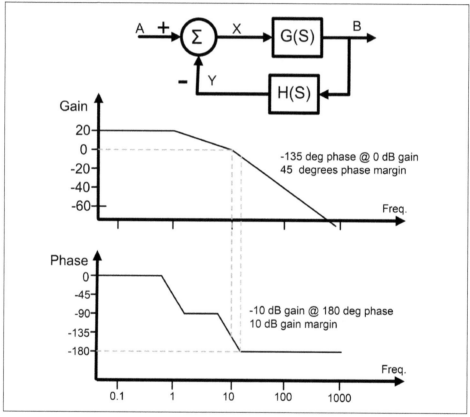

Figure 10-34. Loop stability using gain-phase responses

If additive phase is greater than 180 degrees, at 0 dB gain, the system will oscillate. Two adjustments can be made:

Reduce the phase delays

When possible, reducing the additive phase around the loop is usually desirable, but not always easily done. Frequently, the predominant contribution to phase delay is associated with the DUC. Phase delays can be reduced to create a faster responding loop, so this should be considered if viable.

Reduce the gain

Most often, gain reduction is easy, and can be done to make the control loop stable.

"Transition to Digital Control" on page 372 will show that these analog methods can be readily transferred to digital control.

How much gain and phase margin is necessary is a matter of debate (Figure 10-35). As a starting point, 30 degrees of phase margin and 8 dB of gain margin are suggested. From there, further system testing can determine if adjustments are needed.

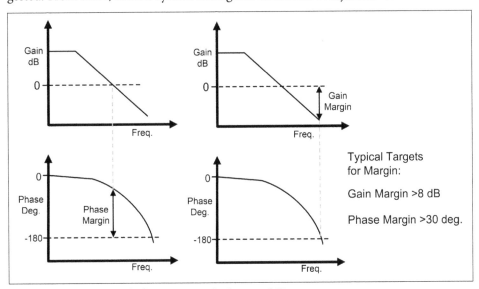

Figure 10-35. Gain and phase margin for loop stability

Bode Plots for Integral and Derivative Response

Several other Bode plots of transfer functions should be understood, since they will be used in digital PID controllers. These are integral, derivative, and time delay Bode plots.

Figure 10-36 shows that integral and derivative functions have a distinct gain and phase response. The integrator at DC has infinite gain that rolls off at 20 dB per decade. The phase response introduces 90 degrees of additive phase. The derivative has the opposite characteristic with 20 dB of gain per decade and 90 degrees of phase lead. As expected, the two functions are essentially equal and opposite.

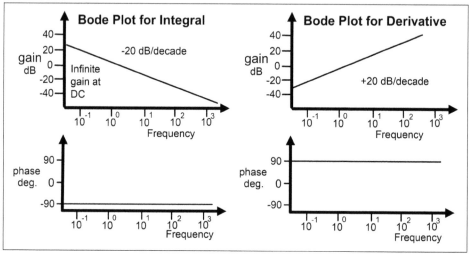

Figure 10-36. Gain and phase response for integral and derivative

Bode Plots of Fixed Time Delays

When developing an additive gain-phase Bode plot for the entire control loop, fixed time delays in the system need to be accounted for. Fixed delays (Figure 10-37) are part of any digital control block, analog-to-digital (ADC) and digital-to-analog (DAC) conversion cycle, or sensor acquisition time.

When plotted on a linear frequency scale (left side of Figure 10-37), any time delay has flat gain, and the phase increases linearly with frequency. A wide frequency range (right side) is used on a log scale Bode plot and shows that a fixed time delay can significantly affect phase margin at high frequencies. The phase equation shown allows creation of a phase response unique to the time delay found in any system. Including fixed time delays in the determination of gain-phase stability analysis is important and needs to be done.

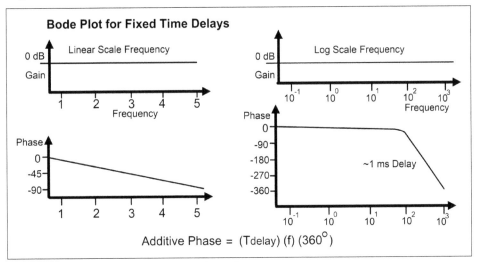

Figure 10-37. Time delays and equivalent phase response

Transition to Digital Control

As a control system is converted from analog to digital, some special items need to be considered:

- Required DUC output accuracy
- Needed DAC resolution and ripple performance
- Sampling rate and clock rate
- Sensor accuracy
- Digital math accuracy

DUC output accuracy and sensor accuracy are common to both analog and digital control. The other items are unique to a digital implementation.

Consider the differences between the systems shown in Figure 10-38. The analog controller uses proportional voltages for all signals. The control algorithm is defined using op-amps to create the math functions. The digital controller functions with digital data streams for all signals. The controller math is implemented in digital logic or a code-defined algorithm loaded into an MCU.

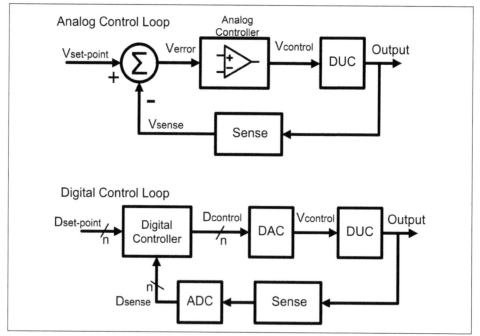

Figure 10-38. Analog versus digital controller comparison

The DAC and ADC are critical to loop performance and need to be considered in the context of driving and sensing DUC performance (FDI: ADC-DAC).

Sampling in time and quantization in amplitude are shown in Figure 10-39. A sampling rate for the sense ADC can be defined using the step response of the DUC. The resolution of the sense ADC will affect the accuracy to which the DUC can be controlled. Both will be examined shortly. First, the resolution of the control DAC and the capabilities of the DUC need to be determined.

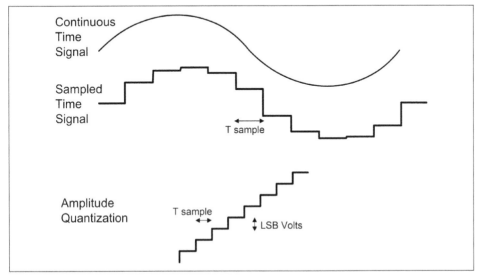

Figure 10-39. Sampled time system

Determine DUC Stability

A quick sanity check of DUC stability will determine suitability for use, as illustrated in Figure 10-40. An input signal that is fixed value and low noise should ideally yield a fixed output response. This could be motor speed, temperature, and so on. For this test, the DUC has a constant environment, without changing mechanical loads (motors) or thermal losses (temperature control) or making other environment changes. Nothing is ideal, and output variance will be seen if the instrumentation has sufficient accuracy. This first analysis is to better understand variance within the DUC.

Figure 10-40 shows an example of a VCO and the stability of the oscillator with a fixed voltage input. The VCO output should be a single frequency, but all oscillators have some inherent phase-frequency noise.

A control loop around a DUC can improve stability, but only within the bandwidth of the control loop. If the DUC has high frequency (HF) variance, a low-bandwidth control system won't compensate for the HF variance.

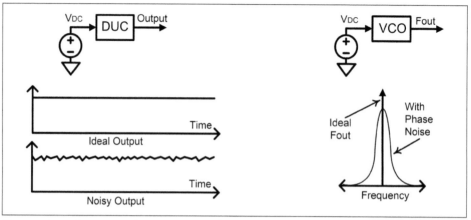

Figure 10-40. Determine DUC output stability

Depending upon the application, understanding open loop DUC stability will be useful knowledge when the DUC is put into a control loop. If a DUC has a large amount of open loop variance, it may be an issue depending upon needed performance.

DAC Performance Requirements

The DC transfer characteristic of the DUC provides the necessary data to define the DAC resolution. Using the VCO example, creating a voltage in versus frequency out plot or data table will be useful. This DC transfer function will be useful in understanding the min–max range of the output and the linearity of the transfer function, and in defining the needed resolution (Figure 10-41) of the input signal to control the DUC.

A test setup (Case A) inputs a series of voltage steps to the DUC and the output frequency is measured. The input should come from a low-noise DC power supply to fully isolate and test just the DUC performance.

The necessary performance specification should be determined at the system level. For the VCO example, the min/max accuracy of the output frequency is needed. Knowing the DC transfer function data for the DUC and the necessary accuracy of its performance, the DAC resolution can be determined.

For the VCO example, the min/max accuracy can be mapped to a min/max input voltage. The min/max voltage can then be used to determine the resolution of the needed DAC. As a rule of thumb, eight, or more, least significant bit (LSB) voltage steps within the min/max voltage window are suggested. Increasing the DAC resolution (more bits) or reducing the amplitude output of the DAC (smaller LSB voltage) are possible solutions.

Case B shows insufficient DAC resolution, but the addition of two additional bits in Case C gives the desired performance.

Multiple LSB steps should fall within the required accuracy window. This gives a margin in the design performance. If the DAC doesn't have a small enough LSB step, the closed loop behavior may produce DAC settings that fall outside the accuracy window.

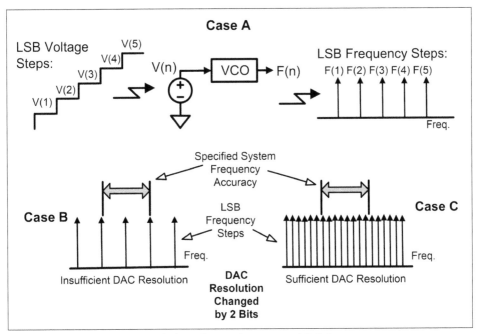

Figure 10-41. Determine needed resolution from control DAC

The DAC performance also needs to be assessed for stability and monotonic behavior (Figure 10-42). See Chapter 7 for techniques to assess and optimize the DAC. As a reminder, a common problem when using pulse width modulation (PWM) DACs is output voltage ripple. For a constant DAC input setting, the output variance must be smaller than the LSB or the LSB performance is meaningless (FDI: ADC-DAC).

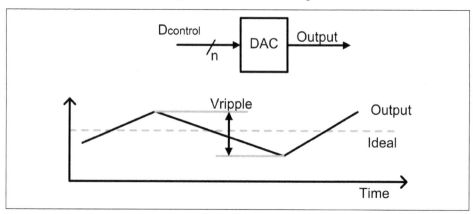

Figure 10-42. Checking ripple performance of the DAC

Accuracy of Control Math

Math accuracy in the controller should be checked. Calculations can be done as floating point or integer. Both are viable, as long as calculation errors are not significant to the computation of the control algorithm.

Most modern controllers use 32-bit math internally, so calculation errors are typically not an issue. If using a legacy MCU with 8- or 16-bit capability, accuracy should be checked, and coding for higher-resolution math may be needed.

ADC Performance Requirements

The ADC provides a digital value from the sense circuit. The sensor and ADC need to measure DUC output with greater repeatability, accuracy, and resolution than the system requires. Without that, it will be difficult (or impossible) to provide precise control (FDI: Sense).

The ADC resolution (number of bits) is determined in a similar fashion as the DAC, namely the resolution needs to be better than the desired output accuracy.

ADC Sampling Rate Determination

Some control textbooks tie the sampling rate to the control loop bandwidth. This can be a problem because the electronics are often designed and built before the overall performance is known. Defining the ADC sampling rate before the hardware is fabricated can be useful. Consequently, using the DUC step response to set a minimum sampling rate provides a solution, as illustrated in Figure 10-43.

The open loop-step response is a slewed output that is dependent upon the time constant of the DUC. Typically, this is a delayed response (lag time) followed by a slew rate-limited transition to the new output value. As shown, the DUC may also exhibit some oscillatory ringing, which is especially common in electromechanical devices.

The process time constant, $T_{process}$, is defined as the time needed to transition to 63% of the final value. Setting the sampling time period such that 10 or more samples occur during this time period is commonly used as a suitable conversion rate.

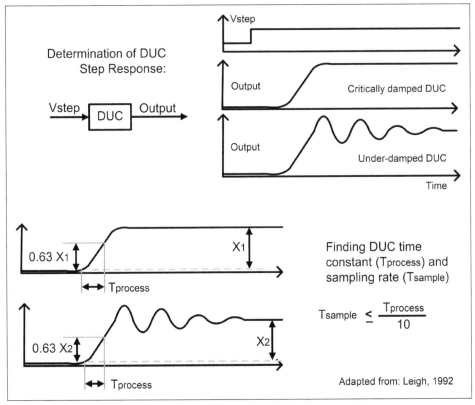

Figure 10-43. Determine sampling rate for control system

Final Selection of ADC and DAC

With the earlier defined bit count and the sampling rate determined, an ADC can be selected. A perfect fit to the numbers may not exist, but an equal or higher resolution and an equal or faster sampling rate should be available. With that, the sampling rate of the selected ADC becomes the sampling rate for the entire digital control loop. Similarly, with the DAC resolution determined, a DAC that functions at the same rate as the sampling clock can be selected.

Dual-Clock Strategy for Improved Phase Margin

Minimizing phase delays around the loop is always desirable, as discussed earlier. Using two synchronized clocks (Figure 10-44), one for ADC sampling and the other for the DSP of the controller algorithm, reduces additive phase in the control loop.

The digital controller is usually capable of faster clocking than the ADC. Consequently, a faster DSP clock allows DAC updates to occur sooner. This reduces additive phase delays in the control loop.

The two clocks must be synchronized to each other, and dividing/counting down the DSP clock to create the sample clock is a solution. Also, setup and hold timing for the digital output of the ADC feeding into the DSP needs to be checked. Many MCU devices have divider counters and multiclock capability built in, so the ADC, clock divider, digital controller, and DAC can often be implemented using a single chip.

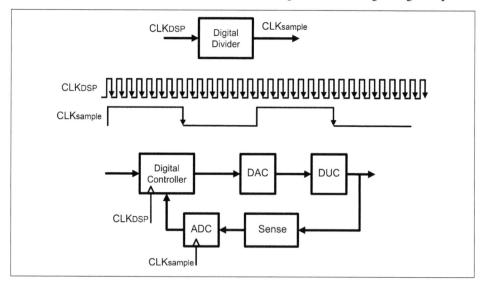

Figure 10-44. Using synchronized clocks for sampling and DSP

Digital Trapezoid Integration

The integration math function is used in many controller structures. In a digital implementation, an approximation by trapezoid integration (Figure 10-45) is suitable.

In its simplest form, integration is the cumulative area under the function curve. The trapezoid area defined by two adjacent samples is easily found by taking the mean value of the two samples and summing the sequential "area" calculations.

This is a simplified view; negative value handling must be included, and integrator range limitations may be needed. Implementation can be code on an MCU, or hardware-defined gates.

The T_{sample} of the horizontal axis is common to all samples, so in use it can be recognized as a common scalar, as long as all T_{sample} periods are the same.

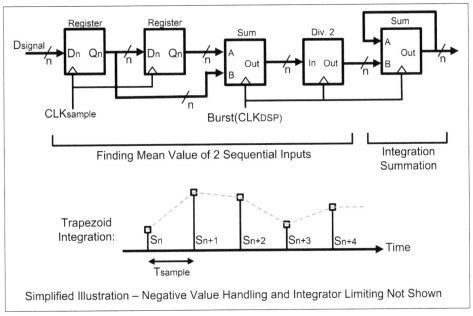

Figure 10-45. Digital integration

Digital Integration: Limit Windup and Avoid Saturation

Integrator windup occurs when a large value is stored within the integration function and drives the system out of the linear response region (analog circuits) or tries to send the DAC beyond its operating range in a digital implementation. Because the integrator stores prior sample integrations, recovery can be slow. Two improvements to the PID controller are suggested in Figure 10-46:

Limit integrator range (Limit A)
> This limit is selected in conjunction with the integrator gain (Ki). If the proportional and derivative outputs are zero, the limit value is selected so that the DAC input is at its maximum. This limits the integrator windup to values within the normal operation range of the DAC. Higher integrator gain indicates a smaller limit value to meet this requirement.

Limit DAC input range (Limit B)
> This limit prevents the DAC from sending a control signal to the DUC that is outside its functional range. Values used here are dependent upon the input signal range of the DUC.

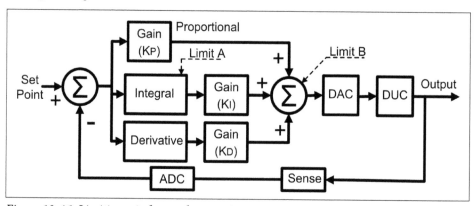

Figure 10-46. Limiting windup and saturation

A digital version of the derivative is also needed.

Digital Derivative by Adjacent Samples

The derivative function, in its simplest form, is the slope of the line at the sample point. In the digital implementation, an approximation is done using the difference between adjacent samples, as illustrated in Figure 10-47.

The value difference between two adjacent samples provides a suitable derivative approximation. This is a simplified view, where coordination of clocks (sample and DSP) has been omitted. Also, the derivative function needs to process both positive and negative value inputs for both positive and negative slope scenarios.

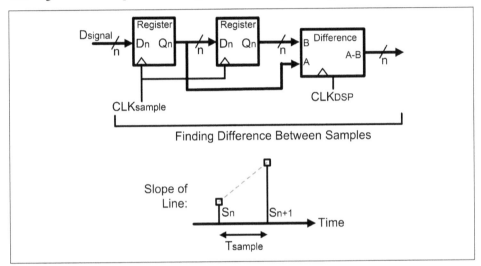

Figure 10-47. Digital derivative

Similar to the integral, the sampling period (T_{sample}) is common to all samples, so in practical use it can be recognized as a common scalar, as long as all time periods are the same.

Additive Time Delays in the DSP

In addition to phase shifts introduced by signal processing, additive time delays in the system need to be accounted for, as shown in Figure 10-48.

Using the two-clock strategy can reduce the additive phase delay of the digital part of the system. In many cases, it is not a significant issue but should be accounted for. The phase delay due to fixed time shifts becomes significant for control loops with fast response times.

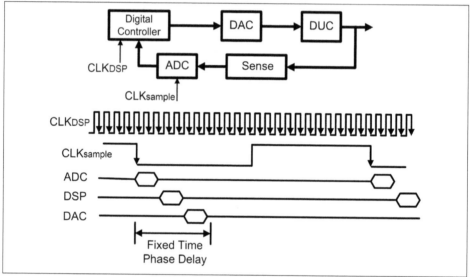

Figure 10-48. Additive delay/phase due to signal processing

The information here on DAC and ADC selection, sampling rate, and phase/gain stability criteria is applicable to most controllers. Some designers will implement a customized controller fitted to the DUC and its required performance. Most designers take a more expedient path and implement a PID control structure.

PID Control Implementation

With the prior information in this chapter, the reader now has the background to implement a digital PID controller. For all implementations, the mathematical structure remains the same, as shown in Figure 10-49.

The PID controller is also called a three-term controller by some authors. They are the same thing.

The PID controller processes the error through three parallel paths: proportional, integral, and derivative. In actuality, the PID controller is often configured without all three paths active. One of the most popular is the PI configuration, with no signal through the derivative path. The different configurations, and which variant to use, will be examined shortly.

PID controllers are widely used in many scenarios. PID control predates electronics, and early variants were mechanical or pneumatic. With the birth of the transistorized operational amplifier circa 1960, PID control was widely implemented using analog electronics. As digital electronics grew during the 1970s–1980s, system-level controllers transitioned to digital emulation of the PID controller. The availability of high-performance and low-cost ADC and DAC devices as integrated circuits has resulted in most PID controllers being digital. Analog implementations still exist and are primarily used within integrated circuits.

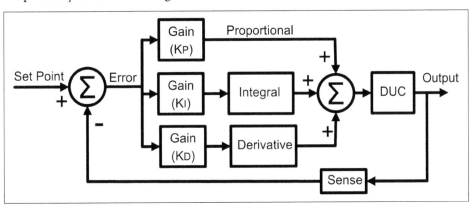

Figure 10-49. Basic PID controller structure

Modern implementations can be done using digital gates or as program code, as illustrated in Figure 10-50. This structure explicitly shows the DAC and ADC with a digitally implemented PID controller. The design methodology explored here is often called "digital emulation of an analog system" in the literature. If the sampling rates and binary resolution are suitable, the performance is quite similar. Digital emulation varies from analog system performance when:

- Sampling rate is too slow
- Samples are improper due to aliasing
- ADC or DAC resolution is insufficient

The procedures defined earlier for converter selection, sampling rate, and math resolution avoid these problems.

Typically, selecting physical logic or code implementation comes down to the needed operation speed. Slower DUC control functions can often be implemented using code downloaded into an MCU. Faster sampling may need a controller implemented in physical logic.

Another approach uses an MCU exclusively dedicated to the feedback and control task. A main controller defines the task and a secondary controller carries out the functions of the control loop (FDI: Architecture).

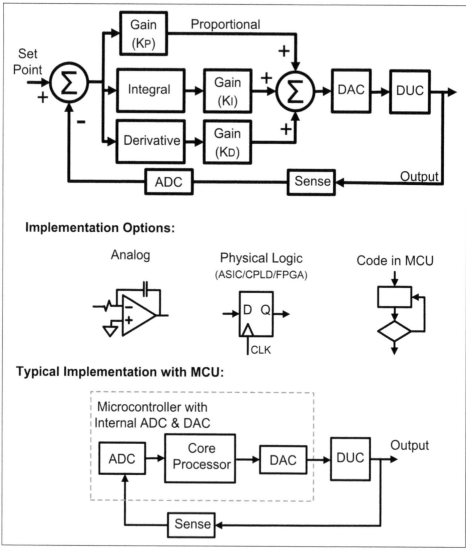

Figure 10-50. PID implementation

Response Variants: P, I, PI, and PID

For the proportional-only response, the gain settings for integration and derivative are set to zero, or are otherwise disabled (Figure 10-51). Due to finite DC gain, the output does not achieve zero error in steady state. Increasing the gain (K_P) reduces steady state errors. However, higher gain causes stability problems, which are indicated by ringing of the output response. With enough gain, the loop will oscillate. Increasing the gain reduces the gain/phase margin.

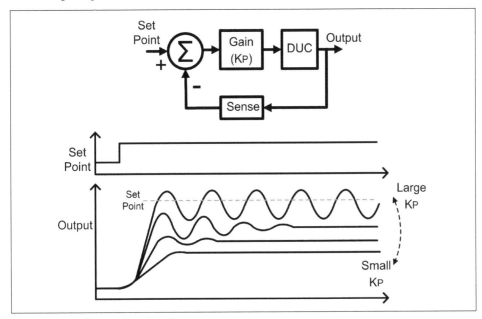

Figure 10-51. Proportional-only response

As gain, K_p, goes up, the gain/phase margins of the system are reduced until the device becomes an oscillator, as shown in the gain-phase plots in Figure 10-52.

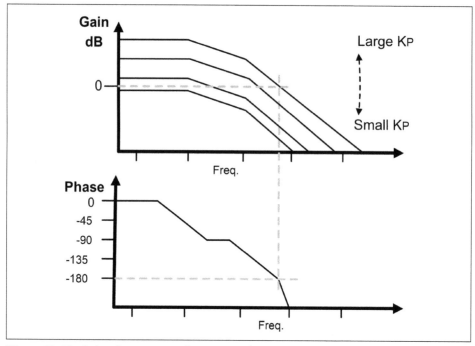

Figure 10-52. Gain-phase margin for increased K_p

To eliminate steady state error requires infinite gain at DC. That requires using an integrator (Figure 10-53). The integral-only response is slow and poor in dealing with rapid changes in either the set point or external loading of the DUC. The integrator response speeds up with increases in gain (K_I), but this will also become unstable as the gain-phase margin is reduced.

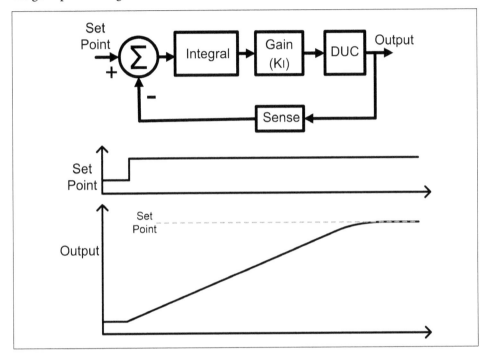

Figure 10-53. Integrator-only response

Both proportional and integral terms can be combined to eliminate steady state error and create a reasonable dynamic response. The PI controller (Figure 10-54) has been widely adopted as a simple but well-performing controller structure.

The PI structure is probably the most widely used controller in a multitude of industrial systems. When properly tuned, it yields negligible steady state error and reasonably quick response to changes in either the set point or DUC loading.

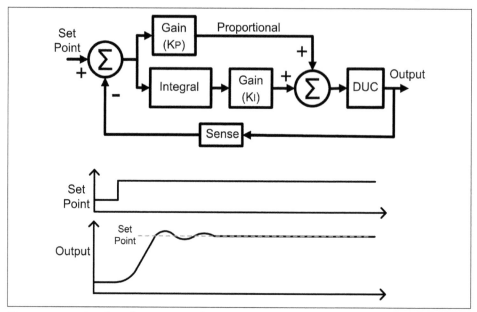

Figure 10-54. Combined proportional/integral (PI) response

If the closed loop PI response exhibits excessive overshoot and ringing, the gain terms can be reduced, resulting in a slower response with less overshoot. An alternative introduces a differential term (Figure 10-55) to reduce both overshoot and settling time.

A common designer strategy implements a PI structure and determines whether the performance is suitable. If the response characteristic includes too much over-shoot and ringing, introducing a differential term frequently provides an acceptable solution.

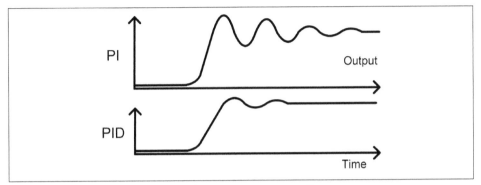

Figure 10-55. Reduced overshoot of PID versus PI controller structure

Typical Effects of Gain Adjustments

Adjusting the gain terms produces predictable performance changes, as itemized in Table 10-1. Any adjustment has to stay within the limitations of gain-phase restraints for stability. A Bode plot for a PID controller can be created through empirical testing or the use of math tools (MATLAB or other). Laplace transforms can be used to get gain-phase equations, but hand analysis of multiple gain iterations is time consuming. The full control loop can be simulated to adjust parameters, but that assumes a DUC model, which is often not available.

There is an easier way!

Table 10-1. Typical effect of raising PID gain coefficients

Gain coefficient	Rise time	Overshoot	Settle time	Steady state error
K_P	Reduce	Increase	Negligible	Reduce
K_I	Reduce	Increase	Increase	Eliminate
K_D	Negligible	Reduce	Reduce	Negligible

Ziegler Nichols Tuning

Simple methods to set PID gain parameters are available. The use of tuning rules to characterize the DUC and set gain coefficients has been available for many years. In this section, two sets of tuning rules will be examined:

- Ziegler–Nichols tuning (ZNT, published in 1942)
- Chien–Hrones–Reswick tuning (CHRT, published in 1952)

Many academic approaches to control system design require an accurate DUC model. Following is one perspective:

> A typical industrial process tends to be poorly defined with, perhaps, at best, a step response available. The specification to be met is not in general easy to translate into servomechanism terms. It is under these typical industrial conditions that the apparently crude route through step-response, Ziegler Nichols procedure and the three-term controller implementation is preferred in practice, although it may appear inferior as a scientific method. (Leigh, 1992)

Tuning rules circumvent the lack of a model by using experimental characterization of the DUC. This expedites the process of controller definition and has been widely adopted.

ZNT is designed to achieve a particular performance profile, as shown in Figure 10-56. The goal of ZNT is to achieve an output that overshoots the steady state operating point, but settles out to 25% of the overshoot in one cycle (t_{period}) of the overshoot oscillation. This target is good for rapid acquisition of the final value, but it may not be suitable for other scenarios.

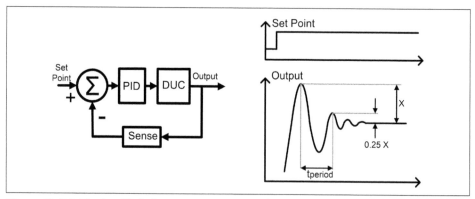

Figure 10-56. Ziegler–Nichols tuning response goal

The first step in implementing ZNT is to characterize the DUC. Two different methods are available: step response and ultimate sensitivity.

When testing the DUC, the step signal should be in the midrange of the DUC input. For example, if the min–max input range of the DUC is from 0 V to 1 V, a step function from 450 mV to 550 mV should be suitable. Using a step input that does not saturate the DUC or go into the nonlinear extremes of operation helps maintain the DUC as a linear device.

All step response plots here have been normalized to the input signal. After the output signal data was recorded, it was divided by the input signal amplitude, creating the plot. For example, if the input step is 0.1 V, the output data is divided by 0.1 V, effectively scaling the plotted values by 10X. This removes the amplitude of the input step from consideration.

For the step response (Figure 10-57), the input is created and the output is recorded. The step response has certain distinct characteristics:

Lag time (t_{lag})
  Indicates the elapsed time between the input of the step and the beginning of the output response. Lag time is part of loop phase delay and can make stable control more difficult to achieve. Consequently, larger amounts of lag time reduce K_P and K_I to help maintain stability.

Rise time (t_{rise})
  Indicates the amount of time it takes the output to slew between the initial and final values.

Output amplitude (A)
  Is proportional to the input step amplitude. This is used with the rise time to determine line slope.

The table at the bottom of Figure 10-57 shows the controller gain coefficients achieved when using the step response plot. Gain values are shown for the P, PI, and PID configurations.

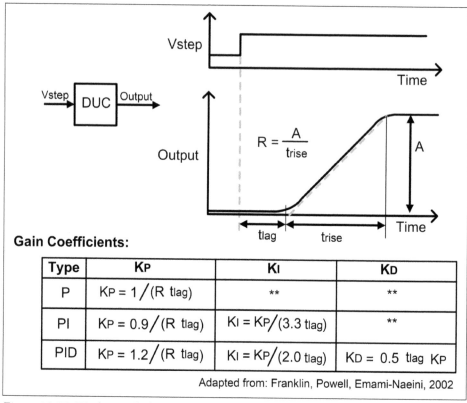

Gain Coefficients:

Type	KP	KI	KD
P	$K_P = 1/(R\ t_{lag})$	**	**
PI	$K_P = 0.9/(R\ t_{lag})$	$K_I = K_P/(3.3\ t_{lag})$	**
PID	$K_P = 1.2/(R\ t_{lag})$	$K_I = K_P/(2.0\ t_{lag})$	$K_D = 0.5\ t_{lag}\ K_P$

Adapted from: Franklin, Powell, Emami-Naeini, 2002

Figure 10-57. Ziegler–Nichols tuning by DUC step response

The "ultimate sensitivity" method of ZNT is implemented with a closed control loop. The closed loop uses a proportional-only controller. Instead of a step response, a loop using only proportional gain is tested. The proportional gain is slowly increased until the loop starts to oscillate. This gain setting is known as the ultimate gain (K_u) and the oscillation period (P_u) is called the ultimate period. These are used to determine PID gain coefficients, as shown in Figure 10-58.

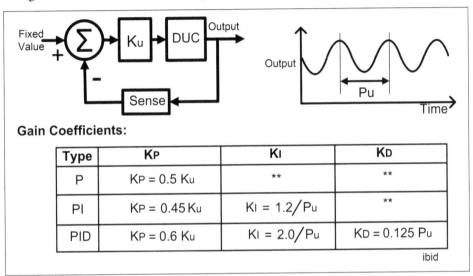

Gain Coefficients:

Type	KP	KI	KD
P	KP = 0.5 Ku	**	**
PI	KP = 0.45 Ku	KI = 1.2/Pu	**
PID	KP = 0.6 Ku	KI = 2.0/Pu	KD = 0.125 Pu

ibid

Figure 10-58. Ziegler–Nichols tuning by ultimate sensitivity methods

ZNT by both step response and ultimate sensitivity are widely used. Selecting between the two often comes down to what method can be done easily and is nondestructive to the DUC.

Chien–Hrones–Reswick Tuning

ZNT creates a system that responds quickly but is often on the edge of stability. A more conservative, more stable tuning method is Chien–Hrones–Reswick tuning (CHRT), which typically has better gain-phase margins (Figure 10-59).

CHRT introduces two overshoot options (none and 20%) in the tuned loop response. Also, CHRT recognizes that changes in set point and load variance have different optimal tuning and consequently provides optimal rules for both.

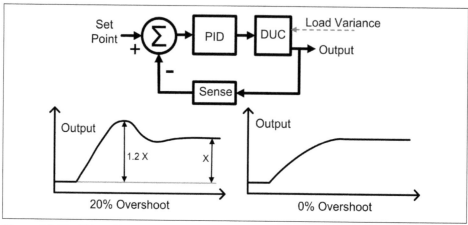

Figure 10-59. Chien–Hrones–Reswick tuning goals

CHRT starts with a step response test (Figure 10-60). CHRT rules require measuring the lag time (L), rise time (T), and gain of the DUC (K_{DUC}). The slope of the response is equal to both (a/L) and (0.63 K_{DUC}/T).

With the step response parameters determined, select for optimal response to changes in set point or loading, and preferred overshoot. Gain coefficients can then be determined from the tables.

Although ZNT methods are more widely known, ZNT results often sit on the edge of stability, especially when system component variance comes into play. Using CHRT parameters tends to produce a more stable system.

A PI configuration controller loaded with CHRT (0% overshoot) gain settings is a good starting point for a new design. Thankfully, with a digital PID controller, changing gain parameters is an easy numeric download or a reflash of memory.

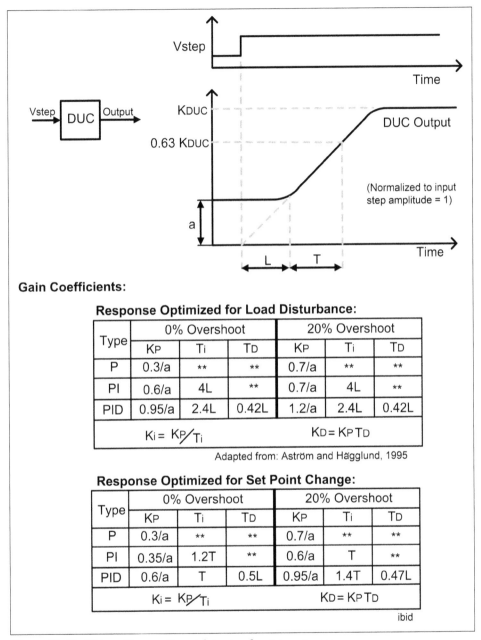

Gain Coefficients:

Response Optimized for Load Disturbance:

Type	0% Overshoot			20% Overshoot		
	K_P	T_i	T_D	K_P	T_i	T_D
P	0.3/a	**	**	0.7/a	**	**
PI	0.6/a	4L	**	0.7/a	4L	**
PID	0.95/a	2.4L	0.42L	1.2/a	2.4L	0.42L
$K_i = K_P/T_i$			$K_D = K_P T_D$			

Adapted from: Aström and Hägglund, 1995

Response Optimized for Set Point Change:

Type	0% Overshoot			20% Overshoot		
	K_P	T_i	T_D	K_P	T_i	T_D
P	0.3/a	**	**	0.7/a	**	**
PI	0.35/a	1.2T	**	0.6/a	T	**
PID	0.6/a	T	0.5L	0.95/a	1.4T	0.47L
$K_i = K_P/T_i$			$K_D = K_P T_D$			

ibid

Figure 10-60. Chien–Hrones–Reswick tuning by step response

Component Variance and Control Tuning

Tuning a control loop with a high amount of precision can quickly become meaningless if parts of the system have large amounts of performance variance. Understanding the sources and magnitude of that performance variance helps to determine whether the variance is tolerable or whether more complex control techniques are needed.

The system depicted in Figure 10-61 comprises several parts:

System clock

    If the clock frequency is defined by a crystal or ceramic resonator, the clock will be stable and repeatable across multiple devices. Accuracy in clock frequency is a problem when the clock is set by an RC time constant or ring oscillator made of digital inverters. Some MCU vendors trim this frequency or provide an adjustment register, but meaningful correction requires an accurate clock elsewhere in the system. It's best to use an accurate clock source from the start.

Sense and ADC

    Sense circuits come in many variants. Things like speed sensors are often digital event counters and their accuracy and repeatability are generally good. Sense circuits using analog measurements (strain gauges and others) need analysis for accuracy and may need calibration for consistent results (FDI: Sense). How sensor variance affects system accuracy needs to be checked. ADC accuracy can have gain, linearity, and offset issues, but converters are available that are both accurate and consistent (FDI: ADC-DAC).

Digital PID controller

    If the clocking rate is accurate and consistent and the math has suitable accuracy, there's little here that can vary. Many papers written about the limitations and accuracy of DSP were published when 8- and 16-bit MCUs were the standard. With the prevalence of fast 32-bit core structures, DSP limitations are rarely an issue.

DAC and driver circuits

    Suitable DAC performance was discussed earlier. If the resolution, min–max range, steady state ripple, and monotonic behavior are suitable, loop accuracy will be controlled by the accuracy of the sensor and controller system. If the DAC output is a string of PWM pulses (FDI: Drive & ADC-DAC), there's very little that can vary from device to device.

    Driver circuits use power transistors and can introduce variance due to changes in the characteristics of the transistors over temperature and from device to device. The power supply associated with the power transistors is usually not an issue, unless the power supply is unregulated or coming directly from a battery.

Then the performance needs to be evaluated at the battery's min–max range. Also, if the device functions while the battery is charging, proper functionality needs to be checked there as well.

DUC

The biggest variance in a control system is usually DUC performance. Electromechanical devices vary with factors such as age, temperature, frictional losses, and load variance extremes. Without calibration or adjustment, the system needs to work with any variance of the DUC, but also across multiple devices due to manufacturing differences. Generally, the DUC is the biggest elephant in the proverbial control room.

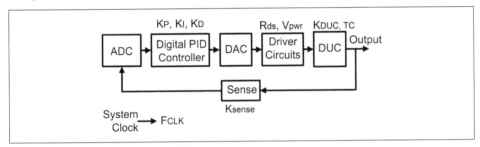

Figure 10-61. Variance of control loop components

Things like high gain in the DUC and increased lag time in the step response can usually be compensated for by slightly reducing PID gain coefficients. The extremes of DUC performance need to be understood to yield satisfactory results. This strategy is not adaptive control, but rather sacrifices some performance to acquire more stability margin.

Adaptive Control Methods

Adaptive control is a general description for a control system whose parameters can be adjusted to improve performance. An adaptive system (Figure 10-62) will usually be an observer of both DUC input and output. DUC performance is monitored, and controller parameters are adjusted to improve performance. This is a very "big picture" view, with the details of adjustment and monitoring unique to the DUC.

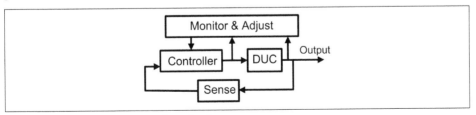

Figure 10-62. Adaptive control methods

Two forms of adaptive control (Figure 10-63) are gain scheduling and self-tuning. With gain scheduling, the DUC output and the characteristics of the environment are observed to determine what parameters should be loaded into the controller. This gain schedule is predetermined for optimal performance under a particular set of conditions. Gain scheduling has been widely used in aviation where airspeed and other dynamics are monitored and parameters are loaded into the controller to adjust flight control characteristics.

Self-tuning—adjusting control parameters based upon performance evaluation—is common to most adaptive control scenarios. One example of self-tuning is adjusting controller gain based upon output response overshoot. Too much overshoot causes the reduction of controller gains. This can compensate for variations in DUC performance. Many other forms of self-tuning exist, with different strategies of system test and adjust, depending upon the nature of the DUC.

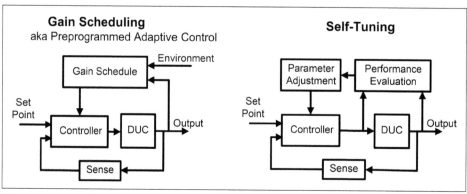

Figure 10-63. Adaptive control: gain scheduling and self-tuning

A method known as Model Reference Adaptive Control (MRAC) uses an ideal reference model as an adjustment reference (Figure 10-64). The reference model is an ideal version of the desired transfer function and includes the equivalent of the controller, DUC, and sense feedback. Typically, the reference model is an equation set or possibly a data table. The ideal model ("Desired Output" in the figure) and the DUC response ("Output" in the figure) are compared, and system adjustments are made to move DUC performance closer to the desired output.

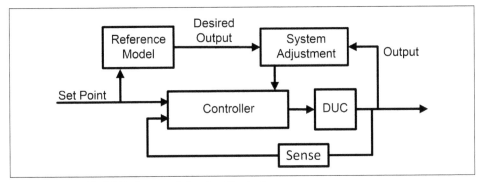

Figure 10-64. Model Reference Adaptive Control

With the availability of fast, low-cost digital processing, many forms of adaptive control with measure and adjust methods have emerged.

Trajectory Control Methods

Trajectory control (Figure 10-65) is another form of adaptive control. Trajectory control changes the dynamics of the loop during the transition, depending on where the system is in the transition.

For the motion control example in Figure 10-65, controller configurations are kept as a set of rules. Whereas gain scheduling swaps different sets of gain parameters into the controller, trajectory control has the capability to change the entire control algorithm. For motion control, rules ("Mode Select" in the figure) are changed based upon real-time location information in comparison to the desired location. The motion rules include desired trajectory data, and controller actions to follow the desired path.

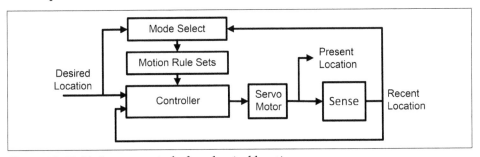

Figure 10-65. Trajectory control of mechanical location

Trajectory control is valuable where DUC performance is predictable and repeatable, such as the servo motion used in a hard disk drive. Test cases can be built and performance optimized for the application. With the freedom to change the control rules during the transition, very aggressive and high-performance results can be achieved.

Developing a target trajectory is shown in Figure 10-66. In this example, mechanical motion from location X to location Y is needed. The target trajectory has been defined and the associated velocity and acceleration needed are shown. A soft acceleration and deceleration are used to minimize jerk and the associated mechanical vibration.

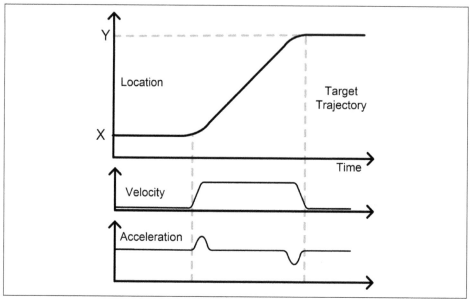

Figure 10-66. Target trajectory

Looking at the desired trajectory, it can be seen that there are four different control regions (Figure 10-67) depending upon where the servo motor (DUC) is in the transition. When a move command is given at (A), Launch rules activate (B), which creates a predefined smooth acceleration protocol until the desired transition velocity is reached. Upon reaching the transition velocity, the system changes to constant velocity Transit rules (C). When the device is a predefined distance from the destination, the system switches to Approach rules (D) for deceleration, using a smooth protocol for slowing down. Approach rules aim for a low velocity when arriving at the destination location. Most of the velocity is gone upon crossing the destination, and the controller changes to Track rules (E).

Transfer between rule sets must not disturb functionality and must maintain a smooth, cohesive operation. This is often referred to as a *bumpless transfer*.

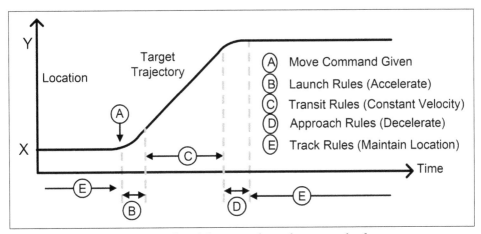

Figure 10-67. Trajectory control with location-dependent control rules

Since the Transit region is constant velocity, it is easy to use this rule set to move different distances, as illustrated in Figure 10-68. Shorter transitions (X to Y2) go directly from Launch to Approach rules. Close-proximity moves (X to Y1) stay in the Track mode. If the system knows present and destination locations, the rule change locations are easy to determine.

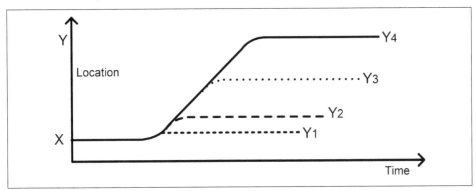

Figure 10-68. Trajectory control: versatility of constant velocity transit

Following are some supplemental comments on the four modes illustrated in Figure 10-69:

Launch mode

The start of rapid motion is a smooth acceleration, which is often run as a predefined sequence of increasing servo motor currents until about 50% of the transit velocity is reached, followed by a decreasing current sequence, arriving at the final transit velocity. In precision-motion electromechanical systems, any velocity changes need to avoid rapid adjustments to minimize mechanical vibration. A launch sequence of increase/decrease currents is often developed through experimental methods and may not be perfectly symmetric due to frictional losses or mechanical bias.

Transit mode

The transit mode is for maintaining a constant velocity until passing a predetermined distance to the final destination, where it switches to approach mode.

Approach mode

The controlled reduction of velocity, without causing mechanical vibration and a suitable low velocity at destination arrival, is the goal here. Velocity at destination is low enough that the track mode can take over in a bumpless transfer.

Track mode

Holding position at the destination (tracking) is optimized for low static error and immunity to disturbances. This mode is not for high-speed transitions, but the need to move quickly is satisfied by the other mode sequences.

In the design phase, a typical strategy is to start with a slow, nonaggressive target trajectory, and progressively speed the system up until desired performance is reached.

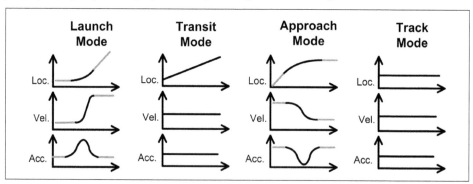

Figure 10-69. Trajectory control: launch, transit, approach, and track

Motion control is the example explored here, but variants of these methods are applicable to other systems. Trajectory control is useful when the DUC response is repeatable and predictable.

A set of rules designed to fit a target trajectory circumvents much of the DUC variance. In this example, the servo motor currents may scale up or down with DUC variance, but the motion and acceleration profiles are ideal definitions and remain constant.

Typically a system can run a calibration routine to determine its current scaling, or parameters can be adjusted while the system is in use. Both methods have been used in high-volume products.

Control system textbooks frequently vaunt disk drive servo performance. Disk drive servo systems have been using digital-based trajectory control since about 1980. A fixed-parameter system would need to always use track mode rules, severely limiting seek time performance.

Implementation of trajectory control is more complex than PID methods but is a good path to high-performance motion control. Recently, robotics applications needing high-performance precision motion have adopted these techniques.

Summary and Conclusions

The goal of this chapter is to explain:

- Control systems, what they do, and how they work
- Bode plots, and gain-phase margins for feedback stability
- Digital emulation of analog control systems
- Functional and performance requirements of a digital controller
- Dual-clock structures to improve phase margin stability
- Behavior and limitations of P, PI, and PID controllers
- Response characteristics of ZNT and CHRT tuning rules
- Component variance in a control loop
- Adaptive control methods
- Trajectory control systems

With that information, readers should be able to:

- Structure sequence control systems and suitable event timing
- Properly "size" a DUC for nonsaturation functionality
- Specify and select suitable DACs/ADCs/sensors
- Determine an appropriate sampling rate
- Define a digital PID structure and select desired tuning rules
- Test a DUC and determine proper gain settings for ZNT and CHRT tuning rules
- Design a PID system and select a configuration (P, PI, or PID)
- Determine whether adaptive or trajectory control methods are needed

Presented here is a small piece of control systems theory. You should have received enough information to get the job done, however. In too many books, control systems are presented in a very abstract manner, and the process outlined here takes more of a "need to know this to get that done" approach.

If a deeper look into the topic is needed, two texts listed in "Further Reading" on page 406 are suggested:

- *Feedback Control of Dynamic Systems*
- *PID Controllers: Theory Design and Tuning*

Both present the topic in a detailed manner, but provide enough support commentary to be readily understandable.

PI and PID control suffices for most control system problems. High-performance methods often aren't necessary, and doing a high-level analysis of optimum control frequently offers little performance improvement, especially when DUC variance comes into play. Selection guidelines provided here for DAC, ADC, sampling rates, and sense accuracy should suffice for any control structure that a designer might implement within a digital controller.

Further Reading

- *Feedback Control of Dynamic Systems, 4th Edition*, by Gene Franklin, J. David Powell, and Abbas Emami-Naeini, 2002, ISBN 0-13-032393-4, Prentice Hall.

- *PID Controllers: Theory, Design and Tuning, 2nd Edition*, by Karl Astrom and Tore Hagglund, 1995, ISBN 1-55617-516-7, Instrument Society of America.

- J.G. Ziegler and N.B. Nichols, "Optimum Settings for Automatic Controllers," Transactions of American Society of Mechanical Engineers, 1942, Vol. 64, 759–768.

- K.L. Chien, J.A. Hrones, and J.B. Reswick, "On the Automatic Control of Generalized Passive Systems," Transactions of the American Society of Mechanical Engineers, 1952, Vol. 74, 175–183.

- *Applied Digital Control, Theory, Design and Implementation, 2nd Edition*, by J.R. Leigh, 1992, ISBN 0-486-45051-1, Prentice-Hall.

- *Adaptive Control, 2nd Edition*, by Karl Astrom and Bjorn Wittenmark, 1995, ISBN 0-201-55866-1, Addison-Wesley.

- *Computer Controlled Systems: Theory and Design, 3rd Edition*, by Karl Astrom and Bjorn Wittenmark, 2011, ISBN-13 978-0-486-48613-0, Prentice-Hall.

- *Real-Time Computer Control: An Introduction, 2nd Edition*, by Stuart Bennett, 1994, ISBN 0-13-764176-1, Prentice-Hall.

- *PID Control Fundamentals*, by Jens Graf, 2016, ISBN 978-15-353-5866-8, Sinus Engineering.

- *Modern Control Engineering, 5th Edition*, by Katsuhiko Ogata, 2015, ISBN 978-93-325-5016-2, Pearson Education Prentice-Hall.

- *Discrete Time Control Systems, 2nd Edition*, by Katsuhiko Ogata, 2015, ISBN 978-93-325-4966-1, Pearson Education Prentice-Hall.

- *Practical PID Control*, by Antonio Visioli, 2006, ISBN-13 978-1-84628-585-1, Springer-Verlag.

- R.K. Oswald, "Design of a disk file head positioning servo," *IBM Journal of Research Development*, November 1974, Vol. 18, 506–512.

- Daniel Abramovitch and Gene Franklin, "A Brief History of Disk Drive Control," *IEEE Control Systems Magazine*, June 2002, 28–42.

- Stuart Bennett, "The Past of PID Control," *Annual Reviews in Control*, 2001, Vol. 25, 43–53.

CHAPTER 11

Schematic to PCB

Prior chapters dealt with various pieces of the system, and the discussion now moves to organizing those pieces into an understandable and documented schematic. From there, the design of a printed circuit board (PCB) is examined, along with what's needed to commercially build a PCB loaded with components.

The PCB design process starts with selecting the components and defining the schematic. Then a detailed bill of materials (BOM) can be created. The selected components require suitable connection footprints for mounting to the PCB. These can be pulled from existing vendor libraries, or for less common devices they can be designed from scratch.

Depending on PCB size and circuit complexity, a suitable set of PCB layers (aka *layer stack-up*) can be defined. A strategy for optimized component placement and connections is explored, with information on many of the "special attention" items necessary for a PCB. These include transmission line design, Kelvin sense connections, low-impedance power decoupling, electromagnetic interference (EMI) and electrostatic discharge (ESD) issues, specialty vias, thermal relief methods, and others.

PCB design, fabrication, and component assembly doesn't follow a single set of fixed rules. Every contract PCB manufacturer (CPM) has somewhat different capabilities and can provide rules suitable to their capabilities. Consequently, frequent referrals are made herein to information needed from the CPM. Both a design rule set and a capabilities sheet are commonly published by every CPM.

PCB design tools (aka EDA or CAD) are discussed in a generic manner to provide a better understanding of design rule checking (DRC), layout versus schematic (LVS) verification, and creation of the needed manufacturing and assembly data files. Vendor-specific EDA tools are not discussed.

Surface mount components are the emphasis here, and designers should minimize the use of through-hole components. This allows automated assembly, lower manufacturing costs, less EMI, and better high-frequency performance. Many CPMs can component-load (aka "stuff") PCBs using automated assembly, even for small (5–10) lots, so even small prototype quantities are built without manual assembly.

PCB Terminology

Understanding some commonly used terminology is useful. Some brief definitions follow, with more details in upcoming sections:

Bill of materials (BOM)
A detailed itemization of all components used to build the PCB.

Bare board
Refers to a PCB without electronic components.

Copper weight
Refers to the thickness of the copper used in the metal layers.

Controlled impedance connection
A connection that uses transmission line methods to optimize interconnect for high-frequency signals.

Cutout
Opening cut into a PCB for mounting or fitting the PCB around an obstruction (Figure 11-1).

Creepage
The distance between two points that follows the shortest path on the PCB (Figure 11-1).

Clearance
The shortest distance between two points on the PCB (Figure 11-1).

Design rule checking (DRC)
Automated layout checking for adherence to the physical design rules of sizing, spacing, and other physical aspects of the design.

EIA (Electronic Industries Alliance)
A trade organization responsible for electronic components standards. The EIA has been absorbed into the ECIA (Electronic Components Industry Association); see *https://www.ecianow.org*.

EDA (electronic design automation) or CAD (computer-aided design)
  Software tools for schematic capture, simulation of circuits, PCB design, verification of design interconnect, verification of physical design rules, and creation of output files used to manufacture/assemble the PCB.

Footprint
  Copper pattern used on PCBs for mounting an electronic component (Figure 11-2), also known as a decal or land pattern in some EDA tools.

IPC
  Industry consortium, originally called the Institute for Printed Circuits, that defines PCB standards and requirements. See *https://www.ipc.org/* for more information.

Gerber files
  A digital file format commonly used to transfer PCB design information. These are sometimes referred to as a PCB manufacturing file set.

Pad
  Singular copper region (aka land) that connects one component pin to the PCB; see Figure 11-2.

Pitch
  Spacing (center to center) between adjacent pins on an integrated circuit (IC); see Figure 11-2.

Layout versus schematic (LVS, aka layout versus netlist)
  Automated checking of a PCB layout for correlation between the physical interconnect and schematic connections.

Layer (conducting, insulating, ground, power)
  Layers of material in a PCB that are laminated together. A conducting layer is typically created from copper and is used for electrical connections. An insulating layer is nonconductive and used to separate the conducting layers. A ground layer is dedicated to a common ground connection for the PCB, and a power layer is dedicated to the distribution of electrical power.

Silkscreen
  Nonconductive comment and documentation layer used on the outside surfaces of the PCB for component labels and other information.

Signal layer
  A metal layer used to define interconnection signals on the PCB.

Surface mount (SMT) component
  Electronic component that doesn't require drilled holes through the PCB to mount (Figure 11-3).

Through-hole component

Component using wires extending from the component body (Figure 11-3). These wires insert into drilled holes. This is also called a leaded component.

Traces and trace routing

Electrical interconnects on a PCB metal layer (Figure 11-3).

Via

A conductive connection that goes through a PCB, instead of across a PCB as a trace does (Figure 11-3). A via is used to change an electrical connection from one layer to another.

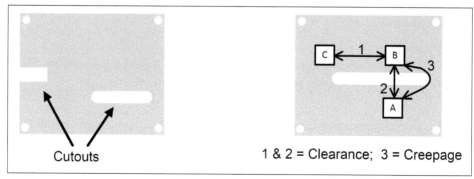

Figure 11-1. Cutout, creepage, and clearance

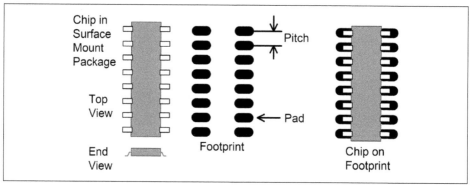

Figure 11-2. Footprint, pitch, and pad

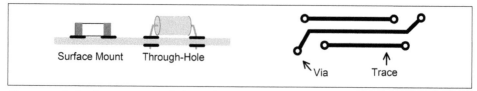

Figure 11-3. Surface mount and through-hole components, vias, traces, and routing

PCB Design (EDA) Tools

EDA tools typically offer the capability to:

- Capture and edit schematics
- Generate a connection netlist
- Create component symbols and footprints
- Define multilayer PCBs and multiple control layers
- Do component placement and interconnect routing
- Check compliance with physical design rules
- Determine whether the physical layout correlates with the circuit schematic
- Generate manufacturing data files (includes metal layers, silkscreen, solder masks, solder paste stencil, automated PCB drilling, BOM creation, and automated component placement)

There are many providers of EDA tools, and these tools are constantly evolving. The discussion here focuses on what needs to be done to create a PCB, and not how to use any specific EDA tool.

Getting Started

As a circuit schematic comes together, certain important features need to be included:

- Externally connected devices need ESD protection.
- Radiated noise needs to be limited to pass EMI regulatory testing.
- Power systems need high-frequency bypass capacitors to ensure power stability.
- Antialias filters (or no stuff placeholders) should be placed at analog-to-digital converter (ADC) inputs.

Suitable circuit solutions for all of these have been discussed in earlier chapters.

Component Selection

All components are vendor supported with specifications and application notes. Always acquire and review this material.

Selecting RLC Components

Selecting discrete components is primarily a strategy of size versus ease of access. Smaller components are necessary for miniaturization but can be difficult to access or rework (Table 11-1). Picking a component package size comes down to available PCB area and whether all the components can fit.

If the PCB area is sufficient, larger devices (0805, 0603 Imperial) are easier to work with and modify without extreme effort. For space-limited situations, smaller components are necessary to fit area constraints. Due to the small dimensions, a "magnifier and tweezers" approach is needed for rework of any surface mount PCB.

Table 11-1. RLC surface mount package

Imperial				Metric			Package
PKG	L (mils)	W (mils)	Power (W)[a]	PKG	L (mm)	W (mm)	
01005	16	8	0.03	0402[b]	0.4	0.2	
0201	20	10	0.05	0603[b]	0.6	0.3	
0402[b]	40	20	0.06	1005	1.0	0.5	
0603[b]	60	30	0.10	1608	1.6	0.8	
0805	80	50	0.125	2012	2.0	1.25	
1206	125	60	0.25	3216	3.2	1.6	
1210	125	100	0.5	3225	3.2	2.5	
1812	180	125	0.75	4532	4.5	3.2	
2010	200	100	0.75	5025	5.0	2.5	5mm
2512	250	125	1.0	6332	6.3	3.2	

[a] Only for resistors.
[b] Metric and imperial 0402 and 0603 are different devices.

Schematic labeling of components (Figure 11-4) should provide enough information to understand the device basics. At a minimum, the component value and reference designator (R7, C11, L25, etc.) are necessary. As schematic space allows, power ratings (resistors), maximum operating voltage (capacitors), device tolerance, and other performance criteria can be added.

Beyond the basics, EDA schematic capture tools typically allow parameter attachments (Table 11-2) that are not visible in the schematic but are linked to each individual component. Things like a manufacturer name, manufacturer part number, definition of footprint, and package size all need to be attached to the component. Each EDA vendor has its own methodology for doing this. Labeling and linking of parameters can also be included as needed.

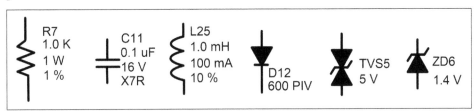

Figure 11-4. Labeling of common discrete components

Table 11-2. RLC device parameters

Resistors	Inductors	Capacitors
Component value	Component value	Component value
Value tolerance (%)	Value tolerance (%)	Value tolerance (%)
Package size	Package size	Package size
Package height	Package height	Package height
Operating temperature range	Operating temperature range	Operating temperature range
Temperature coefficient	Temperature coefficient	Temperature coefficient
Maximum power rating	Maximum current rating	Maximum voltage rating
Material composition	Saturation current	
	Equivalent series resistance	
	Frequency of self-resonance	
	Quality factor at designated frequency	
	Utilizes magnetic shielding	

When selecting resistor values, opt for EIA standard values to keep unit costs down and ensure availability of devices from multiple vendors.

The E24 (5% tolerance) standard values list (Table 11-3) should suffice for most situations. More accurate EIA standard value devices exist (E48, E96, E192), but a price premium will be paid for higher accuracy. Standard value capacitors also use a similar set of values and the same numeric 10 multiplier for larger and smaller values.

Table 11-3. Standard EIA value resistors and capacitors (E24 list)[a]

EIA standard values, 5% tolerance			
1.0	1.1	1.2	1.3
1.5	1.6	1.8	2.0
2.2	2.4	2.7	3.0
3.3	3.6	3.9	4.3
4.7	5.1	5.6	6.2
6.8	7.5	8.2	9.1

[a] Or any 10 multiple of these (i.e., 2.7 is available as 0.27, 2.7, 27, 270, 2.7K, 27K, 270K, etc.)

As discussed in prior chapters, commonly available capacitors include a three-character rating code (Table 11-4) that defines min–max temperature ranges and temperature-dependent variances. More accurate capacitors than these are available, but high-accuracy capacitors come at a higher price and larger size premium.

Table 11-4. Capacitor rating code, class 2 ceramic capacitors

Low temp (°C)	High temp (°C)	Temp variance
X = −55°	4 = 65°	P = +/−10%
Y = −30°	5 = 85°	R = +/−15%
Z = +10°	6 = 105°	S = +/−22%
	7 = 125°	T = +22%, −33%
	8 = 150°	U = +22%, −56%
	9 = 200°	V = +22%, −82%
Temperature versus tolerance, derived from EIA RS-198		

Picking Connectors for Off-Board Wires

When choosing plugs and connectors for off-board wires, consider the following:

- Current and voltage rating
- Size and PCB area used
- Height restrictions
- Number of connections needed
- Polarized connectors to avoid backward plugging
- Insertion life cycle capability and long-term connector life
- Abuse expected from typical or unintended use
- Acceptable shape
- Application environment
- Resistance to dust, vibration, and moisture (IP rating)
- Existing standard requirement (i.e., USB)
- Supply chain and sources (multisource availability preferred)
- Multiple connectors going into the wrong receptacle
- Specialty tooling for the connector (crimping tools)

Most of these parameters are self-evident, but there are some additional issues to consider:

- Many connectors are single source or require specialized tooling to use.
- Designing a connector strategy where it is impossible to plug cables into the wrong location is always preferred.
- Many small connectors are designed for one-time assembly and are not friendly to repetitive replugging (insertion life cycle).

Selecting IC Packages

Frequently, an IC is available with multiple packaging options. Selecting a device package variant is a trade-off between needed board space, ease of access for test and rework, and high-frequency performance.

As shown in Figure 11-5, the area used by an older DIP through-hole component can readily contain five or more smaller area packages. The IC packages with bottom contacts (BGA, uCSP, and QFN) don't offer ready test access to IC pins. The SOIC, SOP, and QFP devices are easier to work with but do require more space. The decision here is again a question of area used versus ease of access.

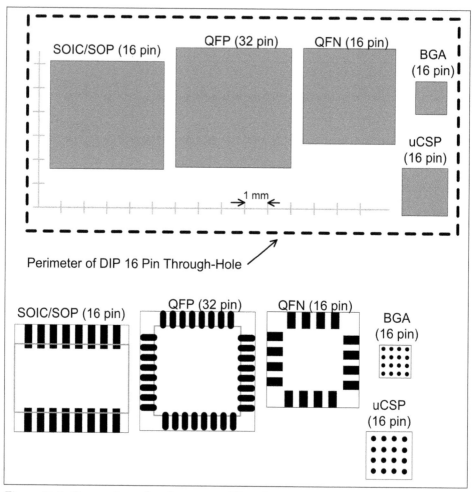

Figure 11-5. Comparison of surface mount IC package options

Checking Component End of Life and High-Quantity Availability

Most RLC components are available from multiple vendors, so multisource availability gives some security that the needed parts can be acquired. Semiconductor devices may have an end of life projected for them, however. Some chips have been selling for 40+ years with no manufacturing termination seen. Other devices may be short lived, or the vendor is promoting a newer variant while phasing out an older (or problematic) chip.

Always check end-of-life projections for any semiconductor device used in the design.

Availability in volume should also be checked for products expected to ship in large quantities. If you have a consumer product shipping millions of devices every month, some vendors may not be able to provide the needed quantities.

Including Test Access and Interface Ports

Typically, a Joint Test Action Group (JTAG) access port to the PCB (Figure 11-6) is needed to download code into the microcontroller (MCU). Most MCU, complex programmable logic device (CPLD), and field-programmable gate array (FPGA) digital devices have a JTAG port, and a plug-in port on the PCB to access the chip is needed. IC vendors provide programming platforms (an IDE) to develop code for their chips, and the JTAG interface is used to load the MCU memory. This is known as flashing the memory or loading the memory.

Access to test signals on the assembled PCB is frequently needed. Test points can take several forms: contact vias, contact pads, or solder-on test points. Ground connections at multiple locations across the PCB are also suggested. High-frequency analog signals may need oscilloscope access, and a buffered coaxial connection can be included or a scope probe access socket (shown in Figure 11-6) can be used. Putting multiple digital signals to a connection header on the PCB is another possibility, allowing easy connection to a logic analyzer.

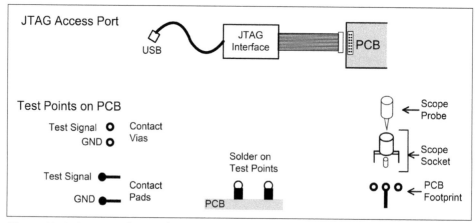

Figure 11-6. Access and control ports

Including features for test, debug, and manufacturing invariably saves time and expense in the long run.

Schematics

Schematics need to be organized, understandable, and well commented. In large organizations, this material will be distributed for use in manufacturing and in field debugging. For smaller design teams, documenting the schematics so that the designer can understand what they did months or years after the fact will be invaluable.

Schematic Sheets and General Organization

The standard American Society of Mechanical Engineers (ASME) drafting sheet is still widely used, although it was created many years ago. Its orientation is mechanical drawing, and it comes from an era of T-squares and manual drafting, when updates were created as markups of the original. A sheet format for electronic schematics is suggested in Figure 11-7.

On the top of the figure is the original ASME drawing sheet. For electrical schematics and modern computer-based design, it's less than optimal. On the bottom is a template better suited for electronics, after removing material that's not relevant to electronic circuits.

A schematic set using a multisheet hierarchy, including extensive commentary and a top-down structure, is easier to understand than a large, single-sheet document with circuits and no explanations.

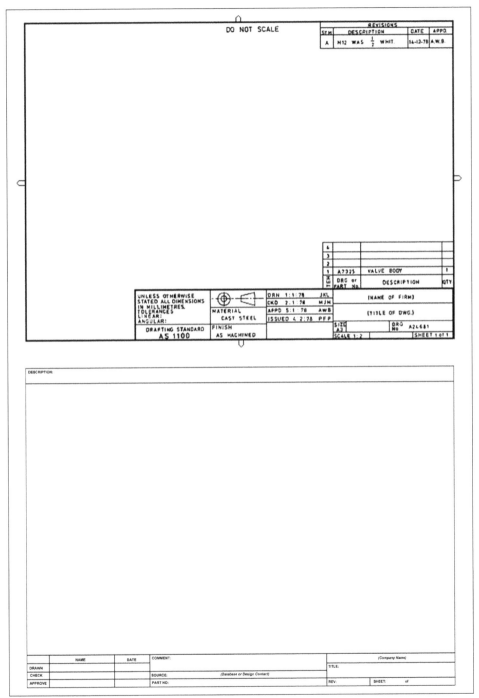

Figure 11-7. Drafting sheet and alternative format

Frequently, the top-level sheet of the schematic may not have actual circuits, and instead a block diagram illustrates the overall structure (Figure 11-8) and is used to explain the sheets that follow, as well as how to find things. In the example shown in Figure 11-8, there are no circuits on the front page. Since it is a large schematic (54 pages), the front page serves as a table of contents and guide to the general hierarchy.

A common convention uses signals and sensors entering from the left side, with controls and outputs exiting to the right. This helps explain the flow of control and signals.

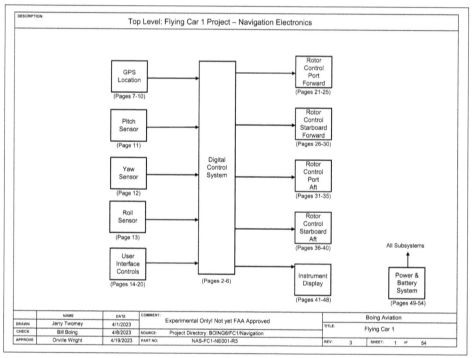

Figure 11-8. Top-level schematic

If circuit components function together, keep them grouped together to aid in understanding. Figure 11-9 provides an example of this. In the figure, Case A shows resistors and transient voltage suppressor (TVS) diodes that serve as ESD protection on the CON12 input connector. Placing those parts close together at the connector is important and should be noted in the schematic. Case B shows a similar situation where R7, R8, and C11 form a filter at the input of the ADC and should be close to the ADC input. Case C has a power connection that has been filtered with a ferrite bead and capacitor to minimize noise, at the point where the 12 V power enters the PCB.

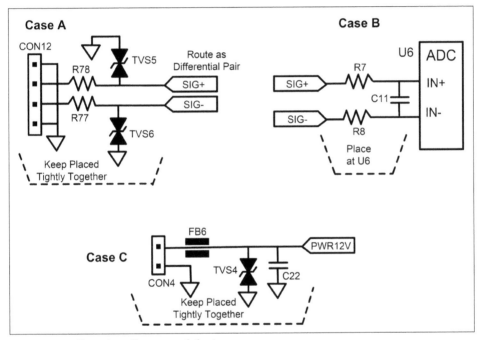

Figure 11-9. Functionally grouped devices

Symbol Organization for Integrated Circuits

IC symbols should also be structured to support the left-to-right flow of signals. For example, the two symbols in Figure 11-10 represent the same MCU. On the right, the symbol is organized by pin location on the chip package, without regard to function or signal flow. On the left, the signals are organized by inputs on the left of the symbol and outputs on the right. This convention makes it easier to understand. All chips should also include a brief functional description to aid in understanding the schematic.

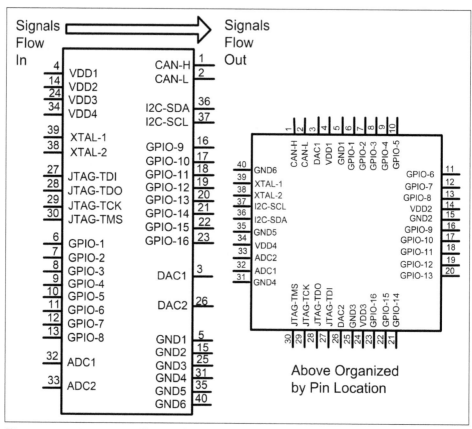

Figure 11-10. Microcontroller symbol showing functional structure

Following are some additional useful schematic techniques:

Connections using net names

When a signal goes to many places in a schematic, using net names for signals instead of a wire/line connection is suggested. This includes things like power, ground, or a clock that goes many places. This minimizes spaghetti wiring, which can be difficult to follow.

Informative net names

PWR12V, ADCIN1, and FANPWR5V all tell an abbreviated story about the signal and connection, which facilitates understanding, whereas NET17, NET45, and NET91 don't contain useful information.

Labeling hierarchy

Signals exiting a circuit section should have a common prefix. For example, signals leaving the power supply could be labeled PWR-5V, PWR-12V, PWR-2p5V, and so forth.

In the flying car example in Figure 11-8, signals leaving the GPS section of the PCB are labeled GPS-CAN-POS, GPS-CAN-NEG, and GPS-CLOCK. (These are a CAN bus and a clock coming from the GPS circuits.) Using this labeling methodology, anything entering a page on the left will have a prefix that instantly tells the reader where it came from. This is especially valuable with a large, multipage schematic. Page numbers embedded in signal names generally don't work due to page shuffling during schematic creation.

Placeholders and "Do Not Populate" Components

Frequently, designers run into situations where a component or circuit may be needed, but a final decision is pending lab testing of the PCB. For such situations, a common technique is to include the PCB locations for the circuits but not to populate those spaces with components. Unpopulated circuits can be swapped in and out with jumper connections or zero-ohm resistors. This is a common technique for first-generation designs.

Provide Generous Commentary

Always provide plentiful written commentary on what's going on within a schematic. Designers know their circuits the best, but the audience for the schematic includes many parties in manufacturing and other groups that need to understand the circuits.

Avoid Ambiguity

If the wires cross, do they connect? Don't leave people guessing; be explicit. For instance, a connection dot at the crossover point tells the story that the signals are connected. Better yet, staggering the wires slightly removes ambiguity from the connection (Figure 11-11).

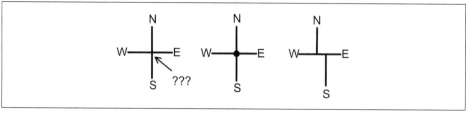

Figure 11-11. Avoiding interconnect ambiguity

Call Out Items Requiring Special Attention

Items requiring special attention in layout and/or component placement and those requiring safety warnings (Figure 11-12) need to be explicitly pointed out. Such items may need isolation or shielding, high-voltage layout spacing, or thermal relief considerations, among others. In addition, noise sources and noise-sensitive circuits should be pointed out, as should high-speed signal paths that require special routing, transmission lines, or path balancing.

Assume nothing, and convey information wherever things need to be dealt with. The circuit designer knows what the signals are on those schematic connections. Everyone else just sees lines on paper unless they are informed otherwise. Again, assume nothing!

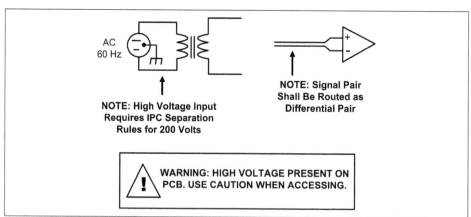

Figure 11-12. Special attention items

Bill of Materials

A bill of materials (BOM) is a detailed itemization of everything included in the PCB build (Table 11-5). Most EDA tools can automatically generate a BOM from the schematic if the necessary data is attached to the components. A careful review of the BOM output is suggested. Human data entry errors occur where several variants of the same device may exist, or a multidesigner schematic may have multiple sources for (essentially) the same thing.

Typically, the BOM format is a spreadsheet, but your CPM may have preferences or require a specific format. Boards with components on both sides will probably require two (top and bottom) BOM lists.

Table 11-5. Typical BOM content

Item #	QTY	Ref Des	Mfg Part #	Mfg	Value	PKG[a]	Comments
1	1	U1	PIC16F690-I/SS	Microchip		20SSOP	PIC MCU, Flash, 4KX14
2	2	R1,R2	ERJ-6ENF4 702V	Panasonic	47K	R0805	1/8 W, 1%
3	3	C1,C2,C3	GRM21BR71C105KA01L	Murata	1.0 uF	C0805	16 V, 10%, X7R
4	1	L1	MLF1608E100KTD00	TDK	10 uH	L0603	10 mA, 1.7 Ω, 10%
5							

[a] Package designations are imperial, not metric.

As with any documentation, make sure the BOM is kept up to date with design revisions and includes project identification, revision version, and date.

Defining Physical, Control, and Data Layers

EDA tools define the physical layers of the PCB but also include control and data layers used for mechanical definition, fabrication information, and component assembly.

The physical layers include the top and bottom (T&B) metal layers, internal metal layers, T&B solder mask, T&B solder paste mask/stencil, T&B silkscreen, board outline, drill drawing, and drill table.

The control and data layers include the assembly layer, courtyard layer, keep-out layer, and component outline layer. The assembly layer is for information and control and is used to automate assembly and aid in component-loading (stuffing) the PCB. The courtyard layer is used to define a perimeter around all the components, creating a component-level keep-out zone so that devices don't encroach on each other. In some EDA tools this is also known as a *boundary* or *bound top/bottom* for the component.

The keep-out layer is used to define regions of the PCB where components and connection traces are not allowed. The component outline layer is used to define the body of the component and aids in automated assembly of the PCB. Neither of these layers are rendered as part of the physical PCB.

Different EDA tools may have somewhat different versions of control and layer structures, but the layers mentioned here are common to most. Many of these layers are used in the creation of a component footprint.

Defining a Component Footprint

A component footprint (Figure 11-13) is also known as a *land pattern* or *decal* within some EDA platforms. Both EDA vendors and component manufacturers develop and maintain component libraries for designer use. Thankfully, the majority of component footprints can be readily pulled from these libraries.

 The accuracy of component libraries is generally reliable, but errors exist, depending upon the source and whether any quality control of the library was done. Always check the footprint against the dimensions in the component specification. Caveat emptor!

Invariably, some custom-made footprints will be needed, and modifying a similar device pulled from a known good library is a good starting point. There's more there than just the solder pads, however.

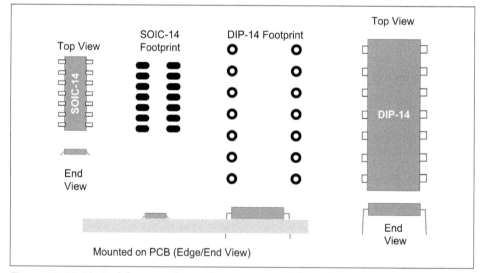

Figure 11-13. Typical footprint for surface mount and through-hole devices

Taking a closer look (Figure 11-14), additional layers are used beyond just the metal layer. For instance, a solder mask is added around all pads, preventing solder from bridging across to an adjacent connection. Also, a solder paste stencil is included to limit where the solder paste is applied to the pad.

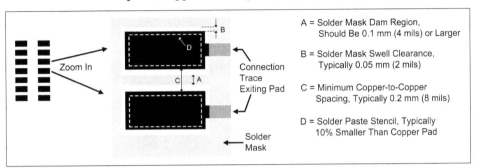

A = Solder Mask Dam Region, Should Be 0.1 mm (4 mils) or Larger

B = Solder Mask Swell Clearance, Typically 0.05 mm (2 mils)

C = Minimum Copper-to-Copper Spacing, Typically 0.2 mm (8 mils)

D = Solder Paste Stencil, Typically 10% Smaller Than Copper Pad

Figure 11-14. Footprint for pad with solder mask and paste mask

In addition to the metal layer, solder mask, and solder paste stencil, several other things are included (Figure 11-15). For instance, a pin 1 indicator and a component reference designator (R1, C5, U15) for all components should be placed in both the assembly layer and the silkscreen layer. Pin number designators for all pads should be included in the control layers, as per the specific EDA tool.

The silkscreen layer provides information for visually locating components using reference designators and a pin 1 marker on the component side of the PCB. Any polarized items (electrolytic capacitors, power supplies, etc.) should include a "+" indicator in the silkscreen layer showing proper component orientation or connector polarity. A component body outline should be included in the silkscreen and assembly layers.

The IPC-7351 document "Generic Requirements for Surface Mount Design and Land Pattern Standard" can provide further insight. EDA tool documentation will also provide guidance unique to the particular tool set.

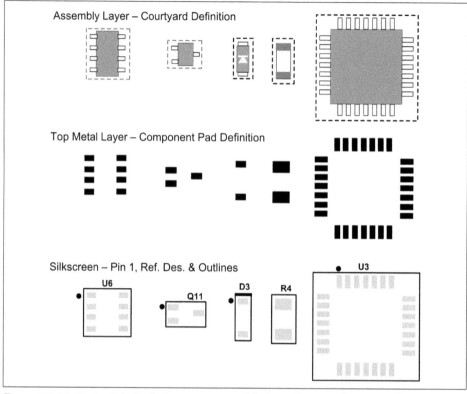

Figure 11-15. Footprint pin designators: assembly, bound top, and courtyard

Mechanical Definition of a PCB

In addition to the electronics, a multitude of mechanical issues need to be dealt with in the design of a PCB.

Metric Versus Imperial Measurements

The metric system is used everywhere in the world except for the United States. Imperial units are still often used in PCBs designed in the United States, largely due to history. Widespread PCB use started in the 1960s to 1980s, largely in the United States. Many early ICs were based upon a pin pitch of 0.1 inch. Many devices designed in that era have survived and are still widely used today. Consequently, dimensions of inches and mils (1 mil = 0.001 inch) will be seen in designs.

PCB designers are encouraged to use metric units for new designs, but they will invariably run into situations where imperial measured devices are present in the design.

The IPC has standardized on a universal grid of 0.05 mm, but many US designers still use imperial (inches and mils) units. Herein, examples are presented in both imperial and metric units. For the moment, designers need to be capable of using both and converting between them. Widespread conversion in the United States from imperial to metric units is slow and has been ongoing for decades.

PCB Mounting

The shape and size of the PCB are typically dictated by the final product enclosure. How the PCB mounts in that enclosure needs to be designed to minimize board flexing, vibration sensitivity, and response to physical shock (Figure 11-16).

First, mounting and spacing in the enclosure must provide sufficient clearance to avoid vibration or flexing that allows the PCB to touch the enclosure. Minimum spacing here is dependent upon regulatory requirements and the voltages involved.

Mounting guidelines will vary depending upon the application and regulatory body that approves products. Military, medical, and aerospace applications will be more stringent than consumer products. As with any electronic design, understanding regulatory requirements needs to happen early in the design process.

IPC-2221 includes some global guidelines and suggests a maximum of 76 mm between mounting points, but also advises that a more stringent mechanical vibration analysis should be performed.

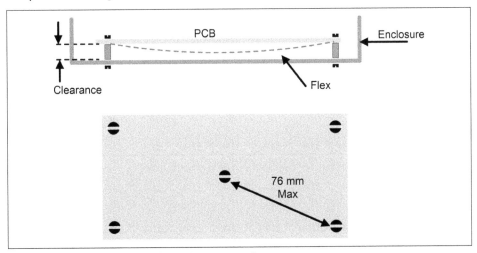

Figure 11-16. Mounting for flex, vibration, and stress

Electrical Grounding Through Mechanical Mounts

Mount holes can either isolate the PCB from the enclosure or be used to connect the PCB ground plane to the enclosure. For a new design, keeping the mounting points electrically isolated gives more freedom in changing the grounding strategy during EMI certification testing.

Drilled Hole Spacing and Keep-Outs

Keep-out regions need to be defined around any drilled through-hole on the PCB and along the perimeter of the PCB (Figure 11-17). A typical PCB is built within a larger panel that is separated into smaller PCBs. A keep-out region of at least 100 mils (2.5 mm) should be established from the PCB edge. This requirement is often called the "copper to edge of PCB clearance."

A keep-out region around all mounting holes should be 30 mils (0.75 mm) or more beyond the region containing the mounting hardware. Any drilled hole that is not part of a via should have an all-layer keep-out region of 6 mils (0.15 mm) or more beyond the outer perimeter of the drilled hole.

IPC-2221 discusses a multitude of spacing rules, and your CPM will provide a list of spacing guidelines. Many CPMs will allow closer spacing, but the preceding numbers are a safe starting point.

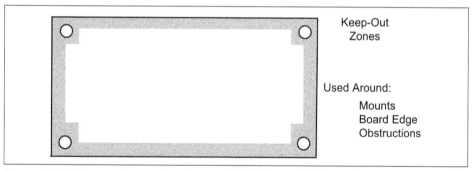

Figure 11-17. Keep-out zones for edges and drilled mount holes

Cables to the PCB

Connectors and cables exiting the PCB need to be secured so that they are not mechanically stressed or vibrating. A common failure in many products occurs when mechanical stress and vibration have loosened a connector and the connection has become intermittent. Securing the cable to the enclosure, in close proximity to the PCB, usually avoids this failure problem.

PCB Alignment References

Modern PCB assembly is done using automated "pick and place" machines to install components. These machines use location data created from the PCB design and are assisted by alignment references on the PCB to determine exact location. The IPC has standardized the *fiducial marks* used, as shown in Figure 11-18.

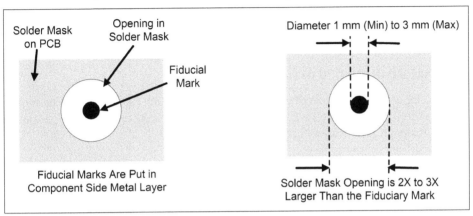

Figure 11-18. PCB alignment references (fiducial marks)

Alignment markers are required at three levels on the PCB (Figure 11-19). PCB panels and individual boards each require three reference points, placed near the corners, to align the pick-place machine. Additionally, fine-pitch IC packages use two markers on the diagonal line of the footprint. Consult with your CPM to verify the suitability of alignment markers.

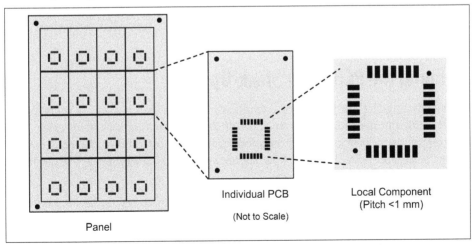

Figure 11-19. Fiducial mark locations: panel, individual PCB, and local component

Conformal Coating

A conformal coating is a thin film coating, typically applied by spraying, that protects the PCB and components from moisture and hostile corrosion environments. Many low-cost products don't use conformal coatings to reduce manufacturing costs. High-reliability products and any device placed in an open environment should consider using a conformal coating to improve reliability and device lifetime. Regulatory requirements of some products may mandate conformal coating for product approval.

Test Fixture Using Bed of Nails

Depending upon the complexity of the PCB, the capability for the PCB to be mounted to a test fixture (Figure 11-20) and electrical connections made through a "bed of nails" (aka pogo pins) may be needed.

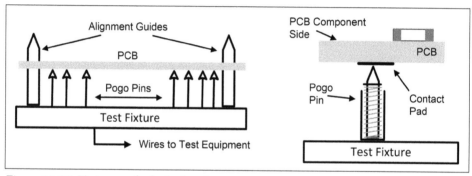

Figure 11-20. Test access by bed-of-nails contacts

A PCB is an electronic design with mechanical aspects that need to be dealt with. Both are part of a successful design.

Defining the PCB Layer Stack-Up

Understanding the fabrication of a multilayer PCB (Figure 11-21) is a good start to defining the needed PCB layer stack-up for a particular design. *Prepreg* is an abbreviation for "previously impregnated," which is most often implemented using woven glass fiber (aka fiberglass) that has been impregnated with resin. This resin can be activated using heat and pressure.

Prepreg is an insulating layer, and as part of a PCB it is often referred to as a substrate or dielectric. The *substrate* term comes from its role as the underlying insulating layer below a conductor. The *dielectric* name comes from the prepreg's role as the insulating layer between the conducting layers.

Fiberglass prepreg, often referred to as FR-4 (fire-retardant glass epoxy laminate), is commonly used. Other specialty materials are used for high-frequency (>1 GHz) devices or high-temperature (>110°C) scenarios. The vast majority of PCBs are created with FR-4.

As shown in Figure 11-21, a copper clad core is created with two sheets of copper and a prepreg layer. The addition of heat and pressure bonds the three together to create a copper clad core. A copper clad core is utilized with two metal layer masks to create unique circuit connections. This is done by removing mask-selected copper regions, creating a core with circuits on both sides. The most common method to do this uses photolithography.

Multiple cores with circuits are bonded together, using additional prepreg layers to create a multilayer stack of circuit connections. The thickness of the multilayer stack-up ranges from 0.2 mm to 6 mm with most CPM services, but there is a strong preference to using a total thickness of 1.57 mm (0.062 inches). This is largely due to legacy use of Bakelite sheets of that thickness. Diverting from this thickness is readily done, but will usually cost more and may limit which CPM can fabricate the board.

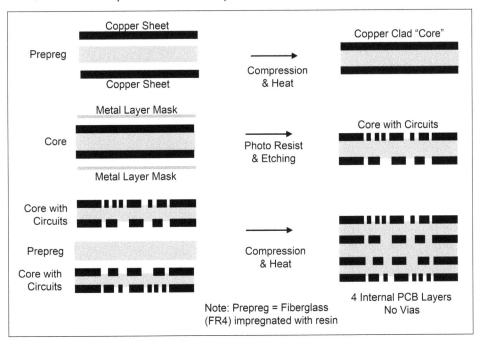

Figure 11-21. Substrate and core layers

The thickness of the copper layers can be selected for a particular application. Copper thickness is commonly referred to by weight. A "one ounce" copper layer is defined as the thickness of 1 oz (by weight) of copper that has been rolled out to an area of 1 square foot. This is about 0.035 mm thick. Different "weight" copper layers are scaled in a similar manner.

For most circuit boards, 1 ounce and ½ ounce copper are commonly used. Specialty boards requiring low circuit resistance or better thermal conductivity may use heavier (thicker) copper layers. Heavier copper options are often used for PCBs with high-current power circuits.

As shown in Figure 11-22, copper resistance changes with temperature. Most circuits won't be affected by this variance unless a precise connection resistance is needed.

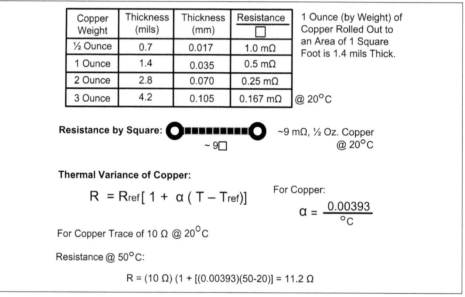

Figure 11-22. Copper weights, thickness, and resistivity

High-density PCBs have been fabricated with over 50 metal layers. Most boards have fewer than 20 metal layers and many have four to eight metal layers. Some commonly used layer stack-ups are explored in Figure 11-23.

In these stack-ups, components are loaded on the top and no components are on the bottom. As shown in the figure, a two-layer PCB is one of the simpler stack-ups used, but it lacks dedicated ground and power planes, so noise on signals, power stability, and ground bounce tends to be a problem. For general-use electronics, a two-layer PCB is discouraged.

The four-layer PCB is the high-volume workhorse of the industry. With a dedicated ground and power layer, it has much better power and ground (P&G) stability and less signal-to-signal coupling due to the P&G layers between the two external signal layers.

As component density goes up and connections get crowded, adding metal layers will help with interconnect space. Typically, metal layers get added in multiples of two due to manufacturing considerations.

Two variants of the six-layer PCB are also shown in Figure 11-23. One adds two internal signal layers and the other adds one signal layer and one additional ground layer. Both are viable depending on how much additional interconnect is needed beyond the four-layer board. The six-layer option A will have some noise coupling between Internal Signal 1 and Internal Signal 2, but using orthogonal signal traces (discussed shortly) between the two layers helps, and digital data connections should survive. Noise coupling may also extend to the Bottom Signal layer as well.

The best noise immunity puts a ground plane between signals to reduce crosstalk. The six-layer option B adds one signal layer and one additional ground layer and keeps ground planes between signal planes.

A useful strategy is to always place a ground layer directly under the top layer, in close proximity to components. Also, the prepreg under the top component signal layer can be kept thin so that any transmission lines on the top surface can use narrower traces and still get the right impedance (see "Specialized Interconnection Methods" on page 455).

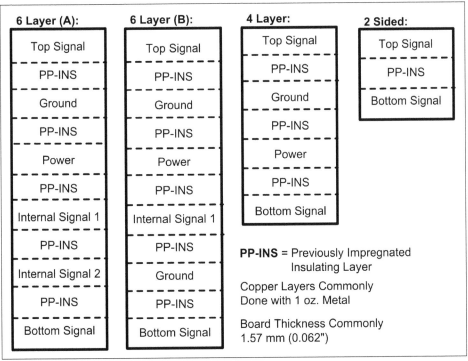

Figure 11-23. Commonly used layer strategies

If two signal layers without components are available, a common strategy is to make the connections of those two layers at 90 degrees to each other, as shown in Figure 11-24. This orthogonal signal routing achieves two things. First, it minimizes the interconnect problems with crossing wires. Connections go up and down on one layer and left and right on the other. A connection between any two points can generally be made with one via in the connection path. Second, it minimizes crosstalk and coupling between the two signal layers. Connections that cross over each other in different layers only have close proximity for the small region of the crossover, instead of lengthy lines on top of each other in the two signal layers.

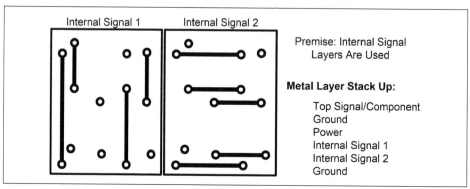

Figure 11-24. Orthogonal signal routing layers

When PCB area is limited and numerous components need to be on the board, designing the board with components on both sides (Figure 11-25) is a possible solution.

Another use of the two-sided components strategy is for noise isolation. Some form of analog signal processing circuit exists on one side and all of the digital signal processing is on the other. The ground plane in between acts as a noise shield, and adding shield covers (Faraday cages) helps keep the two sides isolated from each other (FDI: EMI & ESD).

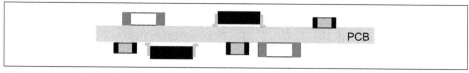

Figure 11-25. Two-sided component builds

A layer stack-up for a two-sided component build is shown in Figure 11-26. This eight-layer stack-up gives a tightly spaced ground under all components and keeps internal connections between two ground planes for minimal EMI. The two internal signal layers should use orthogonal signal routing to minimize crosstalk between the layers. The two ground layers are connected through multiple distributed vias. Most often, a two-sided build is used for a high-density PCB, a design often seen in cell phones.

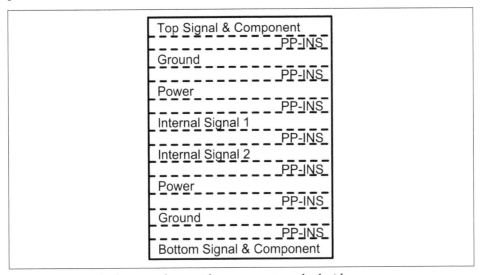

Figure 11-26. Eight-layer stack-up with components on both sides

Interplane Capacitance

Many circuit designers claim that the built-in capacitance between P&G layers is useful as a power supply filter. Taking a quantitative look at the built-in power capacitor shown in Figure 11-27 can be informative.

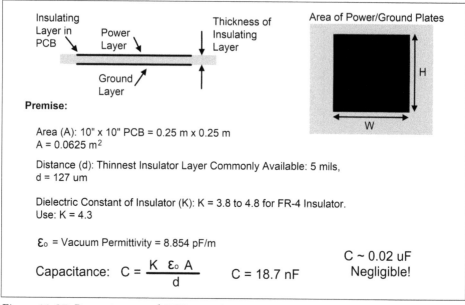

Figure 11-27. Power to ground PCB capacitance

This example assumes a sizable PCB area, a full-sized ground plane and power plane, and a minimal thickness dielectric between them. Doing the math shows a total capacitance between the layers of roughly 0.02 uF for the entire PCB.

That amount of capacitance is insignificant when the PCB should already have a large number of power supply bypass capacitors distributed across the board (FDI: Power). Also, considering that capacitance is spread across a large area, its utility at high frequencies is doubtful due to the distributed network inductance.

Physical Design Rules

Physical design rules should be obtained from the CPM. Following is a sample set of design rules that are checked using EDA software, along with some explanations of the physical design rules (Figure 11-28). This is commonly called *design rule checking* (DRC):

Clearance minimum
    Minimum metal-to-metal separation.

Short circuit
    When two nets with different names are electrically connected.

Unrouted net
    When separate traces have the same net name but are not connected.

Unconnected pin
    When a component pin does not have an electrical connection.

Trace width min/max
    Min/max limits on the width of metal traces.

Corner routing
    Checks for preferred corner routing.

Preferred via
    Checks that all vias meet the specified dimensions of the preferred via.

SMT pad-to-corner minimum
    Defines a minimum distance for traces to exit a pad before turning a corner.

SMT pad neck down ratio
    Ratio of pad width to exit trace from the pad.

SMT pad entry angle
    Restricts the angle at which a trace can connect to a pad.

Solder and paste mask expansion
    Checks the dimensions required for relationships between the solder mask opening and the paste mask.

Power/ground plane via connection
    If desired, inserts a thermal relief connection to the P&G planes. This is a common strategy for components connected to power or ground layers. Doing this allows soldering without a directly connected heat sink created by the P&G planes.

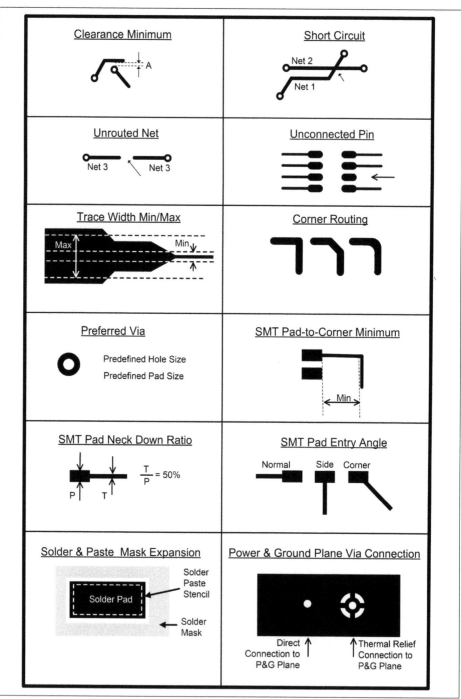

Figure 11-28. Typical physical design rules, part I

Figure 11-29 shows the following additional DRCs:

Power & ground plane clearance
Checks for minimum clearance when any via passes through a P&G plane and is not connected to it

Minimum annular ring
Checks that the copper ring surrounding a drilled hole has a minimum width

Acute angle limit
Checks that a trace has no angles that are less than 90 degrees

Hole size limits (min/max) for vias and through-holes
Make sure that drill sizes are within limits of the drilling system

Hole-to-hole clearance
Determines edge-to-edge spacing of drilled holes

Minimum width solder mask
Makes sure that a wide enough solder mask exists to prevent solder bridges

Silkscreen to solder mask clearance
Keeps silkscreen materials away from edges of solder mask openings

Silkscreen-to-silkscreen clearance
Keeps silkscreen materials from overlapping each other

Board outline clearance
Checks that the perimeter of the PCB remains clear of obstructions

Component clearance
Assures that components are properly separated

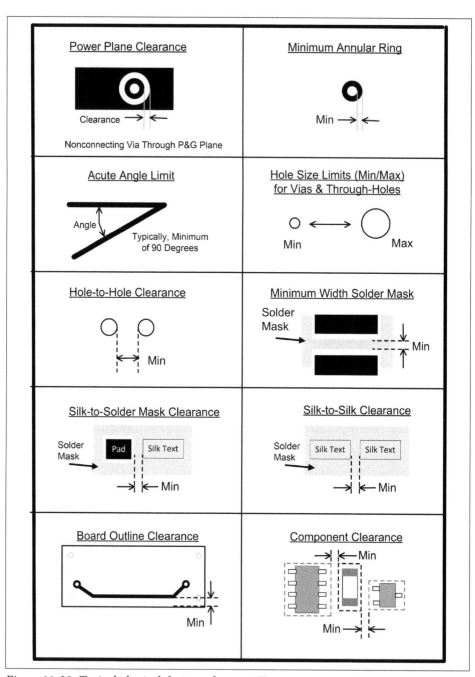

Figure 11-29. Typical physical design rules, part II

High-Voltage Spacing Rules

Voltages greater than 15 V require special attention to PCB interconnect spacing. Under 15 V, the IPC specifies 0.1 mm minimum spacing (Figure 11-30). Common practice is to use a minimum of 0.15 mm to avoid the added costs associated with ultra-fine metal spacing.

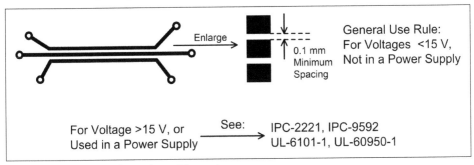

Figure 11-30. Low-voltage (<15 V) spacing

If the PCB has any signal over 15 V or if connection traces are part of a power supply, high-voltage spacing rules come into play (Figure 11-31). For instance, if the voltages on the PCB do not exceed 15 V, using a minimum trace-to-trace separation clearance of 0.15 mm or larger should be suitable. If voltages exist above 15 V or if there is a need for closer spacing, diving into the details of the IPC/IEC/UL regulatory standards is needed.

High-voltage clearance applies laterally, from trace to trace, and also vertically through the PP-INS layer. A common technique for a board with a high-voltage supply is to remove P&G planes under the HV traces to gain additional clearance. If this technique is used, it should be used with caution, on the side or in a corner of the PCB, to keep a solid ground plane under most of the board. Cutouts in the ground plane should be avoided whenever possible (FDI: EMI & ESD).

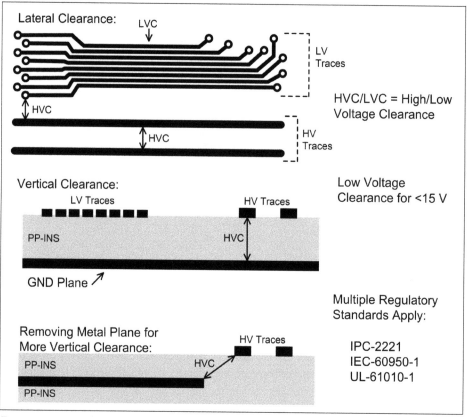

Figure 11-31. High-voltage spacing rules

Component Placement Strategy

When placement of components is well thought out (Figure 11-32), the interconnect should be straightforward. Components are placed in priority order. Location-critical items are placed first, and then other parts are fitted in around them. A typical sequence is shown in Figure 11-32 and outlined as follows:

Define mechanics

This is the physical definition of the PCB and it should include dimensions, mounting locations, and cutouts to define the size, shape, and mounting of the PCB.

Define keep-outs

The perimeter of the PCB should be clear of all components and circuit traces, including internal P&G planes. The keep-out distance off the perimeter of the board is typically 0.1 inch. Keep-out regions should also be established around mounting holes or other mechanical attachment regions or supports.

Place external connectors

Locations of the external cable connectors are often dictated by the mechanical aspects of the system, and there's often not a lot of flexibility on connector locations. Consequently, connectors are the next items to be placed.

Minimize ratline travel and tangle

Most EDA tools can show *ratline connections* where dashed lines (ratlines) show the connection paths that need to be installed as copper traces. Adjust the orientation of the chips, connectors, and other components to minimize crossover connections and the routing paths of the majority of signals relative to each other. If two chips share a dozen connections, placing the two chips in close proximity with an orientation that minimizes the crossover tangle of ratlines is suggested.

Also, FPGA pins and general-purpose input/output (GPIO) ports on MCU devices can often be swapped with a minor code change for the port location. This is useful to untangle connections, minimize vias, and reduce layer changes for signal traces.

If a PCB has high-speed data or analog signal connections, attention should be given to component location and orientation to minimize connection length. Leaving space around analog signal paths and keeping high-speed digital connections away from analog signals is part of the component placement strategy.

Putting chips in neat orderly rows, with all the number one pins in the same orientation, may look organized, but it is not a good idea for high-frequency and low-noise performance. For best high-frequency performance, minimizing the connection length of the high-frequency signals is a top priority (FDI: Digital, FDI: EMI & ESD). If assembly data files and footprints are correct, the pick-place robot can deal with any orientation.

High-priority placements

Location-critical components are placed next. This includes high-frequency power filtering directly at P&G pins, ESD protection at PCB input ports, low-pass filter (LPF) components at ADC inputs, resistor terminations at ends of transmission lines, and all components connected to analog signals. Priority is given to high-frequency signals, noise-sensitive connections, minimal connections for noise sources, and maintaining stable P&G for all chips (FDI: EMI & ESD, FDI: Power).

Other placements

Low-priority connections and components complete the component placement. Low-frequency digital devices, LED controls, and other "loose ends" are included here.

Ground plane

The ground plane should be solid, without cuts or slots, to get optimal EMI and ground stability performance.

Power plane

With multiple chips on the board, different power supply voltages will probably be needed. The power plane can be broken into sections to support the different chip power requirements across the board.

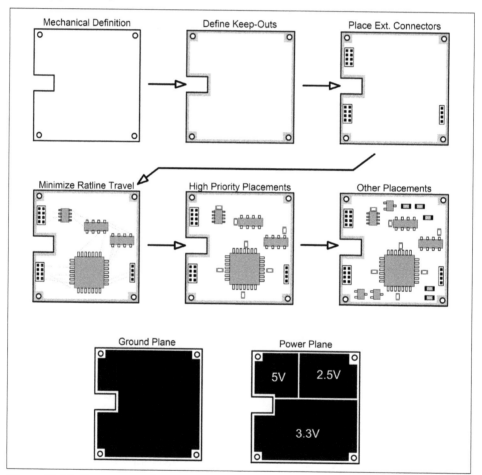

Figure 11-32. Component placement strategy

Following are additional considerations in component placement:

- Keep electrolytic and polystyrene capacitors away from heat sources.
- Whenever possible, bypass caps directly across P&G IC pins, not through vias.

With components optimally placed, interconnect routing of traces should be straight-forward.

General Interconnection Methods

When routing interconnections, designers should be cautious of a multitude of things. Let's explore some of the techniques that will make for a better design.

Easy Estimations of RLC Parasitics

First, consider any nonideal connections. If a connection requires significant current, the resistance of the connection may be an issue. High-frequency signals, both analog and digital, can suffer from capacitance or inductance of the connection. Some rules of thumb (Figure 11-33) can give a rough estimate of RLC characteristics and are useful to determine whether a more careful analysis is needed.

The resistance of a trace can be quickly estimated using the resistance-per-square method. If a 90-degree bend is placed within the trace, these regions are roughly one half square added to the total squares for the straight connections. Knowing the copper weight/thickness, trace length, and trace width quickly yields an approximate resistance.

Inductance is dependent upon the trace width, length, and height above the ground plane. A rigorous electromagnetic simulation tool can give a more accurate number, but for a quick estimate, using 28 to 16 nH per inch will allow a decision about whether or not a more careful analysis is needed.

Capacitance of a trace is a function of the dielectric, distance of the trace over the ground plane, area of the trace, and fringe capacitance of the edges of the trace. Treating the trace capacitance as a two-plate sheet capacitor with no fringe capacitor will yield a quick estimate of the trace capacitance. See Figure 11-33 for the equations and needed numeric values.

Resistance, inductance, and capacitance are estimates that can be made with a calculator and a few minutes of effort. More accurate results can be achieved through more elaborate methods, but these estimates will quickly tell you if closer attention is needed.

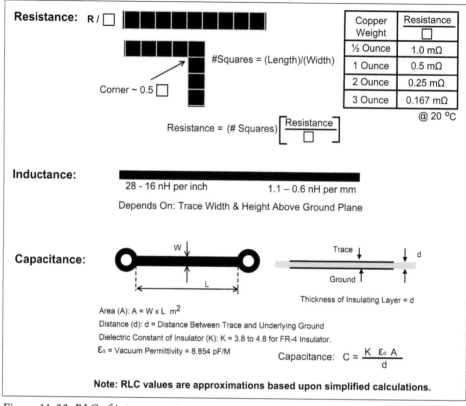

Figure 11-33. RLC of interconnects

Maximum Trace Currents

Another issue to consider is the maximum current for a trace; in other words, how much continuous current through a copper trace is IPC-defined using thermal trace heating, as shown in Table 11-6. Small geometry traces can handle considerable current. But the resistance of the connection and the associated voltage drop should also be estimated because voltage drops across the trace resistance can be significant.

Table 11-6. Maximum continuous current versus trace width (1 oz copper thickness)[a]

Max current (amps)	Trace width (mils)	Trace width (mm)
1	10	0.25
2	30	0.76
3	50	1.27
4	80	2.03
5	110	2.79

[a] Derived from IPC-2152, based upon 10°C thermal elevation. Always check for suitable trace resistance.

Determine Minimum Geometry Trace Requirements

High board density requires the use of minimum geometry traces for interconnect. The narrowest trace necessary (Figure 11-34) on a PCB often comes down to the finest pin pitch IC on the PCB, or the ball-to-ball spacing of the smallest-pitch ball grid array (BGA) packaged IC.

For example, traces that terminate at a surface mount (SMT) pad should be narrower than the pad itself. Design rules often specify a "neck down" of 50% for traces, relative to the pad width, at the pin of the IC. A fine pitch (0.4 mm) quad flat pack (QFP) IC package is shown in Figure 11-34. This requires the capacity to use 0.1 mm traces to connect the IC, and the CPM needs to be capable of this.

The other item that frequently requires narrow traces is the use of fine pitch BGA packages. For the BGA, a narrow trace needs to be able to "escape" the BGA footprint on the PCB between two pads of the BGA footprint. This trace must be narrow enough to go between the BGA pads, and maintain the minimum metal-to-metal spacing the CPM requires.

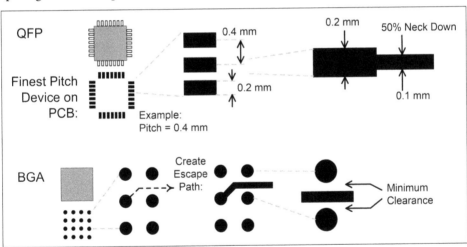

Figure 11-34. Minimum trace width guideline

Acute trace angles also should be avoided. Bends in traces should be 90 degrees or larger (Figure 11-35). Acute angles are a problem during PCB fabrication. The narrow angle can trap chemicals when etching PCB traces and create manufacturing yield problems. Although the 90-degree angle is acceptable to most CPMs, many designers create the 90-degree turn using a pair of 45-degree bends instead. Both methods are viable.

Figure 11-35. Avoiding acute angles in traces

When setting up high-frequency transmission lines, bends and layer changes through vias should be minimized as much as possible.

Solder mask openings around component pads should be slightly larger than the solder pad copper, typically 4 mils bigger on the perimeter. Never allow solder mask openings to overlap or touch each other (Figure 11-36). If two pads don't have solder mask between them, the result in final assembly is often solder bridges between the adjacent pads. A minimum width of the solder mask is typically 4 mils, in what is often called a *solder dam* that prevents solder bridges between pads.

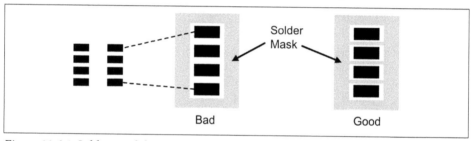

Figure 11-36. Solder mask between solder pads

Vias and Micro-Vias

The drilled via is the primary method used to change connection layers in a PCB. The options available for the drilled via are shown in Figure 11-37. The examples shown in the figure are for a six-layer metal PCB. The concepts are similar for other layer stack-up scenarios. The all-layer via is the most commonly used via, and is usually the lowest-cost manufacturing option because it requires drilling and hole plating of the all-layer stack-up of the PCB.

Both blind and buried vias don't go to all layers and require drilling and plating of selective layers before the entire layer stack is assembled. A blind via accesses either the top or bottom surface of the finished PCB. A buried via has no outer layer access. Use of blind and buried vias is usually a production cost addition. If a PCB can be designed using nothing but all-layer vias, that's typically the lowest-cost method. In Figure 11-37, one all-layer via is shown connected to ground and another all-layer via is shown connected to the power plane.

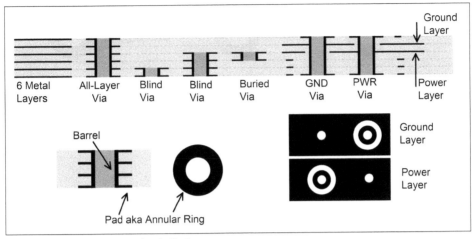

Figure 11-37. Structure of a drilled via

Making a connection through a via will add some inductance to the connection. The values shown in Figure 11-38 are typical via inductance values for a commonly used board thickness. Typically, using a rough estimate of 1 nH per via allows a quick calculation to determine whether a more careful analysis is needed. For circuits under 100 MHz, via inductance is usually not a problem.

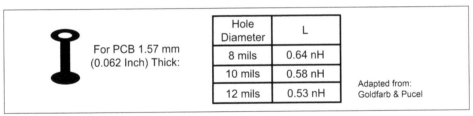

For PCB 1.57 mm (0.062 Inch) Thick:

Hole Diameter	L
8 mils	0.64 nH
10 mils	0.58 nH
12 mils	0.53 nH

Adapted from: Goldfarb & Pucel

Figure 11-38. Inductance of a via connection

Vias should not be placed within a solder pad in most scenarios. Putting a via within a solder pad requires a specialty via that has been filled and plated over. This adds cost to the PCB and not all CPMs have the capability. Except for high-density layouts, this usually isn't needed.

Typically, a via and a component solder pad are connected with a trace and are treated as separate devices on the PCB. Design rules from the CPM usually require a minimum trace length between the via and the solder pad.

Although rare, via connections can stress-crack in some situations. Reliability can be improved using teardrop via connections (Figure 11-39). Most EDA tools have the capability to make teardrop connections, and these should be a straightforward layout addition.

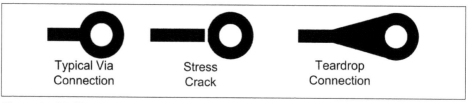

Typical Via
Connection

Stress
Crack

Teardrop
Connection

Figure 11-39. Trace connection to via

For CPM facilities using imperial measurements, Figure 11-40 shows some commonly used via sizes.

Pad Diameter	Drill Diameter	Annular Ring Width	PCB Thickness	Units
(1) 20	10	5	<98	mils
(2) 22	12	5	<118	mils
(3) 18	8	5	<63	mils
(4) 16	6	5	<39	mils

(1) & (2) – General Purpose Widely Used
(3) - Capable of Higher Density, Limits Thickness
(4) - Smallest Via Using Mechanical Drilling

ARW = Annular Ring Width
DD = Drill Diameter
PD = Pad Diameter

Always Ask CPM for Their Preferred Via Geometry List

(Based on CPM Survey)

Figure 11-40. Commonly used drilled via sizes

Metric-sized drilled vias are typically done in the 0.2 mm to 0.6 mm range for the drill diameter and use a pad diameter of twice the drill diameter. Always ask your CPM for their list of preferred sizes for vias.

If multiple manufacturers are needed, a common strategy is to select a group of vendors to work with and create a PCB layout that respects the design limitations of all the vendors.

A drill size smaller than 6 mils (~0.15 mm) requires nonmechanical drilling, which is done with a laser. The micro-via (Figure 11-41) is characterized by laser drilling and a conical drill hole (regular vias are cylindrical due to mechanical drilling), and is primarily used to transition only one layer of the substrate. Multilayer transitions use micro-via stacking or staggered micro-vias. Some CPMs can implement a skip micro-via that transitions more than one substrate layer.

The aspect ratio (depth/diameter) of a micro-via is generally different than that of a drilled via. A drilled via can have an aspect ratio up to 10:1. (The IPC suggests a maximum of 6:1 to 8:1 for manufacturing reliability.) A micro-via typically has an aspect ratio of 1:1 or less. Consequently, a micro-via requires a thin insulating layer. The CPM can provide details of their micro-via capabilities and how it affects the layer stack-up.

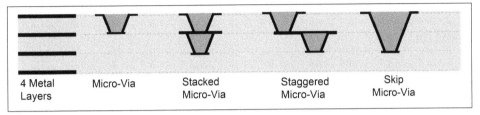

Figure 11-41. Micro-via

High-density component placement requires small component packaging and the micro-via is often necessary to create the needed interconnections. Figure 11-42 is a good example of a high-density BGA package.

As shown, the 8 × 8 BGA can have connections escape the footprint on the component side of the PCB for many of the outer pin locations. However, there are a dozen pads that don't have an escape path on the component layer. This requires a layer change via to create an escape path.

Due to the BGA's fine pitch, a drilled via is too large. A micro-via can be placed between the BGA pads or incorporated directly into the BGA pad. Placing a micro-via in the pad requires a discussion with the CPM about the structure of the via and topping the via with a plated-over metal region for the BGA pad. Different vendors will have different approaches and can advise as to what works best within their capabilities.

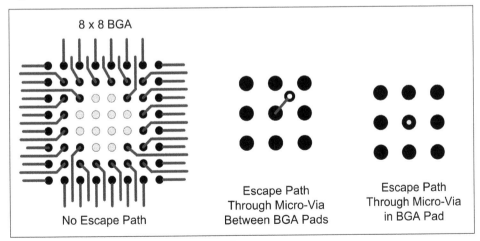

Figure 11-42. Micro-via in component pad

Vias for Thermal Conduction

In addition to making an electrical connection, vias are also used as a thermal conduction path to remove heat from an IC. Specific ICs and power transistors may require heat sinks external to the IC, or they may use the copper of the PCB as a heat sink (FDI: Drive).

Two methods are commonly used to remove heat from an IC (Figure 11-43). The first method places a heat sink directly on the IC to transfer heat to the air. This is a common practice for microprocessors used in many generations of personal computers.

The second method uses the body of the PCB as a thermal heat sink. When using the PCB as a heat sink, the IC package often includes a thermally conductive slug under the chip, which is then mated with a copper square on the component side metal of the PCB. This square has an array of drilled vias in a tightly spaced array that connect to a lower metal layer used to distribute heat. Typical spacing and dimensions are shown in the figure. The vias are sized so that the via plating done in manufacturing creates a solid metal "plug" and thus provides good thermal conductivity to lower metal layers.

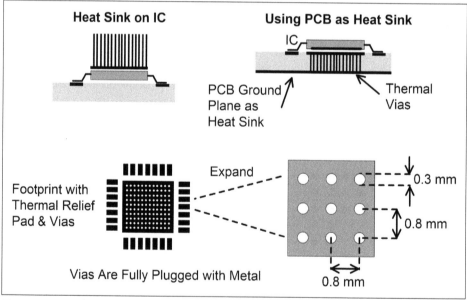

Figure 11-43. Thermal relief vias with heat sink attachment

Spreading the heat out as close as possible to the source is advised. A common layer stack-up includes a ground plane tightly spaced underneath the top metal layer and usually provides a good thermal relief heat sink. Since the vias are drilled through and go to all layers, additional copper squares on the other metal layers are often included.

Specialized Interconnection Methods

When to use a transmission line is dependent upon the connection length and the signal rise/fall time (FDI: Digital). Longer traces and faster rise/fall times indicate a need for a transmission line. Specifics are discussed in Chapter 3.

Differential signals provide better noise immunity, and at lower frequencies they don't need to be an impedance-matched transmission line. The high-frequency transients of low-voltage differential signaling (LVDS) digital signals often require a differential microstrip or stripline with impedance-matched resistive termination.

Differential Signal Routing

Many analog sensors use a differential output (FDI: Sense, FDI: EMI & ESD), and these should be connected using differential signal routing (Figure 11-44) with a grounded perimeter to improve noise immunity. With a ground plane underneath the sensor traces, a perimeter ground is placed around the traces and connected to ground using via connections, a process commonly known as *via stitching*. Most sensors are low frequency and don't need transmission lines. Differential signal routing is sufficient for most sensors.

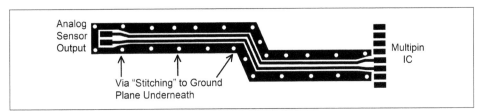

Figure 11-44. Differential signal routing

Microstrip Transmission Lines

Transmission lines come in two variants. The first, a *microstrip*, is placed upon an outer metal layer, with a ground layer underneath it. The microstrip transmission line impedance is defined by the geometry of the connection, including the width of the trace and the distance above the ground plane (Figure 11-45).

Using a thin insulating layer underneath the top component metal layer reduces H and allows for narrower W connections. As shown in Figure 11-45, the typical W/H ratio is about 2.

A CPM can provide the exact transmission line dimensions needed for their PCB products. In addition, the CPM can suggest a suitable layer stack that is transmission line friendly. Moving the transmission line to internal layers changes the geometry and requires a ground plane both above and below the connection.

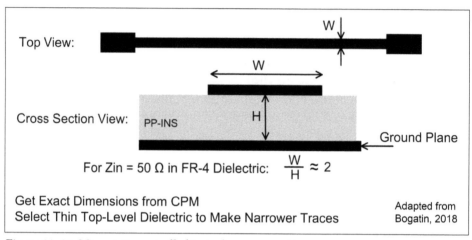

Figure 11-45. Microstrip-controlled impedance connection

Stripline Transmission Lines

This second form of a transmission line is known as a stripline (Figure 11-46). A stripline requires three metal layers to implement with ground planes above and below the connection trace. All three layers can be moved to internal PCB layers, but inductive vias are needed.

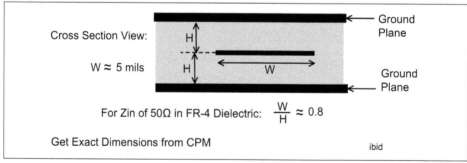

Figure 11-46. Stripline-controlled impedance connection

Differential Microstrips and Striplines

Differential versions of the microstrip and stripline are also available (Figure 11-47). In practice, a differential microstrip is often used for high-speed digital LVDS communication. Slower interfaces (I²C, SMB, SPI, CAN, and others) don't need impedance-matched transmission lines. Differential signal routing with a surround ground is suggested for all differential analog signals. All analog signals should minimize connection length and use surround grounds for noise protection.

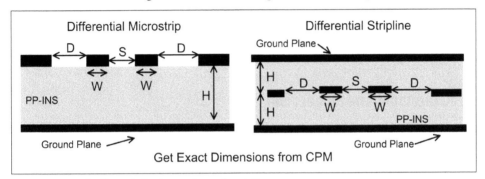

Figure 11-47. Differential microstrip and stripline

Generally, try to keep transmission lines on outer metal layers. This minimizes the number of inductive vias in the high-speed connection. Lower-frequency signals can be routed to internal layers instead.

Kelvin Connections

Kelvin sense connections are often needed when sensing a voltage across a small-value resistance. This is frequently used in power or motor control circuits where a current sensor circuit is needed.

In Figure 11-48, Case A shows a schematic with a small resistor (0.1 Ω) used for current sensing. Case B and Case C show an interconnect that includes voltages across the resistance of the copper trace interconnect (R_{tr} in the Case C image). Case D revises the connections to minimize the interconnect resistance in the measurement by the ADC.

In these examples, the current into the ADC is presumed to be zero. Specialty sense resistors are also available with four pads. These devices use two connections for the high current and two pads to sense the voltage across the resistor. Careful layout of circuit traces is generally sufficient for accurate results with a conventional two-pad resistor.

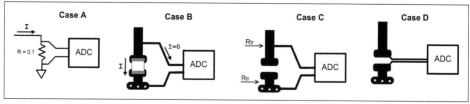

Figure 11-48. Kelvin connections

EMI and ESD Strategies

As discussed in Chapter 6, issues of EMI and ESD are dependent on both circuitry and physical PCB layout.

Solid Ground Plane for Less EMI

According to Kirchoff's current law (KCL), the current in and out of a node must sum to zero. Consequently, any current out of a device will find a way to return to that device, creating what is known as a *current loop*. This becomes an important issue with layout and EMI, as illustrated in Figure 11-49.

Case A in the figure is a simple example of a logic inverter driving another device across the PCB. Current exits the inverter, but the return path is not clearly shown. Case B shows the return path, with current from the power supply exiting the inverter, going down the connect trace and through the receiving inverter, and returning through the ground plane. Case C shows that the current out and the return current are in close proximity to each other. From a distance, the electromagnetic fields created by I_{out} and I_{return} cancel each other out, and any EMI created is minimized.

Case D changes the return path so that it does not tightly follow the output path. Consequently, a current loop transmitting antenna has been created. Case E has a solid ground plane underneath the two inverters and is the physical rendering of the Case C example, with minimized EMI. Case F shows a ground plane that has been notched open and is the physical version of the current loop antenna of Case D.

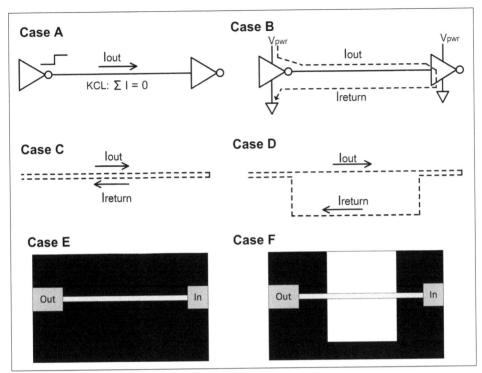

Figure 11-49. Current return paths for signals

The lesson is clear: never notch, slot, or otherwise cut up the ground plane. Instead, keep a solid ground plane for the entire PCB, and EMI from current loops will be minimized. Also, a solid ground plane provides the lowest-impedance ground, and thus minimizes ground bounce (FDI: EMI & ESD).

Flooded Signal Layer Grounds for Less EMI

After routing all interconnects, a common strategy uses flooded grounds on all signal layers (Figure 11-50) to help minimize EMI and cross coupling. Case A in the figure can have cross-coupling problems between the two traces, especially if one connection has high-frequency content and the other is noise sensitive.

Case B separates the two signals and will reduce crosstalk. Case C takes further steps by laying down a grounded region between the signals, with via stitching to the solid ground plane underneath. Any capacitive coupling from the traces will be primarily to ground and not to another signal. Case D takes this one step further and includes a flooded ground surrounding all signals, again with via stitching to the underlying ground plane.

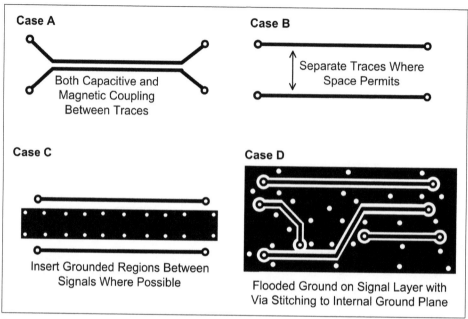

Figure 11-50. Flooded grounds to reduce EMI and cross-coupling

Flooded grounds on all signal layers will reduce crosstalk between signals and radiated EMI (FDI: EMI & ESD).

ESD Interconnect

Placement and routing of ESD protection circuits require special considerations. In the example shown in Figure 11-51, an ESD event at CON4 typically puts thousands of volts briefly on R7, causing a multi-ampere current surge through R7 and TVS4. Because the current surge is brief (<150 ns), a high-power resistor isn't needed, but CON4, R7, and TVS4 should be tightly placed together with minimal impedance connections (FDI: EMI & ESD).

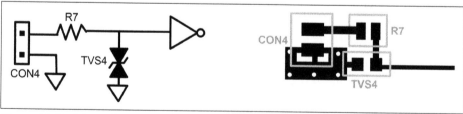

Figure 11-51. Physical routing of ESD protection

High-Frequency Power Bypass Methods

Digital chips put high-current surge demands on the power supply, and bypass capacitors are used to stabilize the voltage. Connection inductance between the IC and the bypass capacitors should be minimized when possible.

As shown in Figure 11-52, C22 has been placed directly across the P&G pins of the IC. Multiple vias to the internal P&G planes are also provided in an effort to reduce connection inductance (FDI: Power).

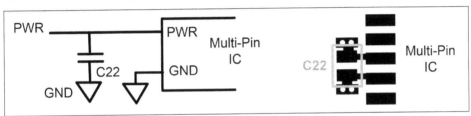

Figure 11-52. Placement of power bypass capacitors for minimized inductance

Features for Manufacture and Assembly

Some items must be included in the PCB and support documentation to satisfy manufacturing requirements. For example, if a PCB has inconsistent copper coverage, it can result in the PCB warping, as shown in Figure 11-53.

Consistent Copper Coverage

To get consistent copper coverage, designers can use flooded grounds on all layers, or they can insert a copper thieving pattern across the regions of each layer that lacks sufficient copper. *Copper thieving* is a repeated pattern of unconnected shapes, frequently squares, used to increase copper coverage without changing PCB functionality.

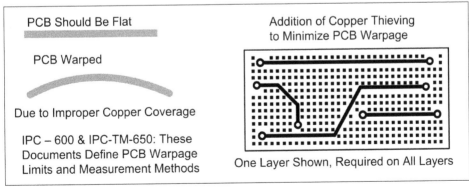

Figure 11-53. PCB deformation and copper thieving

In addition, the perimeter of a PCB or panel of multiple PCBs needs to be free of components, traces, and metal for a 5–10 mm region at the outer perimeter to aid with handling and clamping of the solder paste stencil. The CPM can provide their preferred width.

Panelization and Break-Apart Methods

Typically, the PCB is smaller than the fabricated panel. Consequently, multiple PCBs are put together (Figure 11-54) into a larger panel. A panel with rectangular shaped PCBs is typically created using a V-score break-apart technique. The V-score is created using a router bit designed to cut a 30- or 45-degree score from both sides of the PCB, leaving about 30% of the board thickness. These straight-line scores are easily bent and broken apart after the board has been fabricated and assembled with components.

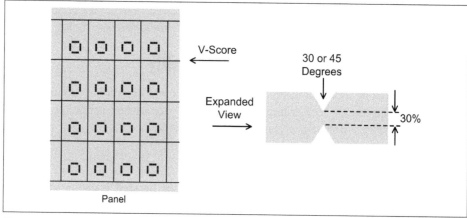

Figure 11-54. PCB panel with V-score break-aparts

Panel cutouts are another method used to separate PCBs. Cutouts are useful when the PCB is a nonrectangular structure without straight edges (Figure 11-55). Using the cutout method allows any PCB shape. The cutouts can follow any path, curved or straight. Some tab material must be left to maintain the structure of the panel. These tab regions can be perforated with a series of small drill holes to create a breakaway tab. This is sometimes referred to as a *mouse bite breakaway tab*.

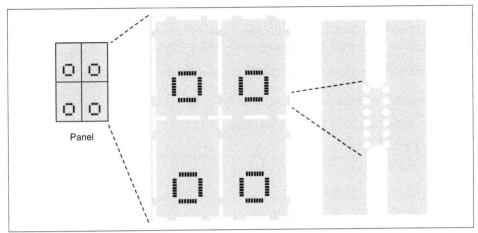

Figure 11-55. PCB panel with cutouts and mouse bite breakaway tabs

Typically, a CPM can handle copper coverage thieving and panelization of the PCB.

 Always carefully inspect what the CPM has done before authorizing actual fabrication.

Fabrication Notes

PCB fabrication notes are provided to the CPM detailing the manufacture and assembly aspects of the build. Fabrication notes must define the following:

Finished board parameters
These include the copper weight for each metal layer, thickness of each internal insulating layer, and finished board thickness.

Surface finish options
These include HASL (Hot Air Solder Leveling with Lead), LF-HASL (Lead-Free Hot Air Solder Leveling, Immersion Tin), ENIG (Electroless Nickel Immersion Gold), Immersion Tin, Immersion Silver, or OSP (Organic Solderability Preservative).

Solder mask color and finish
A green solder mask with a gloss finish is most widely used. A matte finish and alternative colors are available from most CPM sources.

Silkscreen
White silkscreen is most commonly used, but other colors are available.

Panelization information

> This includes a description of the panel size and the desired break-apart strategy. Discuss the preferred panel strategy with your CPM before specifying.

In addition, following is the necessary information to place in the silkscreen layer:

- Traceability information (revision #, serial #, design date, fabrication run)
- Reference designators for all components
- Pin #1 for all chips
- Orientation or polarity markers (diodes, electrolytic/tantalum capacitors)
- Polarity on power connections
- Test point labels
- PCB part number
- Company logo and identification
- Component outlines
- Warning labels ("Contains Lead," "High Voltage," etc.)
- Regulatory labels (UL, CE, FCC, RoHS Compliant, etc.)

 Removal of Harmful Substances (RoHS) initiatives in the electronics industry have (mostly) eliminated the use of lead in PCB fabrication. The CPM can provide information about what surface finish options they are capable of and what is suitable for the PCB and its application environment.

Manufacturing (Gerber) Files

When the PCB design is complete, the design is transferred to the CPM using a set of data files. As of 2023, the most widely used industry standard format for this is a set of Gerber files, created using the RS274X format.

Other formats are in use for PCB definition, namely ODB++, an EDA vendor proprietary database format for the entire PCB; and IPC-2581B, also a database format, that is defined by the IPC coalition. Both of these formats show promise, with IPC-2581B slowly being adopted. However, Gerber files are still the most common method.

Using Gerber files requires that a multitude of individual files be transferred. This includes:

- Metal layers, top and bottom (T&B)
- Silkscreen T&B (sometimes called an *overlay*)
- Solder mask T&B
- Solder paste stencil T&B
- Component map with X-Y component locations T&B
- Assembly drawing T&B
- Internal metal layers (multiple)
- Mechanical definition of PCB shape, size, and cutouts
- Keep-out layer
- Drill drawing
- Drill guide/legend

Although not in the Gerber graphics format, these files are also needed:

- Netlist (ASCII text)
- Bill of materials (BOM spreadsheet)
- Fabrication notes for manufacturing and assembly

Most EDA tools have the capability to generate these files. The BOM should be carefully reviewed for errors. The fabrication notes are manually generated and should be reviewed and agreed to by both the CPM and the designer.

Summary and Conclusions

This chapter takes the circuit design ideas developed earlier and brings them together to create a PCB. Organizing and documenting the design schematic, and selecting components for use, are the first steps. After that, the mechanical and physical aspects of designing a circuit board and how to select a PCB layer structure are covered.

Items needed for interconnect and attaching components are dealt with next. That includes:

- Defining schematic component symbols and PCB footprints
- Restrictions due to physical design rules
- Voltage-dependent interconnect spacing
- Component placements based on critical path performance
- Required trace widths for current handling and high-density interconnect
- Multiple layer-to-layer via methods

Special-attention items include transmission line structures, differential signal routing, Kelvin connections, test and measurement access, JTAG access for programming, EMI and ESD PCB techniques, and others.

Preparing the design for manufacturing includes copper coverage, panelization methods, fiducial mark alignment, Gerber files, and fabrication notes.

With that knowledge, readers should be able to create a schematic, design a PCB, and work with a CPM to produce a finished, component-loaded, and functional PCB.

Further Reading

- *Signal and Power Integrity, Simplified, 3rd Edition*, by Eric Bogatin, 2018, ISBN-13 978-0-13-451341-6, Pearson.

- "PCB Layout Thermal Design Guide," Rohm Semiconductor Document 65AN002E, 2022, *https://fscdn.rohm.com/en/products/databook/applinote/common/pcb_layout_thermal_design_guide_an-e.pdf*.

- Douglas Brooks and Johannes Adam, "PCB Design: A Close Look at Facts and Myths About Thermal Vias," *EDN Magazine*, August 13, 2021, *https://www.edn.com/pcb-design-a-close-look-at-facts-and-myths-about-thermal-vias*.

- Douglas Brooks and Johannes Adam, "Major Design Considerations for Determining PCB Trace Temperature," *EDN Magazine*, April 12, 2021, *https://www.edn.com/major-design-considerations-for-determining-pcb-trace-temperature*.

- Douglas Brooks and Johannes Adam, "Myths and Reality about Via Size and Temperature," *EDN Magazine*, May 20, 2021, *https://www.edn.com/myths-and-reality-about-via-size-and-temperature*.

- M.E. Goldfarb and R.A. Pucel, "Modeling Via Grounds in Microstrip," IEEE Microwave and Guided Wave Letters, June 1991, Vol. 1, No 6.

- AN-958: Board Design Guidelines, Intel Corporation, June 26, 2023, *https://cdrdv2-public.intel.com/677286/an958-683073-677286.pdf*.

- Mikhail Guz, "Practical Considerations for Design and Layout of High Power Digital PCBs," *IEEE Power Electronics Magazine*, June 2021, 53–59.

- Nicholaus Smith, "The Engineers Guide to High Quality PCB Design," *Electronic Design Magazine*, July 17, 2013, *https://www.electronicdesign.com/technologies/embedded/article/21798594/the-engineers-guide-to-highquality-pcb-design*.

- Rush PCB Blog, PCB Design and Fabrication, Rush PCB Inc., *https://rushpcb.com/blog*.

- *The Printed Circuit Designers Guide to… Designing for Reality* by Matt Stevenson, 2022, ISBN 979-8-9856020-8-1, IPC Publishing Group Inc.

- "PCB Design Guidelines for Reduced EMI," Texas Instruments TI Document SZZA009, November 1999, *https://www.ti.com/lit/pdf/szza009*.

- *Electromagnetic Compatibility Engineering* by Henry W. Ott, 2009, ISBN 978-0-470-18930-6, Wiley & Sons.

- *Printed Circuits Handbook, 6th Edition*, by Clyde Coombs, 2008, ISBN 0-07-146734-3, McGraw-Hill.

- *Printed Circuit Boards: Design, Fabrication, Assembly and Testing* by R.S. Khandpur, 2006, ISBN 0-07-146420-4, McGraw-Hill.

- IPC-2221 Generic Standard on Printed Board Design, IPC, *https://www.ipc.org*.

- IPC-7351 Generic Requirements for Surface Mount Design and Land Pattern Standard, IPC, *https://www.ipc.org*.

- "Packaging," guideline to packaging, Texas Instruments, *https://www.ti.com/support-packaging/packaging.html*.

- "Registered Outlines: JEP95," JEDEC database of electronic component packages, JEDEC, *https://www.jedec.org/category/technology-focus-area/jc-10/registered-outlines-jep95*.

CHAPTER 12

Software and Coding

This chapter investigates software- and code-related issues within the context of embedded systems. Topics include:

- What programming languages are used
- When an operating system (OS) is needed
- Understanding the difference between real-time and general-purpose operating systems
- Which real-time operating systems are suitable for use
- Configuring the ports of a microcontroller (MCU)
- Creating device driver code
- Defensive coding for systems functioning under adverse conditions

With coverage of these topics, a designer with coding skills should have the knowledge required to configure and program embedded systems.

Remember, this book focuses on embedded system electronics, not the coding for embedded systems. There are many books on embedded system coding out there, and several good coding references and study textbooks are suggested in "Further Reading" on page 488.

Consequently a tutorial for writing a "Hello World!" program for an embedded system isn't included here.

Coding Languages

The approach to programming an embedded system will usually take one of three paths:

Define a state machine

    For field-programmable gate array (FPGA) and complex programmable logic device (CPLD) programming, the logic structure usually defines a state machine. The state machine applies conditionals to define state changes. For this, the programming uses a hardware description language (HDL), typically Verilog or VHDL.

Use Assembly language

    Simple MCUs (peripheral interface controllers, etc.) and early-generation processors (8051, 6502, and their variants) may require using assembler code. For some of these devices, a compiler may not exist or memory is so limited that compiled programs won't fit. Also, cases where very fast processing or exact control of timing and number of clock cycles is required may need assembler code.

Use compiled code

    The predominant language for controller code is C or one of its variants (C++, C#, Embedded C). Several variants of Python have been developed to fit the limited memory capabilities of embedded systems; examples include MicroPython (*https://micropython.org*) and CircuitPython (*https://circuitpython.org*).

For MCUs, the majority of embedded code development is done in C or a variant thereof. (As of 2023, Python use for MCUs has increased, but C still dominates.) For CPLD or FPGA devices, programming is done using Verilog or VHDL.

Operating Systems

A general-purpose OS controls computer functionality, including task management, file storage management, memory management, and interfacing to peripherals (keyboard, mouse, video, audio, Ethernet, USB, etc.). A real-time operating system (RTOS) adds the capability to respond in a timely manner. A "hard" RTOS is designed to respond in a maximum period of time without exception. A "soft" RTOS is configured so that maximum response time is "mostly" met. The specifics will be unique to each RTOS, and if timing is critical, it needs to be examined to determine suitability.

Since embedded systems respond to sensor inputs and control peripherals, the software must have real-time capability. Whether or not an OS is needed is a question of necessary features.

For example, the following features can be coded and developed:

- Real-time data streaming
- Network interface and security against malware
- Streaming video or audio output
- Multitasking and task scheduling
- Memory management
- Interrupt handling
- Peripherals management
- Error management

However, these features are available within an OS, and if many of them are required, it's often wiser to let an OS take care of them for you. When using an OS, the embedded system code you run is an applications program that gets loaded as part of the OS startup.

About half of all embedded systems don't use an OS. This is commonly known as *bare metal programming*, and it means that the necessary system features are coded into the main MCU program. In simple projects or standalone devices without system networking, bare metal programming is common. As system complexity grows, at some point the need to use an RTOS becomes evident.

Picking an RTOS

Many RTOS options are available. Some are well established, maintained, and supported. Let's examine the ones that are in use and should be viable over the long term.

Embedded Linux

Specialized variants of Linux have been designed for embedded systems. Linux is considered a general-purpose OS and not an RTOS, although there are builds (Linux with a real-time patch) that provide RTOS performance.

Embedded Linux was supposedly optimized for small systems, but it can require heavy hardware resources to run. Typically, 8 to 16 GB of RAM with similar amounts of nonvolatile memory are necessary. Also, don't expect low-power performance or the ability to use a slow clock. The system will probably need at least a 500 MHz clock to provide reasonable response times.

Many builds of embedded Linux (*https://elinux.org*) are available, including the Yocto Project, Buildroot, uClinux, AliOS, OpenWrt, DD-WRT, and the Raspberry PI OS.

Embedded Android

Android was originally based upon a Linux kernel but has been heavily modified, adding features associated with smartphones. Android excels at touch screen interfaces and GUI-based devices. Although the Android source code is available from the Android Open Source Project, actual OS development is internal to Google/Alphabet.

Many cell phone companies have developed Android variants with features unique to their needs. Google developed an embedded systems and Internet of Things (IoT) adaptation called Android Things in 2015, but discontinued support for it in 2020. Various software developers have independently created an embedded OS based upon Android (*https://android.com*), but doing so is a huge undertaking.

FreeRTOS

This OS is optimized for the limited resources of MCU devices and has been configured for over 40 different MCUs. Consequently, it is available for many of the popular controller platforms. Since FreeRTOS comes in controller-specific variants, configuring the OS is much simpler than in the embedded versions of Android and Linux.

Amazon Web Services has taken over maintenance and updates to FreeRTOS (*https://freertos.org*), giving assurance that features, support, and security issues remain up to date. Commercial versions that include support (OpenRTOS) and a version that has passed testing for many of the international safety requirements (SAFERTOS) (*https://oreil.ly/AU58-*) are also available.

QNX

This OS, originally called QUNIX, was designed to be a Unix-similar OS for embedded systems. QNX (*https://openqnx.com*) has existed since the early 1980s and predates many embedded systems, having been used in early industrial control systems. QNX has been widely used in automotive electronics, some Blackberry devices (*https://blackberry.qnx.com/en*), medical devices, industrial automation, and other applications.

Although widely utilized, QNX is being replaced by more modern alternatives.

VxWorks

This OS is proprietary software developed by Wind River Systems. VxWorks (*https://www.windriver.com*) has seen widespread use in aerospace and DoD projects, with many uses in various Mars probes, multiple aircraft, industrial robotics, and others.

INTEGRITY

Green Hills Software (*https://ghs.com*) developed INTEGRITY as a hard RTOS for use on high-performance, multicore processor platforms. The company's claims of reliability and security include certification testing in avionics, medical, railway control, automotive, and other applications. Also included are commonly needed code packages for network security, encryption, wireless interfaces, multiple serial port interfaces, and others.

ARM Mbed OS

ARM Mbed OS (*https://os.mbed.com*) was designed for the 32-bit ARM Cortex-M family of MCU devices with a target market of IoT products. An integrated design environment is available, with multivendor demo boards, hardware development kits, and source code libraries provided.

Zephyr Project

Zephyr (*https://zephyrproject.org*) is designed for the limited resources typical of many MCU implementations. Zephyr is modular and can be built with just the project's necessary features. Multiple communication protocols (Ethernet, WiFi, CAN, USB, Bluetooth, BTLE, etc.) are supported, along with the necessary security support. Processors supported include x86, ARM, ARM64, ARC, MIPS, Altera-Nios, Xtensa, RISC-V, and others.

TI-RTOS

Texas Instruments developed the TI-RTOS and IDE to support its line of MCUs. The TI controller portfolio includes both proprietary and ARM architecture devices. TI-RTOS (*https://www.ti.com/tool/TI-RTOS-MCU*) is compatible across all TI controllers, allowing easy design transfer between devices. Also, these tools function after specifying the target MCU, rather than building the OS.

Nucleus RTOS

Siemens EDA, formerly Mentor Graphics, configured its OS to support multiple processor families, including ARM, NXP, MIPS, TI, PowerPC, Altera-Nios, Xilinx, Atmel, and others. Connectivity functions include USB, Bluetooth, CAN, SPI, I²C, and others. Nucleus RTOS (*https://oreil.ly/HtWZh*) has been used in applications including spacecraft, consumer electronics, cell phones, and medical instrumentation.

Windows 10 IoT

This Microsoft-developed OS (*https://oreil.ly/8Ipr0*) is also known as Windows Embedded and Windows CE. Microsoft also provides cloud-based platforms to manage remote IoT devices using the Azure IoT platform. Processors supported include devices from Broadcom, Intel, Qualcomm, and NXP.

Additional offerings

The aforementioned RTOS offerings are stable and well known. There are many additional offerings, however, including the following:

- Apache NuttX
- Huawei LiteOS
- RTX and RTX64
- ChibiOS/RT
- RTEMS
- RODOS
- LynxOS

- MQX
- PikeOS
- ThreadX
- eCos Contiki
- TinyOS
- RIOT
- RT-Thread

Additional RTOS Considerations

The preceding sections provided an overview of the available RTOSs as of 2023. Some will survive and some will die off. Which OS to use depends on required features, processor compatibility, regulatory requirements, and whether a specific OS is already accepted in the industry sector of the final product.

When choosing, keep in mind the following considerations:

Proprietary or open source
  Licensing and royalties for an OS should be investigated before deciding which path to take. Some OS selections will have financial obligations in perpetuity.

POSIX compliance
  Many RTOS documents will mention whether the OS is compliant with POSIX standards. POSIX stands for Portable Operating System Interface and it is an IEEE standard (IEEE Std 1003.13-2003) that defines how an OS should interface and communicate externally. For any system that is not a standalone application, POSIX compliance is highly preferred. In addition, POSIX compliance should help with design migration to newer product generations when the need arises.

Configuring Ports and Processors

In addition to peripheral interface code, virtually all MCU and microprocessor (MPU) devices have configuration/control registers to initialize the device. This process is often referred to as low-level initialization, preparation for loading a boot routine, or setting up the basic input/output of the system (BIOS). Setup differs among devices, and the paramount advice here is to carefully review device documentation for appropriate settings.

Many MCU, CPLD, FPGA, and digital application-specific integrated circuit (ASIC) devices require the GPIO ports to be configured using digital downloads of the desired configuration. Every chip vendor does this a little differently, but looking at the elements internal to the GPIO port helps to understand how the port is set up.

Shown in Figure 12-1 is a typical I/O port found in many devices. Following is a brief description of its components:

- On the left is the bond pad that connects to the external integrated circuit (IC) pin. An input resistor (shown here as 50 Ω) is often the first item connected. This limits the maximum input current and is useful for protecting the port from electrostatic discharge (ESD).
- ESD protection diode D1 stops input voltage from exceeding:

    $V_{power} + 0.6$ V

- ESD protection diode D2 stops input voltage from going below:

    Ground – 0.6V

- Resistive pull-up (R_{PU}) and pull-down (R_{PD}) are available using EPU and EPD digital control. (For illustration simplicity, these are shown as switches, although the implementation uses transistors.)
- The output drive circuit is shown as a six-transistor array. Positive metal-oxide semiconductor (PMOS) devices P1, P2, and P4 and negative metal-oxide semiconductor (NMOS) devices N1, N2, and N4 are weighted in size, with binary weighted (1, 2, 4) widths. Depending on transistor selection, which is done with the SDS (set drive strength) control, the equivalent transistor width can be 1X through 7X, allowing digitally configurable drive strength for both positive and negative transition speed.

The port can be an input or an output port, depending upon how it is configured, and the option of a tri-state bidirectional port is also available. In addition, some ports may have alternative uses, such as analog-to-digital converter (ADC) input, where a digital control enables the specialty function while disabling the digital GPIO.

The drive strength should be strong enough and fast enough to service the output pins with the expected data rate and printed circuit board (PCB) loads.

 Using either too much drive strength or a light external load creates fast rise/fall edges, which creates more electromagnetic interference (EMI), more signal ringing, and more transmission line reflections. Fast-edged data may require a transmission line configuration on the PCB (FDI: Digital, FDI: PCB).

A common mistake is to use a GPIO port to directly switch a peripheral (LEDs, buzzers, etc.). A buffer circuit is often needed to avoid burning out the port. Always check the maximum current specification of the port before using it for direct drive of a load (FDI: Drive).

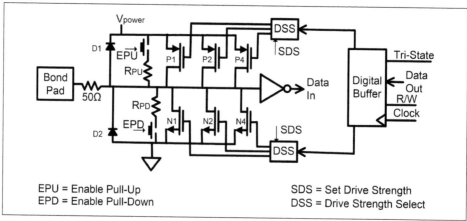

Figure 12-1. GPIO structure for single pin

The suggested strategy is to not rely on default settings for setup. Assume nothing, and explicitly define all port parameters. If a port is not used in the design, configure the unused pin to a defined state that doesn't consume power.

Device Drivers

Herein, a device driver (DD) is a software routine that interacts with and controls a peripheral device. As discussed in prior chapters, driver circuits or circuit drivers are the electronics providing the currents and voltage needed to actuate the device.

Following is a brief look at the nomenclature:

Software, firmware, or code
These three terms are loosely interchanged here. Strictly speaking, source code software (often written in C) is compiled into an executable binary code.

Typically, *software* indicates source or executable code loaded from memory storage, and *firmware* indicates code executed from nonvolatile memory (flash memory or EEPROM) where it is accessed and executed by the host application. Device drivers, as executable binary code, are stored as either software or firmware, depending on the hardware memory structure, and in most cases the soft/firm designation doesn't affect functionality.

Function, procedure, or subroutine
These terms are often used interchangeably, and that's the convention here. In the literature, a function returns a value (or other data) to the routine that called it. A subroutine executes a task and doesn't return anything to the calling routine. Herein the terms *function*, *subroutine*, and *procedure* are used interchangeably.

Problematic Portability

A common problem of most DDs is a lack of easy portability. A lot of software can be moved between devices after being recompiled for the new system. However, any DD requires that the following three things are the same:

- Peripheral performance and driver circuits
- MCU port definitions
- PCB interconnects

A DD taken from one system and loaded into a different system will not function unless the two systems are identical in all related respects. Moving a DD typically requires some parameter redefinition for the DD to function on the new platform.

Peripheral Communication

If the peripheral is directly connected to a dedicated GPIO, writing the DD starts from there. If the peripheral is connected on a serial port (I^2C, SPI, CAN, RS-232, USB, etc.), a serial port protocol needs to be written first.

Frequently, a DD calls another DD to get the job done. Here, designers need to go one level deeper and write a DD for the serial port before writing the peripheral DD. Some serial port procedures are simple to write (I^2C, SPI, RS-232), but others are more complex due to error checking and handshake protocols (CAN, USB). Usually, MCU vendor libraries have drivers for common serial port protocols that can be utilized or adapted to the design (FDI: Digital).

As an example, presume an MCU with an I^2C interface that has multiple devices connected to it. The first step would be to create a routine that communicates over the I^2C serial port, given a target address (read or write) and a data field to send or read back. After that routine is developed, it can be called from any DD that communicates using the I²C interface.

After communication has been created, the desired peripheral behavior defines what's needed in the DD. Here are some examples:

LED device driver
> This driver can be simple, handling on/off functions, or it can be more complex, handling timed light patterns, variable brightness using pulse width modulation (PWM) duty cycles, and fade in/out routines. When connected to an MCU port directly, a DD that turns an LED on/off is about the simplest one out there. For embedded systems, controlling a blinking light is the equivalent of the "Hello World!" program.

Cooling fan
> Early personal computers had on/off drivers for cooling fans. After MPU temperature sensing was introduced (Intel Pentium and newer), the cooling fan driver got more complicated. Newer drivers use temperature data to define the needed fan speed. Also, a fan tachometer was introduced providing rotational speed feedback. For this example, there are drivers for acquiring fan speed, reading the microprocessor temperature, providing variable power to the fan motor, and reducing the clock frequency if the cooling system can't stay below the desired temperature maximum. This example illustrates that device drivers can get complex and often call other device drivers to complete the task.

Initiating Peripheral Communication

Peripheral interaction is initiated through several different methods. For a simple output device, the controller sends the required data to initiate action. For a peripheral that provides data to the MCU, two methods are common: polling and interrupt requests. To *poll* a peripheral, the controller periodically requests a peripheral status update, and that information allows the controller to determine the needed action. The peripheral provides an *interrupt request* when it has data for the host.

Peripheral polling is simple to implement, but it has the weakness of constantly consuming host process time and some uncertainty in the actual time of the peripheral request. Many simple systems use polling because the MCU has a low activity level and loading down the processor with repetitive polling isn't a problem.

Both methods are commonly used, with interrupt requests making more efficient use of the host processing time. High data-throughput requirements may need a direct memory access (DMA) strategy.

Device Driver Features

In addition to basic peripheral communication, the necessary DD features depend heavily on what the device does. In many cases, the capability to switch on/off suffices. However, more complex devices may require considerably more effort. In many cases, a device driver should be capable of the following:

Determine functionality status
Letting the host controller know a peripheral is functioning properly adds to system reliability. Health monitoring of peripheral devices is an important part of high-reliability systems.

Initialization and start-up
The routines necessary to get a device started and into operational mode are frequently different from the normal run routine. Configuring setup of control parameters, turning the power on, moving something to a home position, unlatching an actuator, and using a high-current motor start profile to get a motor spinning are examples of processes handled in start-up routines.

Adjustment and calibration
Examples include calibrating sensors using a known reference stimulus, determining the zero position of an optical encoder, and determining the white/black values from an image scanner output. All of these are commonly implemented (FDI: Sense).

Shutdown procedure
Routines needed for shutdown or deactivation include parking the actuators, rotational braking, locking down an actuator, and other functions needed for an orderly shutdown.

Emergency and failure contingencies
The DD should have methods for recognizing abnormal behavior and executing a safe shutdown while informing the host (FDI: Drive, FDI: Sense).

Understanding how the hardware is supposed to function often serves as a guideline to what features are needed in the DD.

Modularity/Hierarchy for DD Code

For a given peripheral device, put all of the DD code for that device in a single routine. This is especially valuable when debugging or maintaining code. Different control calls to the same device, coming from different blocks of code scattered about the main program, quickly become disorganized and confusing. If multiple code blocks control the peripheral, the system becomes more difficult to understand, debug, and/or update.

For example, say a function named control_motor is created and takes an input argument. Within the function, the input argument is evaluated to determine the desired action; actions of faster, slower, start, and stop would be typical. In this manner, the designer knows all the relevant code is in the control_motor routine, or something called from it.

Testing the DD

As any experienced programmer knows, the code you think you wrote, and what the device actually does, are often two different things. A simple program should be used to test all DD functions with the actual driver circuits and the actual controlled peripheral. By doing this, you can debug the DD in a simplified environment before putting the DD and peripheral into a more complex application program.

Defensive Coding Methods

Defensive coding is defined here as code that deals with nonideal situations, recovers from adverse situations, and remains functional. Embedded systems are often in hostile application environments with noisy signals, data integrity problems, nonideal inputs, unstable power, and unexpected end-user actions. These are all part of the challenges encountered.

If you have ever experienced the Blue Screen of Death or had a system freeze up, the code probably did not use defensive coding or implement self-recovery methods. Reset buttons on the back of the box or pulling the plug to reset the device should *never* be necessary.

Too many embedded systems exist with weak code that can't deal with issues outside of normal operation.

The heart of defensive coding is threefold:

- Things go wrong.
- What can be done to minimize the damage?
- What can be done to correct it without interrupting functionality?

Let's explore some defensive coding methods.

Preprocess Data Inputs (Invalid Data)

Data coming from a sensor or other source always has an expected range of operation. Corrupt data due to noise, a defective sensor, or other issues can be reduced by applying conditionals that improve the probability of the data not being corrupt.

For example, say a power supply sensor monitors a battery, and it is expected to be 1.5 volts when the battery is fresh and 0 volts when the battery is dead. A preprocess algorithm would recognize values from 1.7 to 0 volts and discard data out of that range.

If invalid data occurs, the code needs to be configured to reacquire the data, and if this recurs, the sensor should be flagged as unreliable and an automated indication should specify that service or repair needs to occur.

A common example of this is an automotive "check engine" light. Either something is wrong with the engine's systems, or a sensor is defective. No matter what the root cause is, the sensor data is outside the acceptable range of operation. This is a common strategy in complex system telemetry where many system parameters are monitored and reported on.

Preprocess Data Inputs (Bandwidth Restrictions)

Many sensors monitor things that have physical limits to their transient behavior. Those limits can be used to improve data accuracy. Examples include velocity changes, fluid level changes in a tank, and temperature changes. These things can only change so quickly, depending upon the device physics. A fuel-level indicator in an automobile is a good example. The fuel tank will not transition from full to empty in seconds, so preprocessing the data from the fuel-level sensor should include some averaging so that short-term variants are filtered out.

Depending upon the situation, filtering using a digital signal processing finite impulse response (DSP FIR) filter is also possible.

Preprocess Data (Human Input)

Information coming from a keyboard input or electromechanical push button fits this category. Push buttons require contact de-bounce software (FDI: Essentials, FDI: Sense; see also Ganssle, 2004).

Anything from a keyboard needs to be conditionally examined before use. If the code wants to see a three-digit decimal numeric entered and the end user enters anything else, the system needs to reject the input and prompt the user for a correct entry. A user-friendly prompt that specifies the desired information format is always useful.

Also, the entry may need to be converted from multiple ASCII characters to a single binary equivalent.

Background Reinitialization

Most embedded systems are configured to execute an infinite loop in the code. Figure 12-2, Case A, illustrates this idea.

Case A is a simplified representation of many embedded software setups. The system powers up and resets. The code, executed from nonvolatile memory, configures the ports and internal MCU settings, followed by configuring all the external peripherals. After that, the MCU drops into a loop that services all the system functions. For the majority of applications, this strategy works well.

The weakness of Case A is seen when external peripherals lose their configuration settings due to noise, corrupt power, ESD events, or other "less-than-ideal" events. Volatile memory of the MCU or external/peripheral registers can be corrupted, and a strategy to refresh both configuration and setup data from nonvolatile memory can be useful.

Case B illustrates this idea, with a system that reconfigures and reloads internal MCU setup and port configurations, followed by a peripheral setup refresh, as part of the system's continuous service loop. In most cases, loading port configurations with the same values already there doesn't disrupt system functionality. If the register contents were not corrupt, this process is invisible for most MCUs. If some values were corrupt, the refresh action restores functionality.

An alternative to Case B is refreshing just the peripheral settings, which is useful for a system where data corruption is occurring at the peripherals but not at the MCU.

Case B can function invisibly to the end user, especially for situations where the system is dealing with slow response devices or human interfaces. Other applications may find that repetitive refreshing impairs functionality.

Case C is a variation on this method, where frequent refreshing is disruptive. In this version, a restart occurs when a small proportion of the loop passes through the main control routine. Depending on the system, a refresh of the main control algorithm that occurs every 1,000 iterations might be sufficient.

The system electronics should already have the circuitry to maintain clean power (FDI: Power), guarantee minimal signal noise (FDI: EMI & ESD), and perform error checking of data transfers over cable connections (FDI: Digital, FDI: Architecture). These refresh techniques serve as another layer of protection beyond that available from the physical electronics.

Refreshing is useful for simpler systems. More complex devices typically use device monitoring with a handshaking and verification approach, but that implies the necessary electronics to implement sensing and bidirectional communication of current status and settings.

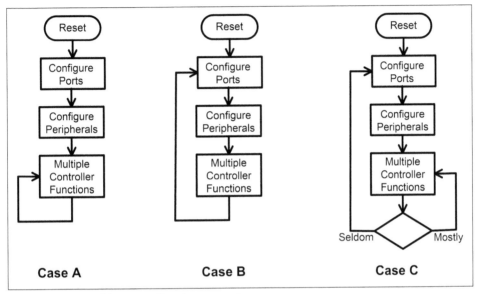

Figure 12-2. Possible reinitialization strategies

All of these methods imply the controller is still functioning in an expected manner and has not hung up internal to the primary service program. If the controller is stuck within the service program, another method of getting things going again is needed.

Watchdog Timers

If the MCU hangs or gets stuck in an infinite loop, using a watchdog timer can be another line of defense (Figure 12-3). Structurally, a watchdog timer is a digital countdown circuit, often included as a feature of the MCU. The watchdog functions independently from the CPU core of the MCU.

Although the specific behavior of a watchdog is unique to each vendor, the basic idea is the same. When the watchdog timer enabled, a digital countdown from a programmed value runs constantly. If the countdown reaches zero, an MCU reset is triggered and a fresh restart occurs.

Within the code of the primary service loop, the capability to reset the counter is used to avoid the forced reset. This is called kicking the can down the road, petting the dog, or feeding the dog. If the system gets lost or hangs up, then feeding the dog doesn't occur, and a reasonable premise is that the system needs a reset.

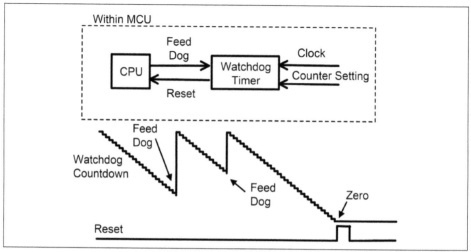

Figure 12-3. Watchdog timer reset

To perform a recovery, some MCUs are structured to do a reset, whereas other devices are structured to execute a block of code that the programmer defines. In both cases, escaping from a nonresponsive control routine is provided without the end user looking for a reset button or pulling the power plug.

Multicontroller Coding

Don't confuse a multicontroller system with a multicore system. Typically, multicore computers and parallel processing computers divide tasks across multiple cores to improve throughput. Many multicore devices share memory and require an OS that can deal with arbitration and efficient task and memory management.

In contrast, a multicontroller system uses multiple MCU devices, and the system is typically separated by what each MCU controls or senses (Figure 12-4).

Multicontroller systems are invaluable in situations where the "local" controllers are tasked with critically timed events that a multitasking controller couldn't service quickly enough. These are also known as distributed systems (FDI: Architecture).

For these systems, manager/local communication is bidirectional, with the manager telling local processors what to do and the local devices dealing with the details of the action and then acknowledging task completion.

Code development for such systems is similar to developing device drivers. Each local controller is black box–tested with the expected commands from the manager and its control/sense of the local peripherals. Once each local controller is responding properly to the expected manager control requests, multiple local devices can be run under manager control and debugged.

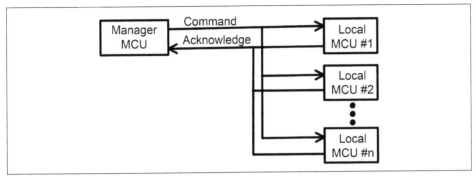

Figure 12-4. Multicontroller system: command and acknowledgment

As with all systems, code should include protective recovery routines if communication is lost or something else goes wrong. Frequently, that's as simple as stopping activity and waiting until a restart request is received from the manager.

Suggestions for Well-Organized Code

Let's explore some ideas to make code easier to understand and debug.

First, breaking a program into separate modules or functions gives structure to the code you are writing. A block of code with a well-defined set of inputs and outputs, plus a well-documented/commented structure, is easier to work with and debug. Keep it simple. Anything complex can usually be broken down into smaller, more easily defined and understood functions.

If the code is modular and each module has a distinct function, present it as physically distinct in the textual rendering of the code. Doing this makes it easier to understand. Do this using whitespace and a consistent labeling and commenting structure (FDI: Oualline, 1997).

Next, add commentary to explain what's going on with both the code and the variables used. Documentation should be sufficient so that you never hear the "what are you doing here?" question. If you have to verbally explain things to a coworker, include the explanation within the code comments. Remember: cave paintings last far longer than conversations around the campfire.

Use a consistent standard of physical structure, such as indentation and whitespace, to show how the code flows and make it easier to read. Also, keep every line 80 characters (or less) in width so that it's easier to read and work with on-screen. For old-school coders who kill trees, the code can be printed out and remain readable.

If you are defining a variable or a routine, do it once and then call the routine (or use the variable) wherever needed. Some designers call this the "Don't Repeat Yourself" principle. For example, when converting degrees from Fahrenheit to Celsius, create a function to do this once and then call the function anyplace it's needed.

Be descriptive and consistent when naming functions and variables. Function and variable names should tell a story. For example:

```
motor_temp_celsius
fuel_level_percent
engine_speed_rpm
```

These are all self-explanatory and don't require hunting down the definition or the units used.

Avoid deeply nested routines. Routines that have many levels of conditional branching can be difficult to follow and debug. Typically there are clearer ways to write things so that what is being done is easier to understand.

Placing all numeric values in a single list makes it much easier to adjust the system parameters. For example, if the max_temp_deg_c is 120, declare it once and then utilize the max_temp_deg_c statement throughout the code. If the temperature requirement changes, then the number only needs to be edited in one location. This is known as a named constant. Here is an example:

```
/* Maximum Celsius temperature for the system is */

   const    int    max_temp_deg_c = 120;

   /*                                            */
```

Trying to find and revise a "120" magic number throughout the preceding code would become messy. This would especially be the case with embedded systems where the physical parameters may be changed many times in the product development cycle. Using a named constant instead of a "magic number" makes things both easier to understand and revise.

For a more detailed discussion of programming style and coding constructs, the book *Practical C Programming* is suggested (see "Further Reading" on page 488).

Summary and Conclusions

Although this is not a software book, certain important topics exist at the border where the code interacts with the electronics. Things like device drivers, configuring the ports, defensive coding, and watchdog timers all deal with the interaction of hardware and software that is at the heart of embedded system design.

Coding languages, and whether an operating system is needed within an embedded system, was discussed, with a limited number of coding language options and a multitude of RTOS offerings available.

Whether or not an OS is needed largely hinges upon how complicated the system is. When there is a need for multitasking, memory management, network communication, and other features, using an OS is often easier than developing the support routines from scratch.

One important takeaway is that the software can be used to improve upon the limitations of hardware errors and the nonideal issues of the application environment. Defensive coding is rarely discussed in programming classes. Errors, noise, end-user abuse, and nonideal responses are foreign to much of the digital world.

In the real world, experienced designers have quickly realized that dealing with those issues is actually a large part of embedded product development.

Further Reading

- *Programming Embedded Systems, 2nd Edition*, by Michael Barr and Anthony Massa, 2006, ISBN-13 978-0596009830, O'Reilly Media.

- *Practical C Programming, 3rd Edition*, by Steve Oualline, 1997, ISBN 1-56592-306-5, O'Reilly Media.

- Jacob Beningo, "How to select your embedded systems operating system: Selection guidelines," Embedded.com, May 31, 2022, *https://www.embedded.com/how-to-select-your-embedded-systems-operating-system-selection-guidelines*.

- Jack Ganssle, "My Favorite Software Debouncers," Embedded.com, June 16, 2004, *https://www.embedded.com/my-favorite-software-debouncers*.

- *Bare Metal C: Embedded Programming for the Real World* by Stephen Oualline, 2022, ISBN-13, 978-1718501621, No Starch Press.

- Stephen Cass, "Top Programming Languages 2022," IEEE Spectrum, August 23, 2022, *https://spectrum.ieee.org/top-programming-languages-2022*.

- *Embedded Android* by Karim Yaghmour, 2013, ISBN–13, 978-1449308292, O'Reilly Media.

- *The Pragmatic Programmer: From Journeyman to Master*, by Andrew Hunt and David Thomas, 2000, ISBN 0-201-61622-X, Addison Wesley Longman Inc.

- Tyler Hoffman, "Defensive Programming – Friend or Foe?," Memfault Inc., December 15, 2020, *https://interrupt.memfault.com/blog/defensive-and-offensive-programming*.

- Chris Coleman, "A Guide to Watchdog Timers for Embedded Systems," Memfault Inc., February 18, 2020, *https://interrupt.memfault.com/blog/firmware-watchdog-best-practices#adding-a-task-watchdog*.

- Leslie Lamport, "Time, Clocks, and the Ordering of Events in a Distributed System," *Communications of the ACM*, July 1978, Vol. 21, No. 7, 558–565.

CHAPTER 13

Special Systems and Applications

Special systems are sectors of the electronics industry that focus on specific products or customer groups. Some sectors have different design priorities or require an entirely different approach to product design due to the application environment or regulatory requirements.

If designers are creating an embedded system for applications in aviation, outer space, military, or medical devices, they are entering a realm of highly regulated products. These devices are qualified and certified only after a suite of required testing. In many cases, the entire design flow including concept, design, testing, and implementation also requires a paper trail of design checks and sign-offs. As well, many sectors also require tight controls on materials, components, and methods utilized in manufacturing.

The upside here is that some devices in these sectors are not constrained by price point. The efforts and expenses incurred to create a regulated product are reflected in a higher price. Other sectors are tightly regulated and require low unit cost, adding challenges to creating a viable product.

Still other sectors, such as automotive and consumer electronics, pay more attention to final cost, recognizing that quality and reliability are important but that final product price often determines successful sales volumes.

The sectors discussed here are ones that I've done product R&D for. This chapter can't make anyone a sector expert, but it should serve as a good starting point. "Further Reading" on page 529 includes many additional sources for deeper discussions of specific areas.

Different Electronics for Different Priorities

Depending upon the product sector, a design can result in some very different outcomes. Consider:

Cost
Quality
Time to Market

Pick Two!

This old engineering proverb expresses the idea that the preceding three things interact during the creation process. This suggests:

- Low cost and quick to market implies low product quality.
- Low cost and high quality implies a long time to design and develop.
- High quality and quick to market implies an expensive product.

Getting all three to happen together can be a bit of a unicorn. But this idea has a certain amount of truth in reality. The important takeaway here is that design priorities interact, and specific industry sectors have different priorities resulting in different outcomes.

Design Priorities

Following are some examples of design priorities and how they affect a product.

Product Cost

This priority dominates consumer electronics, as a low price point is essential to being competitive. Especially true for multisource devices, the tenet "the lowest-cost solution usually sells the best" holds true for consumer commodity electronics.

Quality and Reliability

These two priorities are often discussed together, but they are actually somewhat different. Reliability as a priority tries to optimize or minimize failure by using redundancy in the design, wide design margins on stressed components, and the capability to self-recover from partial system failures.

However, a definition of quality can be a bit evasive. Product design quality often touts these characteristics:

- Performance
- Features
- Reliability

- Conformance
- Ease of use
- Durability

- Ability to service
- Aesthetics

Some areas where high reliability is emphasized include aviation electronics (herein referred to as *avionics*), satellites and spacecraft, military devices, medical devices, and industrial automation. These sectors treat cost as a lower priority.

Power Consumption

Battery-powered devices often dictate the need to minimize power consumption. Bigger batteries are not an option for small form factor electronics like Bluetooth-enabled headsets and cardiac pacemaker implants. Consequently, power-optimized designs reduce or remove anything that consumes energy and minimizes high-energy use powered-up time as much as possible.

Safety

Within nearly all industrialized countries, most products must pass safety testing and regulatory testing. Consumer electronics is probably the most lenient area, with minimal testing of any item without internal AC power. Medical electronics safety requirements are probably the most stringent for both patient and operator safety.

Backward Compatibility

Many product lines require that new products be backward compatible with earlier generations. Backward compatibility has had a heavy design influence on many generations of Intel microprocessors and Windows-based personal computers.

Ruggedness and User Abuse

Military electronics are known for emphasizing protective enclosures, vibration-resilient mechanics and mounts, connectors designed to withstand misuse, and physical resistance to moisture and dirt ingress. In addition, the industrial automation, avionics, and automotive sectors often emphasize systems capable of surviving in challenging application environments.

Capability for Repair

Most consumer electronics are developed without the capability to easily repair or replace components built into the design. Some consumer electronics manufacturers intentionally make their devices difficult to access or repair by third-party repair services. This has exacerbated the e-waste problem with a throwaway mentality being the status quo for purchasers of many consumer devices. Easy system repair tends to be more important for the medical, military, and industrial sectors.

Navigating the Regulatory Maze

Consumer electronics have the easiest path through regulatory testing because the final product is compliance tested, but the design process used to create the product is not regulated. Many sectors regulate the design, qualification, and manufacturing processes, not just the final product. For these, a qualification and quality management process for mechanics, electronics, software, and manufacturing needs to be in place from the start, and then applied throughout the R&D process and for all manufacturing.

Understanding a design flow where the R&D must be both documented and approved is useful. First, a widely published outline of a waterfall design flow is shown in Figure 13-1. This waterfall diagram lacks detail but illustrates several things:

- Review throughout the design
- Verification of final product versus design definition
- Validation of final product versus original user-defined needs

The following discussion uses a medical device as an example, but similar multistage reviews are typical for any device that requires regulatory control of both the design flow and the final product.

The waterfall illustration is simplistic. It was first published in 1956 (Office of Naval Research, 1956) and republished in 1970 by Winston, who coined the "waterfall" moniker for use in software development.

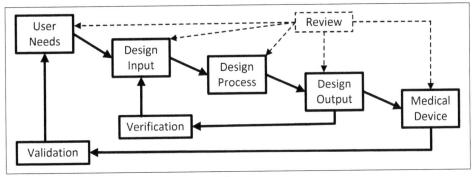

Figure 13-1. Traditional waterfall design flow

A more modern design flow for devices that have mechanical, electrical, and software components is shown in Figure 13-2. This flowchart has been used for medical device development and associated regulatory restrictions many times. It's similar to many design flows that require regulated design.

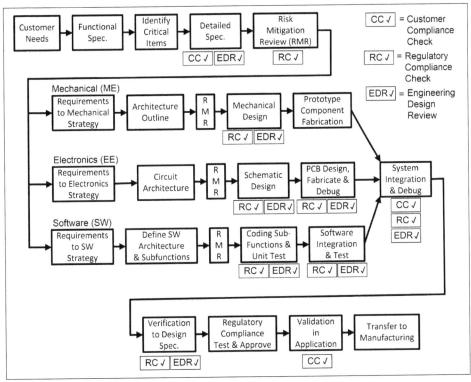

Figure 13-2. Design flow that includes process regulation

Following is a brief explanation of each block in the figure:

Customer Needs

Your customer can be an external client or it can be your company's marketing department responsible for defining the next new product. In any case, understanding what's needed is always the starting point.

Functional Specification

The first pass at product definition is a somewhat more formal way of documenting what the customer needs. Create a document that serves as the starting point for the discussions that will follow. The marketing team often refers to this as a value proposition, and although different, the value proposition and functional specification are frequently merged into a single document.

Identify Critical Items

This is a good point to determine where the binding points are, what's going to be difficult to accomplish, and what resources, time, and money are going to be required. Wherever possible, mitigate the difficult items by using a different method, and sidestep any potential problems.

Detailed Specification

After critical items have been recognized and dealt with, expanding the functional specification into a more detailed set of design requirements is the next step. When completed, an engineering design review is used to get both review and buy-in from the design team. Also, customer approval serves as a sanity check that the specified design goals are aligned with the customer's product vision.

Risk Mitigation Review (RMR)

The first RMR defines all possible risks within the system and determines how they will be dealt with. (Risk analysis and mitigation will be explored shortly.)

Architecture Outline

The mechanical design strategy results in an outline of the mechanical design.

Mechanical RMR

This RMR serves as a check that the original RMR has not changed. There may be no significant changes to the design goals and method. Or some risk items may need to be redefined.

Mechanical Design

Perform detailed design of the mechanical components and prepare those components for fabrication.

Prototype Component Fabrication

Create or obtain the mechanical parts and pieces.

Requirements to Electronics Strategy

Using the design specification, an electronics design strategy is outlined.

Circuit Architecture

A block diagram of the circuit architecture is developed and all major components—microcontroller (MCU), analog-to-digital converters (ADCs), digital-to-analog converters (DACs), and sensors—are selected (FDI: Architecture).

Electronics RMR

Check that the proposed electronics have not changed the original RMR. Adjust as needed.

Schematic Design

This is a detailed design of the system electronics.

PCB Design, Fabrication, and Debug

This is where detailed printed circuit board (PCB) design, bill of materials (BOM) creation, PCB assembly, and debug of the final PCB occur (FDI: PCB).

Requirements to Software Strategy

Using the design specification, a software strategy is outlined.

Define Software Architecture and Subfunctions

Detailing what functional blocks of code are needed, defining their internal functions, and defining all required features and how those software blocks should interact are done.

Software RMR

Check that the proposed software hasn't changed the original RMR. Adjust as needed.

Coding Subfunctions and Unit Test

Detailed code development of the various blocks is done. Block testing is performed, assuring that each block behaves as expected and has the required features.

Software Integration and Test

Build the complete code set, and test out the functionality and interaction of code blocks using emulation and simulation methods or on a preliminary hardware platform.

System Integration and Debug

Bring it all together: mechanics, electronics, and software. Test out functionality, the presence of all desired features, and interaction of electromechanical systems with software control.

Verification to Design Specification

Many regulatory processes require a formal set of tests that verify that product functionality meets the required design specifications.

Regulatory Compliance Test and Approval

Regulatory testing is usually performed by an independent test lab that has been recognized and certified by the regulatory body. Depending on the sector, some testing may be done by the regulatory body, or in-house testing may be accepted by others. Most often, an independent test lab will need to be engaged.

Validation in Application

The device is tested in the expected application environment.

Transfer to Manufacturing

Depending upon the sector, the manufacturing process may also require a regulated and monitored flow of materials and methods.

Sometimes a step backward in the flowchart may be needed if an evaluation, review, or test indicates that something must be redesigned or redefined.

For regulated devices, the overhead associated with the R&D process will generate a number of reports that the regulatory body will want to review. This is illustrated in Figure 13-3. In addition to the documents shown, the regulatory testing lab will generate a compliance test report that describes the tests performed and their results.

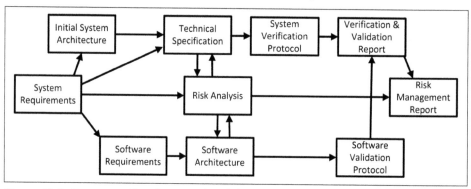

Figure 13-3. Document flow for a regulated design process

Risk Analysis

Depending upon the industry sector, risk analysis is also referred to as *failure modes and effects analysis* (FMEA, used in avionics) and *system safety program requirements* (used for military products). The terms *risk, hazard,* and *failure* are interchangeable in this discussion, and the literature utilizes all three.

As mentioned, regulated R&D processes often include the concepts of risk management and risk mitigation review. Conceptually, the procedure is to identify every known possible failure scenario, hazard, or improper use associated with the device. This list should be extensive, and all involved parties (customer, engineers, manufacturing, typical end users) should be allowed to contribute possible risk/failure/hazard scenarios.

After identifying all possible risks, a classification and mitigation process is performed, as shown in Figure 13-4.

This process allows designers an early opportunity to improve the design by either removing the risk or refining the design to reduce the risk. The outcome of this process should be an improved risk list with a defined level of possible harm and a defined probability of occurrence.

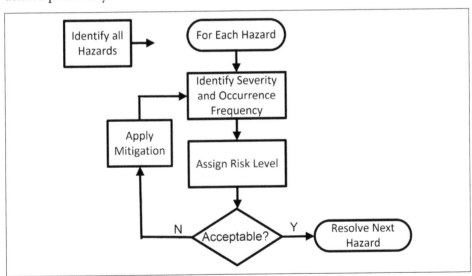

Figure 13-4. Risk management and mitigation

Every defined risk should be assigned a rating for probability of occurrence and a rating for harm severity (Table 13-1). Risks with significant harm severity and high occurrence probability are the items of concern. Risk mitigation through design improvements should attempt to eliminate these issues as much as possible.

Table 13-1. Risk management matrix

		Severity of harm		
		Negligible	Moderate	Significant
Probability of occurrence	**High**	Medium	Medium	High
	Medium	Low	Medium	Medium
	Low	Low	Low	Medium

Even when certain risks can't be eliminated or reduced, a regulatory body will often deem certain higher risks to be acceptable. If the benefit of using the device outweighs the device risks, approval will often be granted while making note of the associated risks. For example, X-ray machines create radiation exposure, which is seen as a risk, but noninvasive medical imaging benefits from its use.

Regulatory documents that deal with risk and hazard management include:

- ISO 14971 – Risk Management for Medical Devices
- ISO 31000 – Risk Management
- MIL-STD 882 – Military Standard: System Safety Management
- MIL-STD-1629 – Military Standard: Procedures for Performing a Failure Mode, Effects, and Criticality Analysis
- SAE J1739 – Potential Failure Mode and Effects Analysis (FMEA) Including Design FMEA, Supplemental FMEA-MSR, and Process FMEA
- SAE 1025 – FMEA/FMECA Global Standard (presently a work in progress but expected to replace both SAE J1739 and MIL-STD-1629)

Although MIL-STD-1629 has not been updated/maintained by the US Department of Defense (DoD), it is still utilized by some. The following maintained standards are most often utilized:

- SAE J1739 is being used for nonmedical devices.
- ISO 14971 is required for medical devices.

Which specific FMEA or risk management standard to use depends on the product sector. (See Oehmen, 2010, in "Further Reading" on page 529 for more on the topic.)

Aviation Electronics (Avionics)

Modern avionics has grown in complexity and capability to the point where automation has replaced flight engineers on modern commercial airplanes. As well, the flight crew now has an extensive array of automated support systems including electronic flight instruments, navigation, communication, automated flight control, multiple information and warning systems, and much more.

Avionics includes many topics and methods. Consequently, the focus here is on just the instruments and systems where embedded controllers are commonly used. Many controllers and computers exist inside most avionics systems.

Design Priorities

The top priority in aviation is safety. That conveys to a high priority for reliable failsafe systems. This includes fault-tolerant systems and hardware, multiple layers of redundancy, and the capability for system reconfiguration. Reconfiguration methods allow the deactivation of failed subsystems and activation of redundant devices to keep the system functional.

Reliability within individual data processing subsystems is often achieved using triplication and voting methods. This uses redundant parallel data processing, where the results of three (or more) processors are compared. If one processor shows a different outcome from the other two, a fault is reported and the majority result is used.

Self-checking processor pairs are another strategy used. If the results from two processors diverge, both are taken off-line while another pair of processors is activated to continue functioning seamlessly. The off-line devices are instructed to execute a self-test and error-check routine before returning to service.

Fault-tolerant coding, error recovery methods, and other defensive coding techniques are also part of avionics safety and reliability.

Special Needs

Avionics requires the capability to:

- Reliably withstand direct lightning strikes for external sensors and lightning-induced electromagnetic effects upon internal systems and instruments
- Tolerate and remain functional over a wide temperature and altitude range
- Deal with restricted magnetic field radiation when in proximity to magnetic navigation instruments

See DO-160 from the RTCA for a comprehensive list.

Regulations, Certifications, and Approvals

The Federal Aviation Administration (FAA) is the regulating body within the United States for aviation systems. However, worldwide consensus on design quality and other regulatory criteria is developed by the RTCA.

For the most part, the FAA has issued advisory documents stating that adherence to the RTCA standards shall suffice for FAA compliance. In addition to RTCA standards, the FAA has issued a number of Technical Standard Orders (FAA-TSO) (*https://oreil.ly/_4V43*) that itemize requirements for individual aviation parts and instruments.

As an example, FAA-TSO-C2d provides FAA requirements for airspeed instruments. Many FAA-TSO documents circle back and refer to information defined within RTCA specifications.

The principle RTCA documents of interest are:

- RTCA DO-160 – Environmental Conditions and Test Procedures for Airborne Equipment (this is aligned with ISO 7137)
- RTCA DO-254 – Design Assurance Guidance for Airborne Electronic Hardware
- RTCA DO-178 – Software Considerations in Airborne Systems and Equipment Certification
- RTCA DO-278 – Guidelines for Communication, Navigation, Surveillance, and Air Traffic Management Systems Software Integrity Assurance (ground-based aviation equipment)
- RTCA DO-248 – Supporting Information for DO-178 and DO-278

When developing an aviation instrument and the associated electronics, following DO-160, DO-254, DO-178, and the FAA-TSO for each device serves as a good starting point.

Satellites and Spacecraft (Astrionics)

Aerospace is a catchall that covers aviation, defense, satellites, and spacecraft. Some design practices, regulations, and requirements spill across these sectors, but many do not. Herein, we parse off the worlds of avionics, military/defense, and satellites/spacecraft as their own disciplines. However, even the "final frontier" needs to be subdivided to understand which preferred methods should be used for embedded electronics.

Is it a spacecraft, a satellite, or something else? A spacecraft is any vehicle that is designed to fly in outer space. A satellite orbits around another body (the sun, the Earth, the moon, or another planet). Consequently, satellites are a subset of spacecraft. A crewed spacecraft includes life support systems, pressurized passenger containment, and (hopefully) the capability to return passengers to their home planet.

This sector can be divided along several different dimensions:

Orbit type
Low Earth (LEO 160–1,500 km altitude), Medium Earth (5,000–20,000 km), Geostationary (GEO all at 35,786 km), Sun Synchronous (600–800 km), and Geostationary Transfer (variable)

Satellite function
Communication (data, video, telecommunication), Earth observation (weather, science research, defense and spy), navigation (GPS), and astronomical observation (Hubble, Webb, Kepler, Spitzer, etc.)

Spacecraft by mission type
Flyby, orbiter, atmospheric, lander, penetrator, rover, observatory, communication, and navigation (see *https://solarsystem.nasa.gov/basics/chapter9-1* (*https://oreil.ly/0Je1w*))

None of these are a good way to determine priorities and methods for spacecraft electronics. Instead, discussing the specific challenges of outer space and how they affect embedded systems should be more appropriate within the context of this chapter.

Radiation

The effects of radiation on integrated circuits can be divided into two areas:

Single Event Effects (SEE) radiation
SEE is caused by an energized particle going through the semiconductor. If a vulnerable transistor is in the path, it can cause the transistor to briefly change state, or it can damage the transistor. In both cases, a logic state error can occur.

Long-Term Total Dose (LTTD) radiation
LTTD can change the performance characteristics of transistors, leading to degradation of the electronics' performance.

Radiation effects are reduced using multiple techniques. Triplication and voting in logic circuits are commonly employed. Circuit architectures that are "radiation hard" utilize circuits where one transistor can be in error but the overall system retains proper functionality. Semiconductor processes that are less vulnerable to radiation are used, including Silicon on Sapphire and Silicon on Insulator devices.

Use of radiation shielding is common, but shielding is minimized because it adds weight and volume to the packaged electronics.

Thermal Extremes

Temperatures created in space can vary over a very wide range, based upon proximity to the sun, devices moving in and out of the shadow of a planet, and others. The dark side and the sun-facing side of a satellite are exposed to extremely different amounts of energy. This environment can create some rapid thermal transients. In addition, due to being in a vacuum, thermal conduction through airflow and the concept of environment temperature don't really apply.

Thermal management techniques are part of most satellites and spacecraft. Early efforts such as passive thermal control mode, better known as the "barbecue roll," would slowly rotate the Apollo spacecraft, exposing all surfaces to both the bright sun and dark space. This would balance out the spacecraft temperature.

Modern satellites use thermal controls with multilayer insulation and other methods. Once a spacecraft is in initial design, energy conduction models will define the resulting temperature range of the internal electronics. That thermal range will be used in the design and qualification of circuitry and components.

Vibration, Shock, and Acceleration

All spacecraft electronics will undergo mechanical stress scenarios, especially those prevalent during launches and return to Earth events. Spacecraft dynamic environments testing is required for all spacecraft and satellite components. The NASA handbook *NASA-HDBK-7008* explores the topic in detail.

Collaboration between electronics and mechanical design will be needed for a vibration- and flexure-resilient mounting strategy. Getting the mounted electronics to pass vibration, shock, and acceleration (VSA) testing prior to integration into the spacecraft is suggested.

Vacuum Environments

Most modern satellites are enclosed, but are not airtight. Consequently, all electronic components selected must be designed and qualified to work in a vacuum. Some components (electrolytic capacitors) will self-destruct in a vacuum, so proper component qualification is important.

Vacuum environments add an additional requirement: namely the electronics must be cooled without using airflow.

These challenges create special considerations when picking materials, components, and PCB methods.

Component Selection and NASA-Approved Parts

Determining whether a part is suitable for space requires expensive and time-consuming qualification testing. Thankfully, NASA, the European Space Agency, and the Japan Aerospace Exploration Agency all maintain parts listings that they deem space suitable and provide information on the device qualification requirements:

NASA
The NASA Electronic Parts and Packaging Program (NEPP) can be found here (*https://nepp.nasa.gov/npsl*).

European Space Agency (ESA)
The "European Space Components Information Exchange System" (ESCIES) can be found here (*https://oreil.ly/MXzxt*) and here (*https://oreil.ly/YPijB*).

Japan Aerospace Exploration Agency (JAXA)
"Common Parts/Materials, Space Use, General Specifications for" Document Number: JAXA-QTS-200F, 2022, can be found here (*https://oreil.ly/AfuoE*) and here (*https://oreil.ly/6OZS_*).

NASA recognizes components based upon MIL-STD qualification testing, as well as certain classes of parts recognized by ESA and JAXA. See the NASA-NEPP site for details and an up-to-date version of its Qualified Parts List Directory.

The following may also be useful:

- MIL-PRF-38535 – Integrated Circuits (Microcircuits) Manufacturing, General Specification for
- MIL-M-38510 – Microcircuits, General Specification for
- MIL-PRF-38534 – Performance Specification, Hybrid Microcircuits, General Specification for

Class K components as per MIL-PRF-38534 are specified by NASA and the DoD as approved for satellite/spacecraft use. Using components not already on the preapproved lists will require considerable effort to gain approval.

PCB Materials and Layout

Certain PCB materials can outgas in a vacuum, especially with large amounts of thermal cycling. Most PCBs for outer space use polyimide laminates.

A useful strategy here is to seek out a contract PCB manufacturer (CPM) that specializes in aerospace electronics and request information on its offerings for outer space–compliant PCBs. A NASA-qualified manufacturer of PCBs can also provide suggested laminate stack-ups and physical design rules for space-compliant circuit boards.

Limited Life of Spacecraft

This discussion of spacecraft and satellites has concentrated upon devices that have high reliability and are optimized to provide long-term survival in outer space. NASA's Voyager program, launched in 1977, was still functioning in 2023. NASA speculates that it may remain functional through 2036.

Typically, a geostationary satellite remains useful for five to fifteen years, with the predominant failure due to propellant running out. Low Earth orbit (LEO) satellites typically last seven to ten years and usually fail due to atmospheric friction slowly eroding their orbit.

Long-term reliability comes at a high cost to implement. There is a different way to do things, however. Consider the idea of mass-produced satellites with a limited life expectation. With that in mind, the satellite system remains functional using multiple devices and a periodic replacement strategy. High-reliability, radiation-hardened, and vacuum-tolerant designs are still needed, but mass production and multiple satellites on a single launch help considerably with cost reduction.

The following notable satellite groups fit the "disposable and mass-produced" category.

Disposable satellites

CubeSats (*https://www.cubesat.org*) are satellites built to a physical size standard (10 cm cube) and launched into LEO. Many early CubeSats were academic research projects and were launched as a secondary payload with other, more traditional satellites.

The physical standard for the CubeSat is defined in ISO 17770, with the "1U" CubeSat being 10 cm × 10 cm × 11.35 cm. Variants on the size are created using multiples of the 1U dimensions, with the 3U version being dimensioned to fit the same space as three stacked 1U devices. This allows multiple devices to be loaded and launched together easily.

The electronics in the CubeSat varies between implementations but has frequently used the PC/104 stackable PCB, which fits within the CubeSat dimensions.

Many early CubeSats were nonfunctional after launch, presumably due to vibration and shock issues. Others died shortly thereafter, presumably due to thermal, radiation, or vacuum issues. Designers were using standard components and a consumer electronics mentality. Later designs were more successful due to the experience gained. As of May 2023, over 2,100 CubeSats have been launched.

Following are some important considerations when developing embedded systems for a CubeSat:

- Use radiation-hardened and vacuum-resilient controller electronics.
- Design mechanical mounts to absorb the mechanical stresses of launch and to protect the integrity of the connectors used between the stacked PC/104 boards.
- Conduct vibration and shock tests of the assembled unit.
- Conduct thermal cycling tests of the assembled unit.
- Conduct vacuum chamber functionality and survival tests.

That small list of items should improve the success and survival rate of future CubeSat launches. There are numerous interest groups and academic resources that support CubeSat R&D online.

Mass-produced satellites

Mass-produced satellites are starting to be widely used in LEO applications. The intention is not extreme reliability and long life, but instead the use of redundancy with multiple devices in orbit. The periodic replenishment of satellites is part of the system strategy. Using this approach, satellites can fail, or they can burn up when their orbits degenerate, while the system remains functional.

The Starlink (*https://www.starlink.com*) satellite constellation from SpaceX is the most prominent example, with over 5,000 Starlink satellites deployed as of 2023 and more planned. Although much bigger than the CubeSat, the Starlink also uses a standard form factor, allowing 60 satellites to be stacked together for a single launch.

Satellite cost should drop considerably by using mass-production methods, but specifics of Starlink's internal satellite design, component selection, and other technical data are not publicly available as of this writing.

Other operational satellite constellations are Orbcomm, Globalstar, and Iridium. Other companies and government agencies are in various stages of R&D.

Regulations, Certifications, and Approvals

Regulatory requirements for satellites and spacecraft require ITU clearance for orbital placement. Both the FCC in the United States and the ITU will need application and approval for frequency and communication allocations. Adhering to NASA technical standards should result in electronics that can survive both launch and environmental challenges of outer space:

- NASA technical standards can be found at *https://standards.nasa.gov/nasa-technical-standards*.

- Spectrum allocation for satellite communication comes from the International Telecommunication Union (ITU). See *https://www.itu.int/en/Pages/default.aspx*.

- Orbital locations for geostationary satellites and orbital characteristics for non-geostationary satellites are allocated by the ITU.

- The FCC licenses satellite communication for use in the United States. This is covered in the code of federal regulations: CFR, Title 47, Chapter I, Subchapter B, Part 25, Satellite Communication (25.101–25.702).

- The International Aerospace Quality Group sets quality standards for the supply chains associated with the aerospace industry. It has implemented a quality standard, AS9100, that is aligned with the aerospace industry while retaining many of the principles associated with the ISO 9001 quality standard. See *https://iaqg.org*.

Military Electronics

Military Standard (MIL-STD) electronics are devices used anywhere within the infrastructure and support of the DoD, military facilities, and field operations. Other countries have similar requirements, with many NATO countries aligned with US DoD requirements via the NATO Standardization Agreement (see *https://www.nato.int*).

The US DoD recognizes and purchases electronics in many forms, including (summarized from MIL-STD-196G, the "Joint Electronics Type Designation System"):

- Radios
- Radar
- Data processing
- Flight control
- Navigation
- Weapons control
- Electronic countermeasures
- Radiation detection
- Infrared devices
- Lasers
- Meteorological

- Magnetic amplifier and detection
- Wire communications
- Televisions
- Fiber optic systems
- Noise detection
- Underwater sound
- Training equipment
- Satellites
- Robotics
- Maintenance/support

Design Priorities and Unique Requirements

Most MIL-STD devices are subjected to an extensive set of environmental and handling challenges, as outlined in MIL-STD-810. This includes:

- Altitude
- Humidity
- Shock
- Salt fog
- Temperature
- Vibration

- Radiated EMI
- Conducted EMI
- Operating temperature
- Storage temperature
- Rapid decompression
- Acoustic noise level

This results in device enclosures that are rugged, with the capability to withstand abuse and remain functional. Military-compliant enclosures are defined in MIL-STD-108.

Internal mounting of components and circuit boards is engineered to survive shock and vibration testing. External electrical connectors that are designed for this abusive environment utilize protective covers, moisture resistance, and larger physical sizes to provide better mechanical durability.

Examples of military circular connectors in common use can be found in:

- MIL-DTL-38999
- MIL-DTL-26482

- MIL-DTL-5015
- MIL-DTL-26500

A strategy of explicitly defined rugged durability is common to most MIL-STD devices. There are exceptions for specialty scenarios where weight or size takes a higher precedence.

Electronic designs that emphasize easy internal component replacement and repair access are usually well received in DoD-conducted design reviews. There have been documents (MIL-STD-1389D, MIL-M-28787D) that emphasize modular electronics in the past, but these no longer seem to be followed or updated.

Regulations, Certifications, and Approvals

The DoD used to maintain a list of all military standards (AD-A273-295, Index of Specifications and Standards) where the list of document titles spanned over 900 pages. List maintenance was discontinued in 1993. The number of documents is huge, and the applicable material for the project at hand needs to be determined early in the R&D process.

When developing a military product, asking the product customer to itemize any and all applicable documents, regulations, and requirements is the most direct path to understanding what must be considered in the design. Asking the customer to define the compliance list up front is probably the safest way to define the required compliance documents.

Some of these documents should be on the customer-requested compliance list:

- MIL-STD-167-1A – Test Methods for Mechanical Vibration of Shipboard Equipment
- MIL-STD-196 – Joint Electronics Type Designation System (JETDS)
- MIL-STD-202 – Electronic and Electrical Component Parts Test Methods
- MIL-STD-348 – Radio Frequency (RF) Connector Interfaces
- MIL-STD-461 – Requirements for the Control of Electromagnetic Interference Characteristics of Subsystems and Equipment
- MIL-STD-464 – Electromagnetic Environmental Effects Requirements for Systems
- MIL-STD-498 – Software Development and Documentation
- MIL-STD-750-2 –Test Methods for Semiconductor Devices
- MIL-STD-810 – Test Methods for Determining the Environmental Effects on Equipment
- MIL-STD-882 – Standard Practice for System Safety
- MIL-STD-883 – Test Method Standard for Microcircuits
- MIL-STD-1397 – Input/Output Interfaces, Standard Digital Data
- MIL-STD-1553 – A Digital Communications Bus
- MIL-PRF-38534 – General Specification for Hybrid Microcircuits
- MIL-PRF-38535 – General Specification for Integrated Circuits (Microcircuits) Manufacturing

This is not a complete list, but these are some of the more commonly referenced standards. Many specific documents are also available that explicitly cover a particular product or group of products. Certified testing labs are widely available that can test to the requested MIL-STD standards.

Designing a MIL-STD product can be a frustrating exercise in determining which compliance documents apply and then reaching an agreed-upon design specification. That process is made more difficult when two different documents define two somewhat different requirements. If your customer is an active advocate of the process, it makes things much easier. If all else fails, consult MIL-C-44072C for some tasty insight into the problem.

Medical Devices

Medical instrumentation has expanded from monitoring and diagnostic devices to a plethora of devices for diagnosis, minimally invasive observation/surgery, emergency therapeutic devices, automated care and monitoring, robotic surgery, and implanted devices. As well, the human–machine interface is becoming populated by more devices, with direct brain–machine interfacing techniques evolving rapidly.

There are two major groups of medically related electronics. As per the IEC, medical electrical equipment is defined as "electrical equipment having an applied part or transferring energy to or from the patient or detecting such energy to or from the patient…." These devices are regulated under the umbrella of the IEC 60601-1-1 requirements. Example devices include:

- EEG monitors
- ECG monitors
- Infusion IV pumps
- Blood glucose monitors
- MRI systems
- Ultrasound systems
- Vital sign monitors
- Cardiac pacemaker implants
- Nerve stimulus implants
- Neural pacing implants
- Defibrillators and AEDs
- Cochlear implants

However, any "lab test" instrument that does not have direct patient contact is controlled under a different set of regulatory rules: IEC 61010. Examples of these include:

- Centrifuges
- Chromatography systems
- Autoclaves
- Microscopes
- Hematology analyzers
- Blood gas analyzers
- Cell counters
- Lab incubators

The discussion here centers on devices that directly attach to the patient. These medical devices are tightly regulated in the design flow, risk management, device test and qualification, materials used, and manufacturing processes. Here the focus is on the design and regulatory qualification of those devices.

Starting with the regulatory requirements allows us to determine the special needs of medical systems.

Regulations, Certifications, and Approvals

Most countries use a device approval process that is driven by IEC/ISO requirements. Some countries have accepted this approach without restrictions. Others developed their own variants of or addenda to the IEC/ISO documents. Following the IEC/ISO documents will get the job mostly done, but country-specific variants should be checked for anything that might affect the design. The relevant documents are:

- IEC 60601-1-11 – Part 1 Medical Electrical Equipment – General Requirements for Basic Safety and Essential Performance.
- IEC 60601-2-2 – Part 2 Medical Electrical Equipment – Particular Requirements for Basic Safety and Essential Performance.
- IEC 60601-2-XX – Specific requirements for individual devices; there are about 80 specialty documents as of 2023, and these are updated and added to as need becomes evident. For example, IEC 60601-2-26 covers specific requirements of EEG machines.
- ISO 13485 – Medical Devices – Quality Management Systems.
- ISO 14971 – Medical Devices – Application of Risk Management to Medical Devices.
- IEC 62304 Medical Devices Software – Software Life Cycle Processes.

Herein, *IEC 60601* refers to the preceding suite of tests and requirements.

In many cases, these documents lay out specific testing or functional requirements, or they require utilization of procedures specified in other documents. For example, IEC 60601-1-2 requires that the device under test (DUT) pass these tests:

Radiated Emissions (IEC CISPR 11)
    EMI transmitted into the air, also defined in FCC Title 47, CFR Part 15 B

Power Line Current Harmonics (IEC 61000-3-2)
    Current loads created by the DUT as seen in the spectral content of the current on AC power

Power Line Flicker (IEC 61000-3-3) and Voltage Dips and Short Interruption Immunity (IEC 61000-4-11)
    Power dips of various magnitudes and periods, including a 5-second AC power cutoff

Electrostatic Discharge Immunity (IEC 61000-4-2)
    Electrostatic discharge (ESD) events up to +/−8 KV applied to all external contact points of the DUT

Radio Frequency Immunity (IEC 61000-4-3)
    DUT is placed in controlled environment with RF injected into the test chamber; multiple frequencies and modulation sources are applied

Electrical Fast Transient Burst Immunity (IEC 61000-4-4)
    Fast transients applied to the AC mains input; similar to a noisy power supply created by a motor on the same mains power lines

Power Line Surge Immunity (IEC 61000-4-5)
    A multitude of different high-voltage transients injected into the AC power wires

Power Frequency Magnetic Field (IEC 61000-4-8)
    DUT is placed in low-frequency (50 Hz and 60 Hz) magnetic field

Specific details of each test can be found in the documents cited.

It is important to note that medical devices need to survive and continue to properly function through these tests. Surviving without damage is not sufficient. Medical devices should have:

- No errors, no system resets, no false alarms, and no display glitches
- No component failures
- No changes in programmable parameters
- No reset to factory defaults
- No change of operating mode

Clean Functionality Throughout EMC Tests

This is explicitly spelled out in the IEC 60601 documentation, which states the following:

- No cessation or interruption of any intended operation
- No initiation of any unintended operation, including unintended or uncontrolled motion
- No error of a displayed numerical value sufficiently large to affect diagnosis or treatment
- No noise on a waveform in which the noise is indistinguishable from physiologically produced signals or the noise interferes with interpretation of physiologically produced signals
- No artifact or distortion in an image in which the artifact is indistinguishable from physiologically produced signals or the distortion interferes with interpretation of physiologically produced signals
- No failure of automatic diagnosis or treatment equipment

Techniques for ESD protection, power supply filtering, defensive coding, error detection and correction, and others outlined elsewhere in this book are all valuable tools for passing these tests. Use of a "mostly digital" approach with ADCs/DACs closely placed to their utilization point and EMI/ESD methods outlined in prior chapters should allow successful approval on most of these tests.

Also, for many devices, an external AC/DC power supply can be used, where the AC/DC supply has already been certified as IEC 60601 compliant. This will simplify the compliance safety examination process.

Special Needs

Requirements that present unique challenges for medical devices include:

- Remaining functional through ESD events
- Power supply dips
- Patient isolation from the medical instrument

ESD functionality

This requires both hardware mitigation (FDI: EMC & ESD) and defensive coding methods (FDI: Code) to prevent or recover from the system being disrupted. If all access connections are properly circuit protected and suitable recovery and protection routines are present in the code, this often suffices.

Suggestion: obtain an ESD test gun and perform in-house ESD testing before taking the device to a certified compliance test facility. This allows the iterative process of fixing ESD sensitivity issues. Typical testing may reveal issues that need software tweaks, parameter adjustments to ESD circuits, moving of internal cable paths, or other mechanical changes.

Power dips

Power dips are part of required EMC testing. This includes assorted power dips and transients. The most challenging of these is the 5-second power drop out, shown in Figure 13-5. In this test, essentially the power plug gets pulled for 5 seconds, and the medical device needs to keep functioning through the power loss. There are two solutions here:

- Specify that the medical instrument be qualified with an external Uninterruptable Power Supply (UPS) as part of the system.
- Build a battery and charging system into the medical device.

Both solutions are viable. The internal battery doesn't need to be huge, just large enough to get through the 5-second power drop out. However, for portable instruments, it is suggested that the battery capacity be sufficient to allow patient transport with active instruments.

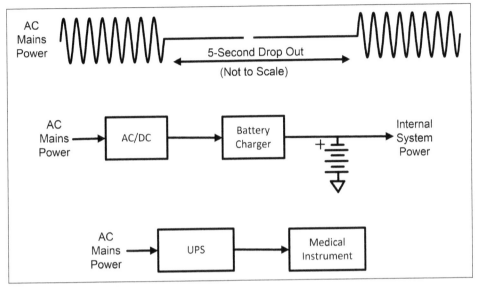

Figure 13-5. Power dip test

Patient isolation safety

Patient isolation safety requires that no medical instrument can create a scenario where "significant" current flows through the patient. Let's get a better understanding of the problem, as illustrated in Figure 13-6.

On the left of the figure is a typical scenario for an instrument attached to a patient. The voltage difference between the patient and the instrument is not well defined or easily controlled. This is represented as two arbitrary voltages (V_1, V_2) where the values can be unknown, or even variable, as the patient and instruments are moved and handled.

On the right of the figure, a simplified circuit representation is shown, with hypothetical resistances associated with the instrument and a human body model (HBM).

The regulatory restriction here is that the maximum total current into the HBM can't exceed 10 uA DC, or 100 uA AC under normal conditions.

Figure 13-6 is an example of the "short answer" guideline. The numbers vary depending upon classification of devices and allowed fault scenarios. For a more detailed answer, IEC 60601-1 Section 8.7, Protection against Electrical Hazards, deals with the details of allowable leakage currents into the patient (from IEC 60601-1, 3rd edition, 2005; newer versions may be available).

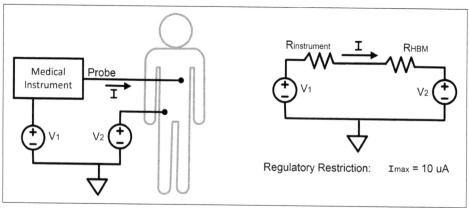

Figure 13-6. Current through the patient

With little capability to control the voltages or grounding associated with the patient and medical instrument, eliminating the possibility of current leakage requires removing any DC connections between the instrument and the patient. Figure 13-7 illustrates the methods used.

As shown, DC paths are opened using a DC-to-DC power converter and digital isolation couplers. The DC-to-DC converter has no DC path from input to output, and the output side will readily "float" by adapting to any voltage created at the patient. Internally, the device uses an isolation transformer and switching circuits to transfer energy using electromagnetics.

The digital isolation couplers allow the passing of data, while also eliminating any DC path between input and output. Internally, the most common method used for data isolation is an LED and phototransistor to transfer data by optical methods.

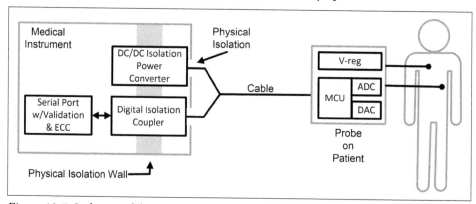

Figure 13-7. Isolation of the patient connection

Seek out devices that are described as meeting medical isolation requirements, where the required maximum voltages across the devices, physical creepage, and separation requirements have been met.

In addition to the DC path isolation components, well-defined physical separation, dielectric/insulation materials, and creepage parameters (FDI: PCB) for isolation gaps must be met. For devices that intentionally apply energy to the patient (defibrillators, electroshock, cauterization, ablation devices, etc.), specialty requirements are provided within the suite of IEC 60601-1-XX documents.

Regulatory Requirements for Software and Firmware

There are two different approaches to meeting regulatory requirements for code-related parts of the system:

- If the code is frozen prior to regulatory testing and will not be modified during the product's lifetime, a black-box approach can be used, where the medical instrument is tested and qualified without an active software management strategy in place. This eases regulatory testing but requires requalification if the code is changed.
- If the code is more complex, or if an updatable operating system is used and is expected to be revised or modified during the product's lifetime, a software lifecycle process (IEC 62304) in conjunction with a quality management system (ISO 13485) need to be utilized.

Typically, use of Verilog/FPGA/CPLD, digital ASICs, and bare metal MCU code can follow frozen prequalification testing, and then application of IEC 60601 testing without software management can be used.

Complex software with multiple developers, multiple code revisions, multiple code sub-blocks, and updating operating systems needs to meet the IEC 62304 process, in addition to the mechanics and electronics complying with IEC 60601.

Further information on software management in a programmable electrical medical system (PEMS) can be found in the IEC 60601-1 documentation; see Annex H, PEMS Structure, PEMS Development, Life-Cycle and Documentation.

Medical electronics must meet regulatory requirements. Without compliance, it's illegal to use them in the United States. No compliance means not usable even on an experimental basis. Compliance requirements can heavily affect the design of mechanics, electronics software, and manufacturing as well as component selection. Designers need to be somewhat literate in the regulatory issues or have a resource that understands electronics–compliance interplay.

Automotive

Automotive electronics have gone through some radical changes. The original electrical system of the Ford Model T had a magneto and four ignition coils to fire the engine sparkplugs, but that was it. Modern use of electronic control units (ECUs) has skyrocketed, with some autos using up to 150 of them within the vehicle.

Typical Electronic Control Units

The Clemson University Vehicular Electronics Laboratory (*https://oreil.ly/7ubI_*) has assembled a comprehensive list of where modern automobiles use embedded control electronics:

- Accident recorder
- Active aerodynamics
- Active cabin noise suppression
- Active exhaust noise suppression
- Active suspension
- Active vibration control
- Active yaw control
- Adaptive cruise control
- Adaptive front lighting
- Airbag deployment
- Antilock braking
- Auto-dimming mirrors
- Autonomous emergency braking
- Battery management
- Blind spot detection
- Environment controls
- Communication systems
- Convertible top control
- Cylinder deactivation
- Driver alertness monitoring
- Electronic power steering
- Electronic seat control
- Electronic stability control
- Electronic throttle control
- Electronic toll collection
- Electronic valve timing
- Engine control
- Entertainment system
- Event data recorder
- Head-up displays
- Hill hold control
- Idle stop-start
- Instrument cluster
- Intelligent turn signals
- Interior lighting
- Lane departure warning
- Lane keeping assist
- Navigation
- Night vision systems
- On-board diagnostics
- Parental controls
- Parking systems
- Precrash safety
- Rear-view camera
- Regenerative braking
- Remote keyless entry

- Security systems
- Short-range communication
- Tire pressure monitoring
- Traction control
- Traffic sign recognition
- Transmission control
- Windshield wiper control

Most of these devices are widely used in cars now on the market. As the self-driving automobile becomes more refined, expect more additions.

Design Priorities and Special Needs

The ECU appears in many forms throughout a modern automobile. These devices are challenged with demanding environments that include temperature extremes, moisture and corrosion, mechanical vibration and shock, variable/noisy power, and others. High volumes of devices are required, and unit cost needs to be kept low to keep final product prices competitive.

Problems and suggested solutions for automotive ECU devices include the following:

Dirty power: A noisy and variable power supply should be expected.
   Possible solutions: Use of filters, RF chokes, and internal voltage regulators to clean up the incoming power supply

Flipped polarity power: Accidental polarity power reversal should be nondestructive.
   Possible solutions: Use of a series-connected diode in a power connection to open a reverse-powered circuit; an internal voltage regulator that is designed to tolerate inverted power; a power plug connection that mechanically is impossible to reverse-plug

Output abuse: Any ECU output should be able to survive being externally forced to ground or forced to the power rail without destruction.
   Possible solutions: Current-limiting resistance in series with outputs or PTC thermal fuses to protect connections

Input abuse: All inputs should be able to survive forced inputs to ground and power, and be ESD event tolerant without destruction.
   Possible solutions: ESD protection circuits, and selecting appropriate ICs that already have input protection

Dropped power recovery: Devices should self-recover and restart upon power up, or after brief power outages.
   Possible solutions: Suitable defensive coding methods, use of MCU devices that have brownout detection and recovery capability

Mechanical consideration of vibration, shock, impact, moisture/corrosion, connector abuse, and enclosure abuse should also be included. See the consolidated test list that follows.

Regulations, Certifications, and Approvals

Regulatory requirements for automobile safety come from many sources.

Vehicular safety and performance

The following regulatory requirements largely focus on the safety and requirements of the total vehicle. The ISO 26262 series covers safety and quality of electrical/electronics in cars, but lacks specifics on testing criteria:

United Nations
World Forum for Harmonization of Vehicle Regulations.

European Union
The Automotive Regulatory Guide of the European Automobile Manufacturers' Association (ACEA) is making attempts to harmonize the many different sets of regulations used across Europe.

International Standards Organization
ISO 26262-1 through ISO 26262-12 establish safety and risk criteria for electrical and electronic devices in automobiles.

US Federal Motor Vehicle Safety Standards
Created and codified as federal regulations created by the National Highway Traffic Safety Administration.

Electronic components internal to ECUs

The Automotive Electronics Council (AEC) deals with the internal electronics of ECU modules. From its mission statement:

> The AEC Component Technical Committee is the standardization body for establishing standards for reliable, high quality electronic components. Components meeting these specifications are suitable for use in the harsh automotive environment without additional component-level qualification testing.

The AEC was originally formed by General Motors, Ford, and Chrysler. Since its inception in 1992, many component vendors and automakers have joined the council (see *http://www.aecouncil.com*).

AEC documents focus on the individual components utilized within an ECU. This includes chips, discrete transistors, sensors, and multichip packages. However, the AEC doesn't address the testing and qualification of the ECU modules (see *http://www.aecouncil.com/AECDocuments.html*).

Coalition for a software-defined ECU platform

The AUTOSAR (AUTomotive Open System ARchitecture) coalition is trying to standardize a software-based framework for ECU devices. The methods it is promoting are focused on the ECU as software-defined devices but not the physical testing or regulatory approval of ECU devices (see *https://www.autosar.org*).

Society of Automotive Engineers

The SAE offers a wealth of documentation and standards for various aspects of electrical and electronic devices within automobiles (see *https://www.sae.org/standards*).

Testing and regulatory requirements for ECU modules

A set of globally accepted test criteria for the certification of all ECUs was not available as of this writing. A number of specific automaker documents that define acceptance testing do exist. Following is a consolidated test list based on these documents:

- Ford Motor Company: CETP 00.00-E-412
- General Motors (GM): GMW3172
- Volkswagen: VW80000

Automotive ECU consolidated test list

Temperature
- Low temperature exposure/operation
- High temperature exposure/operation
- Powered thermal cycle
- Thermal shock resistance
- Thermal shock endurance
- High temperature endurance
- High temperature degradation
- Powered temperature cycle
- Thermal shock/water splash
- Humid heat cyclic
- Humid heat cyclic (with frost)
- Humid heat constant
- High/low temperature storage

Moisture
- Water/fluids ingress
- Salt mist atmosphere
- Water freeze
- Humidity test
- Condensation test
- High-pressure cleaning
- Corrosion degradation

Mechanical
- Powered vibration
- Audible noise
- Mechanical shock
- Vibration with thermal cycling
- Crush for housing
- Connector installation abuse

- Free fall
- Stone impact test
- Dust test
- Vibration test
- Endurance shock test
- Mechanical endurance

Other tests
- Dust
- Chemical resistance
- Durability and mechanical wear
- Parasitic current
- Battery voltage dropout
- Sinusoidal superimposed voltage
- Pulse superimposed voltage
- Crank pulse capability and durability
- Battery line transients
- Long-term overvoltage
- Transient overvoltage
- Jump start
- Load dump
- Superimposed alternating voltage
- Slow decrease/increase supply voltage

- Slow decrease, quick increase supply voltage
- Reset behavior
- Short interruptions
- Start pulses
- Connector interruption
- Reverse polarity
- Ground offset
- Short circuit in signal circuit and load circuits
- Insulation resistance
- Closed-circuit current
- Dielectric strength
- Backfeeds
- Overcurrents

Since there is no universally recognized testing standard, the customer needs to define the required tests and methods. Specific automakers have developed their own suite of test criteria.

Consumer Electronics

Consumer electronics covers a broad swath of devices purchased for personal use. Examples include cell phones, tablets, laptops, personal computers, televisions, handheld remotes, electronic games, calculators, DVD players, digital video recorders, audio sound systems, and countless others.

Design Priorities

Consumer product design is driven by low unit cost and capability for high-volume manufacturing. Visual appeal and ease of use are also important to help drive sales volumes.

Cost

Cost-optimized design has to be a high priority. Something can be a technical success, but if it's too expensive, it won't sell. This is especially true if similar products are available at a lower cost.

The cost of the Apple Lisa computer is an excellent example. For its era, Lisa was ahead of the technology curve with a graphical user interface, while the IBM PC of the same era used DOS driven from the command line. Lisa initially sold for $10,000 in 1983, about the same price as a new Ford Mustang. Due to the high price, Lisa did not sell well.

Volume manufacturing

The capability for high-quantity manufacturing must be part of the design. High-capacity supply chains need to be available for all components. PCBs should be surface mount with automated builds available. Electronic circuits should avoid anything that needs to be tuned or manually adjusted. Mechanical parts need to be simple to assemble and use minimal components.

Visual appeal

Appearance is a large part of any successful consumer product.

Ease of use

Devices need to be easy to understand and simple to interact with. Forcing users through large amounts of training to use a device is never well received. Consumer products do best with a "plug and play" setup and a "point and click" interface for the customer. As one example, the USB interface is widely popular due to its capability to plug and play a computer peripheral without configuration and setup efforts. Prior devices required manual configuration that often stumped nontechnical users.

Repair capability

This is not a priority for consumer electronics. The capability to repair or replace internal components often leads to a bulkier device or a price increase. The desire for compact size and low unit cost has created a throwaway, rather than repair, mentality.

Reliability

Improved reliability that increases product cost or enlarges a portable device can actually hurt device sales. Limited life reliability (3–5 years) seems to be the status quo, with newer product offerings replacing older devices instead.

Better reliability can grow brand loyalty with a *very small portion* of the customer base. For consumer products, however, periodically getting a newer replacement device is typical.

Backward compatibility

The capability for a user to utilize a new device quickly motivates backward compatibility of many devices. Consider how many devices are "new and improved" but still somewhat similar to their predecessors. Radical redesign can alienate some existing customers. This also holds true with the needed capability to use older software applications on newer computers.

Special Interest Groups, Technology Coalitions, and Technical Standards

A Special Interest Group (SIG) is an organization formed to define and promote a particular device, feature, or method. Companies have learned that compatibility and interoperability of electronic devices is an important product marketing feature. Consequently, cooperation among competitors is seen as beneficial to all parties. The consumer electronics space includes many such organizations that steer a common approach to a commonly needed goal.

Products that "play well with others" are typically the outcome of technical standards or technology coalitions. A technical standard explicitly outlines the exact performance of a device. One example is Ethernet, which is based upon the IEEE 802.3 technical standard. This ensures that all implementations of Ethernet can communicate with each other. Technical standards exist for WiFi, Bluetooth, cellular phone communication, and countless others.

A coalition of technology companies is a common strategy used to define a product, a media format, or a method to achieve something. By developing and presenting a unified goal, consumer acceptance is often better or more quickly achieved. Historical examples include Sony and Phillips forming a coalition to back the compact disc.

The VHS tape format was backed by JVC, Matsushita (aka Panasonic), RCA, Quasar, General Electric, Samsung, Zenith, Sharp, and others.

As for modern examples, the Wi-Fi Alliance exists to promote a standardized WiFi product line, define newer generations, and provide vendor certification of compatibility. The MIPI Alliance promotes compatibility for items internally used in the creation of cell phones. The Bluetooth Special Interest Group promotes technical standardization of Bluetooth wireless technology.

Many examples of coalitions and special interest groups steer consumer electronics toward widely accepted products.

Regulations, Certifications, and Approvals

Regulatory requirements for consumer products are actually quite simple. Certain requirements are legally required; others are considered necessary for marketability.

FCC EMI certification

Limiting radiated EMI is required for all products. In the United States, FCC-enforced regulations of "Unintentional Radiators" will require approval from a certified testing facility. Similar EMI requirements also exist in most other countries (FDI: EMI & ESD).

Underwriters Laboratories safety certification

The Underwriters Laboratories (UL) performs testing and defines safety standards for most commercial and consumer products. Although UL testing is not always legally required, comprehensive safety testing ensures safe operation and is valuable as a marketing tool. Additionally, many vendors concerned with liability issues refuse to sell certain products unless they have received UL certification.

Following is a sampling of the many UL safety standards:

- UL 60065: Standard for Audio, Video and Similar Electronic Apparatus – Safety Requirements
- UL 2097: Reference Standard for Double Insulation Systems for Use in Electronic Equipment
- UL 1642: Lithium Batteries
- UL 2054: Household and Commercial Batteries

For more information, see *https://www.ul.com* and *https://www.shopulstandards.com*.

CE mark

Qualifying a product for CE (Conformité Européenne) certification is necessary to sell in the European Union (EU) countries. There are presently 24 different product directives for various product types and features. Multiple directives may apply to a single product, and any applicable directive needs to be validated by a certified testing facility. Further information can be found on the US Department of Commerce on EU Legislation and CE Marking website (*https://oreil.ly/OYjwk*).

Restriction of Hazardous Substances

Restriction of Hazardous Substances (RoHS) certification is required in the European Union. Within the United States, only California requires RoHS certification, although other states are starting to regulate RoHS issues and electronic waste requirements. The list of countries requiring RoHS compliance is growing, and any new products should be compliant to allow widespread distribution.

RoHS verifies that electronics don't contain certain designated hazardous materials and chemicals in excess of defined thresholds. Most RoHS restrictions are linked to or a variation of the European Union's Restriction of the Use of Certain Hazardous Substances (RoHS) Directive 2011/65/EU.

Registration, Evaluation, Authorization, and Restriction of Chemicals (REACH)

REACH regulations come from the European Union and restrict potentially dangerous substances in all consumer products. Compliance is required for EU sales. Further information can be found at *https://environment.ec.europa.eu/topics/chemicals/reach-regulation_en#*.

Regulatory testing and compliance certification is widely available through certified commercial testing laboratories. These labs provide test and compliance reports used when applying for government approvals.

In addition, these labs usually offer consultations and guidance on which regulatory requirements need to be met and what must be done to obtain compliance certification.

Engaging a compliance test facility early in the design process is suggested. Using a lab representative as part of the design review process avoids costly surprises when the device finally enters the lab for compliance testing.

Industrial Automation

Industrial automation (IA) is used for many manufacturing and assembly line systems. In addition, warehouse facilities and material handling systems are frequently automated. IA is where robotic manufacturing was born and continues to expand. As this sector continues to grow, more factories will continue to produce more goods, at a lower cost, while using fewer people and more automation.

The top design priorities of IA are reliability and safety. Failure of a high-volume production line can cost millions of dollars per day. Consequently, reliability takes precedence over cost for automation equipment.

Use of robotics and other machines can be very hazardous to human workers on a production floor. Therefore, facilities with a mix of automation and humans on the production floor need to carefully consider safety issues.

Industrial robotic safety is explored in the following documents:

- ANSI R15.06: Industrial Robots and Robot Systems – Safety Requirements
- ISO 10218-1: Robots and Robotic Devices, Safety Requirements for Industrial Robots, Part 1: Robots
- ISO 10218-2: Robots and Robotic Devices, Safety requirements for Industrial Robots, Part 2: Robot Systems and Integration

Oddly enough, OSHA has not yet issued specific rules for industrial robotics, but points out its applicable standards at *https://www.osha.gov/robotics/standards*.

Most documents on IA will mention programmable logic controllers (PLCs) and Supervisory Control and Data Acquisition (SCADA) systems. Looking at Figure 13-8, a typical PLC should appear familiar. The internal parts of a PLC are very similar to many embedded MCU systems, using inputs, outputs, serial ports, and so on. Essentially, a PLC is an embedded system that has been bundled into an enclosure with power, various interfaces, and support systems.

Instead of building an embedded control system from scratch, the IA designer obtains PLC devices with the appropriate features and can then focus on the installation and programming of the system being automated.

Many variants on the PLC exist, but the essentials are the same. Depending on features, the devices can be quite expensive, whereas an embedded MCU can often be implemented for a few dollars.

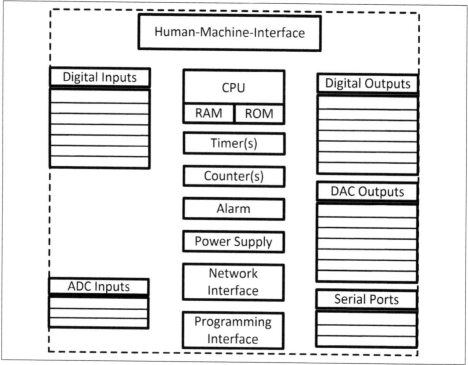

Figure 13-8. Typical programmable logic controller

A typical SCADA system is a hierarchy of networked computers and automation systems. Looking at Figure 13-9, the bottom two rows (On Machine Sequence Manager and MCU Dedicated Function Controller) are similar to the distributed controller systems discussed earlier. The sequence manager can be implemented with an MCU as discussed before, or using a PLC module. One level up, the System Supervisor may be implemented using a PLC or a dedicated computer.

At a higher level, the On-Site Manager Supervisor is typically a dedicated computer that requests services from subordinates while also receiving status and production data from those subordinates. This computer typically consolidates data and provides the central control computer with materials requests and production output reports.

The central control computer monitors subordinate output production, provides desired production volumes and schedules to subordinates, coordinates product shipping, generates external material orders, and informs other company systems of financial transactions needed for the materials produced.

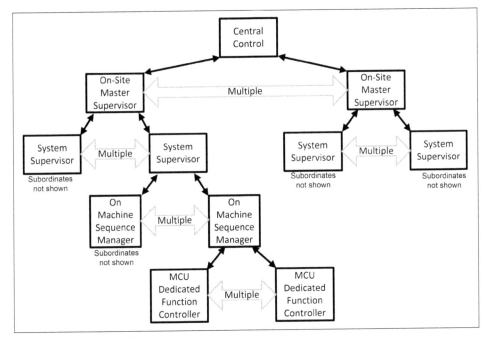

Figure 13-9. Typical SCADA system

Embedded controllers exist at the bottom two tiers of the SCADA system shown in the figure. Apart from that, industrial automation is beyond the scope of this book. Several useful books are recommended in "Further Reading" on page 529 for those interested in exploring this valuable and growing discipline.

Summary and Conclusions

Depending upon the product sector, a suitable design often has different priorities and results in different outcomes of the design process. Regulatory controls frequently shape the design methods, process flow, materials used, and manufacturing methods.

For highly regulated categories such as medical, avionics, and military, reviews and records of the planning, specification, design, testing, and manufacturing processes are often required. Qualification, approval, and quality management systems are used for design planning, device specification, materials and components, the manufacturing process, and the final product.

Always obtain up-to-date documents that define the regulatory requirements for the project. Specific performance values may have changed, or additional requirements may have been added. Any parameters cited here may need an update.

A clear definition of the design specification, which regulatory controls are required, and what qualification testing will be done should be outlined and agreed on. If not finalized, getting customer consensus on those important details should happen before design starts. Without those items nailed down, backtracking, redesign, and wasted time are to be expected.

Frequently, regulatory compliance is mandatory to sell a product, and there are countless examples of organizations that developed something and had their product launch stopped because they addressed regulatory issues too late in the development process.

Further Reading

Waterfall Model
- Symposium on Advanced Programming Methods for Digital Computers, 1956, United States, Navy Mathematical Computing Advisory Panel (29 June 1956), Washington, DC, Office of Naval Research, Dept. of the Navy (Original waterfall model).
- Winston W. Royce, "Managing the Development of Large Software Systems," Proceedings of IEEE WESCON, 1970.

Risk Management
- Josef Oehmen, Mohammad Ben-Daya, Warren Seering, Muhammad Al-Salamah, "Risk Management in Product Design: Current State, Conceptual Model and Future Research," ASME, 2010, *https://asmedigitalcollection.asme.org/IDETC-CIE/proceedings-abstract/IDETC-CIE2010/44090/1033/340428*.
- ISO 31000, Risk Management – Principles and Guidelines, International Organization for Standardization, *https://www.iso.org/iso-31000-risk-management.html*.
- ISO 31010: Risk Management – Risk Assessment Techniques, International Organization for Standardization, *https://www.iso.org/standard/72140.html*.
- E.A. Silk and P.H. Dash, "Risk Management in Space Systems Design and Technology Development," Proceedings of the Institution of Mechanical Engineers, Part G, *Journal of Aerospace Engineering*, 2008, 222(6), 907–913.

Avionics
- *Advanced Avionics Handbook*, FAA-H-8083-6, 2012, Federal Aviation Administration, 2012, ISBN 978-1-61608-533-9, FAA with Skyhorse Publishing.
- *Digital Avionics Handbook, 2nd Edition*, Cary Spitzer (Ed.), 2007, ISBN 0-8493-8438-9, Taylor & Francis, CRC Press.
- *Digital Avionics Systems* by Cary Spitzer, 1987, ISBN 0-13-211517-4, Prentice Hall.

Satellites and Spacecraft
- Embedded.com staff, "The Challenges and Evolution of CubeSat Electronics," Embedded.com, 2017, *https://www.embedded.com/the-challenges-and-evolution-of-cubesat-electronics*.
- Lina Tran, "How NASA Prepares Spacecraft for the Harsh Radiation of Space," NASA's Goddard Space Flight Center, June 10, 2019, *https://www.nasa.gov/feature/how-nasa-prepares-spacecraft-for-the-harsh-radiation-of-space*.
- Thermal Systems in the Mars Reconnaissance Orbiter, NASA, *https://mars.nasa.gov/mro/mission/spacecraft/parts/thermal*.

- P. Aggarwal, "Dynamic (Vibration) Testing – Certification of Aerospace Systems," NASA-Marshall Space Flight Center, *https://ntrs.nasa.gov/api/citations/20110002782/downloads/20110002782.pdf.*

- D. Kern and T. Scharton, "NASA Handbook for Spacecraft Structural Dynamics Testing," 2005, *https://ntrs.nasa.gov/api/citations/20050180670/downloads/20050180670.pdf.*

- "Spacecraft Dynamic Environments Testing – NASA Technical Handbook," NASA-HDBK-7008, 2014.

Industrial Automation
- *Exploring SCADA Systems* by Charles Vance, 2023, ISBN 9798394567124, Mountaintop Publishing.

- *Getting Factory Automation Right (the first time)* by Edwin Zimmerman, 2001, ISBN 0-87263-526-0, Society of Manufacturing Engineers.

- *Programmable Logic Controllers, 3rd Edition,* by Frank Petruzella, 2005, ISBN 0-07-829852-0, McGraw-Hill.

- *Automation, Production Systems, and Computer-Integrated Manufacturing, 3rd Edition,* by Mikell Groover, 2008, ISBN 978-81-203-3418-2, Pearson Prentice Hall.

Creating Great Products

Life's too short to build something nobody wants.

—Ash Maurya

The goal of this book was to provide the essential information to design embedded system electronics that are reliable and problem free. In closing, knowing what to design and how to determine the likely success of potential products is worthwhile knowledge.

What makes a great product? For consumer products, high-volume sales are a good indicator of success. Devices that sell by the millions can't be argued with. However, many products are never destined for high-volume sales. In the military, medical, avionics, and satellite sectors, the volumes are smaller than in the consumer products sector. Nonetheless, many devices in those sectors are valuable and useful.

The consumer electronics market is flooded with multiple competitors and innumerable offerings. Consequently, pricing is cutthroat and product life spans are short, making it difficult to compete. As an alternative, there are many niche markets where there is much less competition and product life spans are friendlier to smaller organizations. Smaller markets with fewer competitors can still yield successful sales.

Yes, this is an "Introduction to Marketing" discussion, but designers should know some of the basics. Understanding the common characteristics of successful products is useful knowledge.

Create Products That Solve Problems or Fulfill a Need

Customers don't care about your solution; they care about their problems.
—Dave McClure

Creating a product that solves a problem or fulfills a need is a good path to success. If the problem is common to many people, the potential customers are there. Frequently, the customer won't have the creative insight to see the solution to their problem. But they will be very aware of their problem and should be receptive if shown a suitable solution.

The world of kitchen gadgets is filled with products that claim to solve cooking problems. Some, such as can openers and bottle openers, cheese graters, cutting boards, and meat thermometers, are valuable problem solvers that have sold consistently for many years.

The smartphone is very successful, with over 1 billion phones sold annually (2015–2023) largely because it is a platform that solves many problems and makes life easier for the owner. Consider the features available:

- Internet browser
- Maps and GPS navigation
- Texting and messaging
- Music player
- Flashlight
- Still and video camera
- Voice recorder
- Email access
- Gaming platform

- Calculator
- Fitness monitor
- Exercise coach
- Calendar and reminders
- News reader
- Contact and telephone list
- Online shopping
- Language translators
- Video player

This "telephone" solves many problems, and makes life easier for the owner. Solve a problem, fulfill a need, or make something easier to do, and the potential for a great product is there.

Identify the Target Market

No matter what the product idea is, initial research into product viability should determine what the target market is. This can be divided in many different ways. Here are some examples:

- Age groups (toddlers, teens, young adults, Gen Xers, boomers)
- Specific interests (cyclists, campers, skiers, boaters, musicians, parents)
- Industry sectors (avionics, automotive, industrial automation, military)

This is certainly not a complete list. Many possible markets can be defined with different customer bases. Information about potential customers in the target market will usually help define both the product and desired features thereof.

Identify What the Customer Wants

Many opinions exist on understanding what a customer wants:

> If I had asked people what they wanted, they would have said faster horses.
> —Henry Ford

> People don't know what they want until you show it to them.
> —Steve Jobs

The original "customer is always right" adage came from a businessperson who focused on customer service. For new products, ease of use, customer-facing interfaces, and physical presentation are all items where the end user should be carefully listened to.

But Henry Ford's "faster horses" comment illustrates that customers have a good understanding of what their pain points are, yet they often won't understand the innovation and new methods that can be utilized to improve their situation.

Steve Jobs once made the claim to never rely on market research, but the truth is that Apple performs extensive market research during product definition and development (Vascellaro, 2012). Jobs, like Ford, understood that customers often don't understand the merits of new methods until they see the final product.

Understanding human needs is half the job of meeting them.
    —Adlai Stevenson

Understanding potential customer needs is often the starting point to product definition. Conducting market research is valuable in developing a successful product. Some ideas for market research include:

- Seeking user input early and often.
- Finding early adopters to try your product and listening to their feedback.
- Conducting user interviews.
- Getting feedback on the proposed price point.
- Addressing the needs of the end user, not the wants of the design team. The best path forward is to create consensus of all parties.
- Getting prototypes reviewed by end users and refining the design based upon their feedback.

Examine Competing Products

Competitor research will help guide product definition. What features are offered, what price point they are sold at, and sales volume are all valuable pieces of information.

To be competitive, first examine currently available products. Then expect competitors to launch next-generation devices with better features and lower cost. Projections of competitors' future product offerings can then be used to define a new product design. Product definition is about competing with future products, not with what's on the shelf now.

Also, determine what customers like and dislike about competitors' products. Objectionable features in a competitor's product can be an opportunity to be exploited. Any new-product definition should improve on the positive features and eliminate the negative features of the competition. At product launch, the marketing department will use the competitor's shortcomings as a springboard to market the "new and improved" product.

Finding ways to offer a better product and undercut a competitor's price point is a good way to gain market share. However, "me too" products based upon competitor offerings rarely do well.

Define the Value Proposition

In marketing, a value proposition outlines the principal features and aspects of the product—things like a functional description of the product, what it does, who it's for, and where it gets used. Also, what makes the product unique, what competitive products already exist, a target market description, and details regarding the market size are usually included. A value proposition serves as a starting point to get a better understanding of what's wanted and who wants it. Also, it is often used to start the R&D team on hammering out the details.

Determine Viable Pricing

Depending on the sector, product cost can be a low priority, as it is in the military, industrial automation, and avionics sectors, or it can be the predominant consideration in making a purchase. Price sensitivity is typically highest for consumer electronics.

Early versions of products are often too expensive for consumers. Things like computers, wireless phones, washing machines, clothes dryers, and microwaves were initially too expensive for consumers. Expensive "commercial" versions sold for business use, but consumers stayed away due to price. Once design and manufacturing refinements brought the prices down, consumer sales volume went up.

For consumer products, pricing should be part of market research. An important question to ask potential customers is "What would you pay for this?" If a customer consensus on price is lower than the desired selling price, the markup or manufacturing cost needs to be lowered.

Determine a Properly Timed Market Window

The viable time window for a product needs to be estimated, and the difficult question "Can we make the window?" needs an answer.

An example can be seen with digital video converters for analog TV. In 2009, the FCC stopped broadcasting analog television signals. Briefly, there was heavy demand for a converter box to transform the digital broadcast video to the older analog format. This was for millions of analog CRT televisions still in use. Within a few months, this market disappeared, as most analog televisions had received the needed converter. New video displays were shipped with built-in digital video capability, so only the older analog sets needed them. This was a narrowly timed market window, but several vendors briefly sold lots of converter boxes.

A product can be too early to market, and the support technology won't be ready or it will be too expensive or too radical for customers to be comfortable with. A product can be too late to market, and the market will already be flooded with competitors. Market timing can be fickle.

Establish Coalitions and Strategic Partners

Strategic partnerships with organizations that agree to use your future product can be invaluable. For example, automobiles use many different electronic control units (ECUs). As a strategic partner with an automaker, designing an ECU for its next-generation car gets you a customer, a valuable feedback resource, a knowledge source to help define the ECU, and guidance on viable unit cost.

A strategic partnership as an OEM supplier can be a win-win relationship. Collaboration opportunities can be found in many places. Semiconductor manufacturers have formed coalitions to develop application-dedicated chip sets for cell phones, computers, set-top boxes, and other products.

Focus on Ease of Use

As the product definition is put together, remember that if it's not easy to use and understand, it won't sell well as a consumer product. A "plug and play" mentality should be the goal. Avoid cumbersome setup and configuration whenever possible. The user should be able to take the product out of the box and use it.

This includes:

- Documentation and setup guides that are simple and clear
- Physical layout of buttons and displays for intuitive use
- Easily understood GUI, menus, and selection options
- Easy and simple navigation of controls
- A simple return to start/home (Home)
- A simple way to go back a step (Undo)
- Safety checks ("Are you sure you want to delete that?")
- Prompts and help messages to aid decisions
- Interfacing with the device that is consistent for all actions

Also, if an industry-standard interface is already widely adopted, use it. Going to new methods may alienate some users. It may also violate regulatory requirements for some devices. Going away from what users are familiar with requires a compelling reason.

Test out an early prototype with end users from your target market. Ask the users to use your product while you observe problems encountered and ask for their impressions and suggestions. Usability testing by people working on the project is not suggested due to their pre-exposure to the product.

If you are focusing on medical devices, doctors and nurses should be your usability testers. Avionics instruments should be user-tested by pilots. Find your end users, get them to test-drive the product, and listen to all their suggestions.

Determine the Needed Resources

Getting the right mix of people on a design team is important. Most electromechanical products require designers for electronics, mechanical hardware, and software. Also needed are individuals to deal with marketing issues, regulatory compliance, materials procurement, lab testing and fabrication, and manufacturing issues.

In some roles there can be some multitasking, but be realistic about what individuals can do beyond their principal skill set. Make sure you can get the right talent set to get it done.

With a team in place, make sure members buy in to the design targets and communicate well with one another. Changes in market, electronics, software, and mechanical design can interact and cause disruption.

Get Design Specification Consensus

Product requirements and parameters need to be well defined. Nailing down the specifics of the product expedites the design process, especially when multiple designers are involved. The entire team should give input when defining the design specification so that consensus and buy-in are in place. Typically, changes made late in the design are more problematic than those made early on. Consequently, getting team consensus on the design specification is crucial.

A design specification can be updated or revised during the design process as problems are identified or must-have features are added. However, implementing changes during the design process will affect schedules and resources.

Minimal Design and Feature Creep

Use a minimal design approach to solve the problem. Keep it simple. Building a product that's crammed with features may look good in a sales brochure, but it takes time and money to make it real. Products with many extra features frequently run into time delays and other avoidable problems getting to production.

Focus on the design, execution, and manufacturing of a product with the necessary features that meet the initial device definition. Changing the product direction midstream is often a symptom of a poorly managed project. Changing the design or adding features should require very compelling reasons and should not be agreed to lightly.

Feature creep and schedule slips can be deadly to new products, especially in cash-limited and time-strapped startup environments. Always ask the question: "Is this feature necessary or nice to have?"

Identify Obstacles Early

No matter what obstacles may come your way, stay focused, stay positive, and you will overcome them.

    —Author unknown

It's extremely important to assess which issues in a design are easily executed and which ones pose problems. Technical challenges, availability of specialized materials, and single-source components are all red flags. Worse yet is the need for components that are not yet on the market, requiring a custom-built solution from an outside vendor.

Ideally, all components have multisource availability and all technical challenges are seen as straightforward to resolve. Frequently, that's not the case. Wherever possible, remove limited-availability components or features to minimize supply chain risks. High-risk but must-have technical features command immediate attention from the design team. If it turns into a showstopper, it's best to know ASAP.

External source components that are not yet available need to be carefully managed with vendor availability and quantity assessments over a defined timeline. Dependency on external suppliers of specialty items has crushed many startups that ran out of money while waiting for something a vendor could not deliver. If at all possible, design these components out, or seek multiple external vendors early in the design cycle. This reduces your product's risks if one external vendor fails to deliver.

Get User Feedback on Prototype Builds

Getting user feedback on prototype builds is an invaluable part of a successful design. For example, rapid-turnaround printed circuit board (PCB) builds can be done in a few days, if needed. Getting a PCB fabricated and assembled quickly will usually incur rush fees, but these are often worth the extra expense. This has the additional bonus of providing actual electronics that software designers can run their code on.

Quick turnaround of prototypes for mechanical parts has become easier with the availability of 3D printing and automated machining based upon CAD files. High-quality metal components are also available from 3D printers in addition to the more common printers that produce plastic parts.

Prototype builds for designers are useful for assessing design progress. In addition, rapid prototyping allows target market customers to see, handle, and evaluate the product. As discussed earlier, their feedback as an end user helps the design team better understand what works well and what needs to be improved.

Additionally, if the project team is looking for funding, being able to do a prototype "show and tell" demonstration is a valuable selling point for potential investors.

Make It Easy to Manufacture

Except in some large companies, the fabrication of parts will be contracted out to external vendors. Setting up and understanding the supply chain for all parts needs to be wrung out during the design cycle.

Realize that rapid prototyping works well for small lots, but 3D printing is not cost or time effective for high-volume orders. Molds for plastic injection–molded components will need to be designed and fabricated prior to volume production. Plastic injection molds are typically machined steel, and their fabrication can take one to three months depending upon complexity and vendor. Plan ahead, and start a dialogue early with mold manufacturers.

You want readily available parts and an easily done assembly procedure. Make the design easy to manufacture. Minimize the custom parts where possible, as off-the-shelf components cost less and are easier to procure. Carefully manage supply chains of all custom components, and understand the volume capability of your support vendors.

Summary and Conclusions

Product marketing is a complicated topic. What's been discussed here touches upon some important points, but an in-depth analysis would require multiple books.

Having a great product idea is the starting point. Doing market research comes next, as you better understand what the target market is and what the customer wants. This helps to develop the idea into a product proposal.

Products that solve a problem for the customer are generally well received, especially if they are easy to use and have a realistic price. Verify the solution viability with early prototype evaluations by target customers.

Understand where the competing products are now, and estimate where the newer products will be when your product is ready for market. This allows a better definition of a competitive product and its features.

Better yet, recognize the need for a product where none yet exists, thus giving you a head start on your competitors. Your success will create that competition.

Remember:

> *Any sufficiently advanced technology is indistinguishable from magic.*
> —Arthur C. Clark

Thanks for reading!

Now go make some magic!

Jerry Twomey
San Diego, October 2023
effectiveelectrons.com

The End :)

Further Reading

- *Running Lean: Iterate from Plan A to a Plan That Works, 3rd Edition,* by Ash Maurya, ISBN 978-1-098-10877-9, O'Reilly Media.

- *The Art of the Start 2.0: The Time-Tested Battle-Hardened Guide for Anyone Starting Anything* by Guy Kawasaki, 2015, ISBN 978-0-241-18726-5, Penguin Random House.

- *Lean B2B: Build Products Businesses Want* by Etienne Garbugli, 2014, ISBN 978-1495296604, Create Space Independent Publishing Platform.

- Jessica Vascellaro, "Turns Out Apple Conducts Market Research After All," *The Wall Street Journal,* July 26, 2012.

- *The 1-Page Marketing Plan: Get More Customers, Make More Money, and Stand Out from the Crowd* by Allan Dib, 2018 & 2021, ISBN 978-1-989025-01-7, Successwise & Macmillan.

Glossary of Acronyms

Variants of commonly used acronyms (i.e., CLK, SCLK, SCL) are included herein, due to various industry standards that utilize different naming conventions.

0–9

3D three dimensional

A

AC	alternating current
ADC	analog-to-digital converter
AEC	aluminum electrolytic capacitor
AEC	Automotive Electronics Council
AED	automated external defibrillator
AFE	analog front end
AGM	absorbed glass mat
Ah	amp-hour
ANSI	American National Standards Institute
ANT	Adaptive Network Topology
ARM	Advanced RISC Machine or Acorn RISC Machine
ASCII	American Standard Code for Information Interchange
ASIC	application-specific integrated circuit
ASME	American Society of Mechanical Engineers
AUTOSAR	Automotive Open System Architecture
AWG	American Wire Gauge

B

BDCM	brushed DC motor
BGA	ball grid array
BIOS	basic input/output system
BLDC	brushless DC (re: motors)
BJT	bipolar junction transistor
BMS	battery management system
BOM	bill of materials
BPF	band pass filter
BT	Bluetooth
BTLE	Bluetooth Low Energy
BW	bandwidth

C

CAD	computer-aided design
CAN	Controller Area Network
CD	compact disc
CE	Conformité Européene or European Conformity
CHRT	Chien—Hrones—Reswick tuning
CISC	Complex Instruction Set Computer
CJ	cold junction (re: thermocouples)
CLK	clock
CMOS	complementary metal-oxide semiconductor
CMRR	common-mode rejection ratio
CPLD	complex programmable logic device
CPM	contract PCB manufacturer
CPU	central processing unit
CSI	Camera Serial Interface
CSP	chip-scale packaging

D

DAC	digital-to-analog converter
DC	direct current
DCU	digital control unit
DD	device driver
DIP	dual inline package
DMA	direct memory access

DNL	differential nonlinearity
DoD	Department of Defense
DOS	disk operating system
DRC	design rule check
DSP	digital signal processing
DUC	device under control
DUT	device under test
DVD	digital video disc or digital versatile disc
DVI	Digital Visual Interface

E

ECG	electrocardiogram
ECL	emitter-coupled logic
ECU	electronic control unit
EDA	electronic design automation
EEG	electroencephalogram
EEPROM	electrically erasable programmable read-only memory
EIA	Electronic Industries Alliance
ELV	extra-low voltage
EM	electromagnetic
EMC	electromagnetic compatibility
EMI	electromagnetic interference
ENIG	electroless nickel immersion gold
EPC	Equivalent Parallel Capacitance
ESA	European Space Agency
ESD	electrostatic discharge
ESL	Equivalent Series Inductance
ESR	Equivalent Series Resistance
EU	European Union
E-waste	electronic waste

F

FAA	Federal Aviation Administration
FAA-TSO	Federal Aviation Administration – Technical Standard Orders
FB	ferrite bead
FB	feedback
FC	Faraday cage
FCC	Federal Communication Commission

FDA	Food and Drug Administration
FDI	further discussed in
FET	field-effect transistor
FMEA	failure modes and effects analysis
FPGA	field-programmable gate array

G

GaNFET	gallium nitride field-effect transistor (aka GaN)
GEO	geostationary orbit
GPIO	general-purpose input/output
GPS	global positioning system

H

HASL	Hot Air Solder Leveling with Lead
HBM	human body model (re: ESD)
HDD	hard disk drive
HDL	hardware description language
HDMI	High-Definition Multimedia Interface
HDTV	high-definition television
HF	high frequency
HL-DS	high-low differential signaling (re: CAN bus)
HPF	high-pass filter
HR	high resolution
HV	high voltage
HW	hardware

I

I^2C or I2C	Inter-Integrated Circuit
I^2S or I2S	Inter-IC Sound
IA	industrial automation
IC	integrated circuit
IDE	integrated design environment
IEC	International Electrotechnical Commission
IEEE	Institute of Electrical and Electronic Engineers
IGBT	insulated-gate bipolar transistor
IIR	infinite impulse response
INL	integral nonlinearity

IP	ingress protection (re: IEC 60529)
IPC	Institute for Printed Circuits (re: IPC.org)
IRN	input referred noise
ISM	industrial, scientific, and medical (re: ISM radio band)
ISO	International Organization for Standardization
ITU	International Telecommunication Union

J

JAXA	Japan Aerospace Exploration Agency
JFET	junction field-effect transistor
JTAG	Joint Test Action Group

K

KCL	Kirchoff's Current Law
KVL	Kirchoff's Voltage Law

L

L	inductance or inductor
LAN	local area network
LC	inductor – capacitor
LCD	liquid crystal display
LDO	low dropout (re: voltage regulator)
LED	light-emitting diode
LEO	low Earth orbit
LF	low frequency
LF-HASL	Lead-Free Hot Air Solder Leveling
Li-ion	lithium ion
LiPo	lithium ion polymer
LNA	low-noise amplifier
LPF	low-pass filter
LSB	least significant bit
LTAC	long-term average current
LTTD	Long-Term Total Dose (radiation)
LVDS	low-voltage differential signaling
LVS	layout versus schematic

M

MAC	Media Access Control
MCU	microcontroller
MEMS	microelectromechanical systems
MIPI	Mobile Industry Processor Interface
MIPS	million instructions per second
MIPS	microprocessor without interlocked pipeline stages
MISO	Manager In, Subordinate Out (re: SPI)
MLCC	multilayer ceramic capacitor
MM	machine model (re: ESD)
MOSFET	metal-oxide-semiconductor field-effect transistor
MOSI	Manager Out, Subordinate In (re: SPI)
MPU	microprocessor
MRAC	Model Reference Adaptive Control
MRI	magnetic resonance imaging
MSB	most significant bit

N

NASA	National Aeronautics and Space Administration
NEPP	NASA Electronic Parts and Packaging Program
NFET	negative field-effect transistor
NiCd/NiCad	nickel cadmium
NiMH	nickel metal hydride
NMOS	negative metal-oxide semiconductor
NPN	negative-positive-negative (re: bipolar transistor)
NRZ	non-return-to-zero
NTC	negative temperature coefficient

O

OATS	Open Area Test Site
OLED	organic light-emitting diode
Op-amp	operational amplifier
OS	operating system
OSHA	Occupational Safety and Health Administration
OSP	organic solderability preservative
OTS	off the shelf

P

P&G	power and ground
P2P	point to point
PA	power amplifier
PAL	programmable array logic
PAN	personal area network
PC	personal computer
PCB	printed circuit board
PCBA	printed circuit board assembly
PCIe	Peripheral Component Interconnect Express
PCM	pulse code modulation
PEMS	programmable electrical medical system
PFET	positive field-effect transistor
PGA	programmable gain amplifier
PI	proportional-integral (re: feedback control)
PIC	peripheral interface controller or programmable intelligent computer
PID	proportional integral derivative
PLC	programmable logic controller
PLD	programmable logic device
PLL	phase locked loop
PMOS	positive metal-oxide semiconductor
PN	positive-negative (re: diodes)
PNP	positive-negative-positive bipolar transistor
POSIX	Portable Operating System Interface
PP-INS	previously impregnated insulator
PPTC/PTC	polymeric positive temperature coefficient
PTC	positive temperature coefficient
PWM	pulse width modulation

Q

Q	charge in coulombs
Q	quality factor
QFN	quad flat no lead
QFP	quad flat pack

R

R	resistor or resistance
R&D	research and development
RAM	random access memory
RC	resistor capacitor
RF	radio frequency
RFI	radio frequency interference
RISC	Reduced Instruction Set Computer
RLC	resistor inductor capacitor
RLL	run length limited
RMR	risk mitigation review
RoHS	Removal of Harmful Substances
ROM	read-only memory
RPM	revolutions per minute
RTC	real-time clock
RTCA	Radio Technical Commission for Aeronautics
RTD	resistance temperature detector
RTL	resistor transistor logic
RTOS	real-time operating system

S

S&S	satellites and spacecraft
SAE	Society of Automotive Engineers
SATA	Serial Advanced Technology Attachment
SCADA	Supervisory Control and Data Acquisition
SCL	serial clock (re: I^2C)
SCLK	serial clock (re: SPI)
SDA	serial data (re: I^2C)
SDI	Serial Digital Interface
SEE	Single Event Effects (radiation)
SerCom	serial communication
SerDes	serializer/deserializer
SFDR	spurious free dynamic range
SG	strain gauge
SiC	silicon carbide
SIG	Special Interest Group
SLA	sealed lead acid
SLA-AGM	sealed lead-acid absorbent glass mat

SMB	System Management Bus (aka SMBus)
SMPS	switched-mode power supply
SMT	surface mount technology
SMT-MLCC	surface mount technology multilayer ceramic capacitor
SNR	signal-to-noise ratio
SoC	system on a chip
SOIC	small outline integrated circuit
SOP	small outline package
SPI	Serial Peripheral Interface
SPICE	Simulation Program with Integrated Circuit Emphasis
SRF	self-resonant frequency
SSB	solid state battery
SSD	solid state disk
SSSN	Slow, Soft, Short Net
SW	software

T

T&B	top and bottom (re: PCB)
T&M	test and measurement
TCC	timer counter circuit
TIA	Telecommunications Industry Association
TLT	transmission line and termination
TTL	transistor–transistor logic
TVS	transient voltage suppressor
TX	transmitter

U

UART	universal asynchronous receiver transmitter
UL	Underwriters Laboratory
USB	Universal Serial Bus

V

VAC	volts alternating current
VCM	voice coil motor
VCO	voltage-controlled oscillator
VHDL	VHSIC Hardware Description Language
VHSIC	Very High Speed Integrated Circuits Program

VIRLC	voltage, current, resistance, inductance, or capacitance
VoIP	Voice over Internet Protocol
VSA	vibration, shock, and acceleration

W

| WDT | watchdog timer |

X

| X86 | Intel microprocessors, 80x86 family |

Z

| ZNT | Ziegler–Nichols tuning |

Index

About the Author

Jerry Twomey has designed many consumer, medical, aerospace, and commercial products. This includes: high-speed data communication, satellite chipsets, medical instrumentation, cell phones, RF devices, and others. His focus is nondigital electronics, namely those areas where the biggest challenges are. With a breadth of experience designing ICs, PCBs and systems, he has developed the design techniques necessary for reliable electronics.

Jerry has authored multiple trade magazine publications that provide solutions for common electronic design problems. That series of articles became the catalyst for a book that provides a clear-cut methodology to design problem-free systems.

Beyond design, Jerry teaches seminars in embedded systems, IC design, EMC, and medical technology. Positions held include: Chair IEEE SD Solid State Circuits, Lecturer UCSD ECE Graduate Studies, Chair IEEE SD Microwave Theory and Techniques, Reviewer IEEE Journal of Solid State Circuits, Reviewer IEEE Journal of Microwave Theory and Techniques, and Instructor UCSD Extension.

Beyond electronics, Jerry enjoys racing and cruising sailboats, restoring classic cars and other hands-on projects, downhill skiing, hiking and camping, yoga, music, and margaritas. After being raised in Massachusetts, migrating to Silicon Valley, and moving to San Diego, he's seen the sunset from both sides of the country.

Colophon

The animal on the cover of *Applied Embedded Electronics* is a yellow-footed rock wallaby (*Petrogale xanthopus*), an Australian marsupial found in western New South Wales, eastern South Australia, and portions of Queensland.

With a robust body, accompanied by a bushy tail, this wallaby sports grizzled grey or brown fur that provides effective camouflage in its rocky environments. Notably, its paws and tail base are adorned with vibrant yellow markings, lending the species its name.

Found in rugged, rocky landscapes like cliffs, gorges, and rocky outcrops, the yellow-footed rock wallaby exhibits exceptional agility in navigating challenging terrains. Their diet mainly consists of grasses, supplemented by various herbs, shrubs, and leaves.

These creatures are crepuscular, meaning they are most active during dawn and dusk, and will seek refuge in rocky crevices or caves during the day to escape the heat. Socially inclined, they form small groups, or mobs, typically led by a dominant male with several females and their offspring. Female wallabies have a pouch where they nurture and carry their underdeveloped young, known as joeys, for an extended period.

Classified as "Near Threatened" by the IUCN, the yellow-footed rock wallaby faces threats such as habitat loss, competition with introduced species, and predation. Many of the animals on O'Reilly covers are endangered; all of them are important to the world.

The cover illustration is by Karen Montgomery, based on an antique line engraving from *Natural History of Animals*. The cover fonts are Gilroy Semibold and Guardian Sans. The text font is Adobe Minion Pro; the heading font is Adobe Myriad Condensed; and the code font is Dalton Maag's Ubuntu Mono.

Milton Keynes UK
Ingram Content Group UK Ltd.
UKHW051203051024
449280UK00003B/6